Characteristic Functions and Models of Nonself-Adjoint Operators

Mathematics and Its Applications

Managing Editor:

M. HAZEWINKEL

Centre for Mathematics and Computer Science, Amsterdam, The Netherlands

Characteristic Functions and Models of Nonself-Adjoint Operators

by

A. Kuzhel

Department of Mathematics,
Simferopol State University,
Simferopol, Ukraine

KLUWER ACADEMIC PUBLISHERS

DORDRECHT / BOSTON / LONDON

A C.I.P. Catalogue record for this book is available from the Library of Congress.

ISBN-13: 978-94-010-6566-5   e-ISBN-13: 978-94-009-0183-4
DOI: 10.1007/978-94-009-0183-4

Published by Kluwer Academic Publishers,
P.O. Box 17, 3300 AA Dordrecht, The Netherlands.

Kluwer Academic Publishers incorporates
the publishing programmes of
D. Reidel, Martinus Nijhoff, Dr W. Junk and MTP Press.

Sold and distributed in the U.S.A. and Canada
by Kluwer Academic Publishers,
101 Philip Drive, Norwell, MA 02061, U.S.A.

In all other countries, sold and distributed
by Kluwer Academic Publishers Group,
P.O. Box 322, 3300 AH Dordrecht, The Netherlands.

Printed on acid-free paper

CONTENTS

APPENDIX 3. LAX–PHILLIPS ABSTRACT SCATTERING SCHEME

INTRODUCTION

Over the last decades, the study of nonself-adjoint or nonunitary operators has been mainly based on the method of characteristic functions and on methods of model construction or dilatation for corresponding operator classes.

The characteristic function is a mathematical object (a matrix or an operator) associated with a class of nonself-adjoint (or nonunitary) operators that describes the spectral properties of the operators from this class. It may happen that characteristic functions are simpler than the corresponding operators; in this case one can significantly simplify the problem under investigation for these operators.

For given characteristic function of an operator A, we construct, in explicit form, an operator that serves as a model $\bar{A}$ of the operator A in a certain linear space (to some extent this resembles the construction of diagonal and triangular matrices' unitary equivalent or similar, to certain matrix classes). The study of this model operator may give much information about the original operator (its spectrum, the completeness of the system of root subspaces, etc.).

In this book, we consider various classes of linear (generally speaking, unbounded) operators, construct and study their characteristic functions and models.

We also present a detailed study of contractions and dissipative operators (in particular, from the viewpoint of their triangulation).

The methods and results presented in this book have found important applications not only in the investigation of various classes of linear nonself-adjoint or nonunitary operators, but also in scattering theory, in the theory of nonstationary random processes, in theoretical radiophysics, and in the other fields of modern mathematics and mathematical physics.

In what follows, we use standard terminology (see, e.g., Akhiezer and Glazman [1, 2] and Dunford and Schwartz [1, 2]) with some necessary refinements and modifications.

The sign ■ denotes the end of proof.

1. REGULAR EXTENSIONS OF HERMITIAN OPERATORS

1. Principal Concepts

1.1. Terminology and Notation

A linear operator A acting in a Hilbert space $\mathcal{H}$ is called *Hermitian* if

$$(Ax, y) = (x, Ay) \quad (\forall \, \{x, y\} \subset \mathfrak{D}_A).$$

An Hermitian operator A is called symmetric if $\overline{\mathfrak{D}_A} = \mathcal{H}$.

The defect space $\mathfrak{N}_\lambda$ of an Hermitian operator A is defined by the equality

$$\mathfrak{N}_\lambda = \mathcal{H} \ominus \Delta_A (\lambda),$$

where $\Delta_A (\lambda) = \Delta_{A-\lambda I}$ and $\Delta_T = T\mathfrak{D}_T$ is the range of the operator T.

We also use the following definitions:

$\langle \mathcal{M}_1, \mathcal{M}_2, \ldots, \mathcal{M}_g \rangle$ denotes the linear span of sets from $\mathcal{H}$ (in particular, of lineals; the term 'lineal' means a 'linear subspace');

$L \vee M = \overline{\langle L, M \rangle}$ is the closure of the linear span of the sets (lineals) L and M from $\mathcal{H}$;

$\mathcal{B}[\mathcal{H}]$ is the algebra of all linear bounded operators acting in $\mathcal{H}$;

$N(\mathcal{H})$ is the set of all closed operators densely defined in $\mathcal{H}$; an arbitrary operator A from $N(\mathcal{H})$ is called an *N-operator;*

$\operatorname{Ker} T = \{x \in \mathfrak{D}_T \mid Tx = 0\}$ is the kernel of an operator T;

$\rho (T)$ is the resolvent set of the operator T;

$\sigma (T)$ is the spectrum of the operator T;

3

$\sigma_p(T)$, $\sigma_c(T)$, and $\sigma_r(T)$ are, respectively, the point, continuous, and residual spectra of the operator T.

If $\mathcal{M}$ is a set of complex numbers, then $\mathcal{M}^* := \{\bar{\lambda} \in \mathbb{C} \mid \lambda \in \mathcal{M}\}$.

1.2. Hermitian Domain of a Linear Operator

Assume that a linear operator A acts in a Hilbert space. The lineal

$$\mathcal{G}_A = \{x \in \mathcal{D}_A \mid (Ax, y) = (x, Ay), \ \forall \, y \in \mathcal{D}_A\} \tag{1.1}$$

is called the *Hermitian domain* of A, and the operator $A_0 = A \mid_{\mathcal{G}_A}$ is called the Hermitian part of A.

It follows from (1.1) that the operator A_0 is Hermitian. In addition, if A is closed, then, as can be easily seen, the operator A_0 is closed, as well.

Proposition 1.1. *Let* $A \in N(\mathcal{H})$. *Then*

$$\mathcal{G}_A = \text{Ker}\,(A - A^*). \tag{1.2}$$

Proof. If $x \in \mathcal{G}_A$, then $(Ay, x) = (y, Ax)$ $(\forall \, y \in \mathcal{D}_A)$. Consequently, $x \in \mathcal{D}_{A^*}$ and $A^*x = Ax$, i.e., $x \in \text{Ker}\,(A - A^*)$. Conversely, if $A^*x = Ax$, then

$$(Ax, y) = (A^*x, y) = (x, Ay) \quad (\forall \, y \in \mathcal{D}_A)$$

and, hence, $x \in \mathcal{G}_A$.

■

Proposition 1.2. *The Hermitian parts of N-operators* A *and* A^* *are equal.*

Proof. By virtue of Proposition 1.1, we have

$$\mathcal{G}_{A^*} = \text{Ker}\,(A^* - A^{**}) = \text{Ker}\,(A - A^*) = \mathcal{G}_A.$$

Moreover,

$$(A^*)_0 = A^* \mid_{\mathcal{G}_{A^*}} = A \mid_{\mathcal{G}_A} = A_0.$$

■

Proposition 1.3. *If $A \in \mathcal{B}[\mathcal{H}]$, then $\mathcal{G}_A$ is a subspace of $\mathcal{H}$.*

The proof follows from (1.2).

∎

Lemma 1.4. *The defect space $\mathfrak{N}_{\bar{\lambda}}$ of the Hermitian part A_0 of an operator A is connected with $\mathcal{G}_A$ by the relation*

$$(A - \lambda I)(\mathfrak{N}_{\bar{\lambda}} \cap \mathcal{D}_A) = \mathcal{G}_A^{\perp} \cap \Delta_{A-\lambda I}. \tag{1.3}$$

Proof. Let $x \in \mathfrak{N}_{\bar{\lambda}} \cap \mathcal{D}_A$. Then

$$(y, (A - \lambda I)x) = ((A - \bar{\lambda} I)y, x) = 0 \qquad (\forall\, y \in \mathcal{G}_A).$$

Consequently, the vector $u = (A - \lambda I)x$ belongs to $\mathcal{G}^{\perp}$. Or, which is the same,

$$(A - \lambda I)(\mathfrak{N}_{\bar{\lambda}} \cap \mathcal{D}_A) \subset \mathcal{G}_A^{\perp} \cap \Delta_{A-\lambda I}.$$

Conversely, if $u \in \mathcal{G}_A^{\perp} \cap \Delta_{A-\lambda I}$, then $u = (A - \lambda I)x$, where $x \in \mathcal{D}_A$ and $((A - \bar{\lambda} I)y, x) = (y, u) = 0$ for any $y \in \mathcal{G}_A$. Consequently, $x \in \mathfrak{N}_{\bar{\lambda}} \cap \mathcal{D}_A$. But in this case, $u \in (A - \lambda I)(\mathfrak{N}_{\bar{\lambda}} \cap \mathcal{D}_A)$, and equality (1.3) is proved.

∎

Corollary 1.5. *If the lineals $\mathfrak{N}_{\bar{\lambda}}$ and $\mathcal{D}_A$ are linearly independent, then the lineals $\mathcal{G}_A^{\perp}$ and $\Delta_{A-\lambda I}$ are also independent. Conversely, if $\lambda \,\bar{\in}\, \sigma_p(A)$ and the lineals $\mathcal{G}_A^{\perp}$ and $\Delta_{A-\lambda I}$ are linearly independent, then the same is true for the lineals $\mathfrak{N}_{\bar{\lambda}}$ and $\mathcal{D}_A$.*

The proof follows from (1.3).

∎

Corollary 1.6. *If $\overline{\mathcal{G}}_A = \mathcal{H}$ and $\lambda \,\bar{\in}\, \sigma_p(A)$, then the lineals $\mathfrak{N}_{\bar{\lambda}}$ and $\mathcal{D}_A$ are linearly independent.*

The proof follows from (1.3).

∎

The following statement is a specific case of the previous corollary:
If the operator A is symmetric, then the lineals $\mathfrak{N}_{\bar{\lambda}}$ and $\mathcal{D}_A$ are linearly independent for any nonreal λ.

Corollary 1.7. *Let A be a closed operator and $\lambda \in \rho(A)$. Then $\mathfrak{N}_{\bar{\lambda}} \cap \mathcal{D}_A = R_\lambda \mathcal{G}_A^\perp$, where $R_\lambda = (A - \lambda I)^{-1}$.*

The proof follows from (1.3).

∎

1.3. Regular Extensions

An operator A is called a regular extension of an Hermitian operator H if $H \subset A_0$, where A_0 is the Hermitian part of the operator A. Clearly, an arbitrary operator A, is a regular extension of its Hermitian part A_0.

We denote the collection of all regular extensions of an Hermitian operator H by $\mathcal{P}(H)$. In particular, $\mathcal{P}(H)$ contains self-adjoint extensions of the operator H (if such extensions exist).

Proposition 1.8. *An operator A is a regular extension of the operator H if and only if $H \subset A$ and*

$$(Hx, y) = (x, Ay) \qquad (\forall\, x \in \mathcal{D}_H, \quad \forall\, y \in \mathcal{D}_A).$$

The proof directly follows from the condition $H \subset A$ and relations (1.1) and (1.3).

∎

The criterion given by Proposition 1.8 is convenient for many applications.

Note that extensions of an Hermitian operator H are not necessarily regular. Indeed, assume that $\mathcal{H} = \langle e_1, e_2 \rangle$, $\mathcal{D}_H = \langle e_1 \rangle$, and $He_1 = e_2$. An arbitrary extension A of an operator H has the form

$$Ae_1 = e_2, \quad Ae_2 = ae_1 + be_2,$$

where a, b are complex numbers. But it is easy to verify that the operator A belongs to $\mathcal{P}(H)$ only for $a = 1$.

Let an N-operator A belong to $\mathcal{P}(H)$. Then, by virtue of the equality $(A^*)_0 = A_0$, the operator A^* also belongs to $\mathcal{P}(H)$. Since $A_0 \subset A$ and $A_0 \subset A^*$, we have

$$H \subset A, \qquad H \subset A^*. \tag{1.4}$$

In view of Proposition 1.1, the converse statement is also true: If an N-operator A satisfies condition (1.4), then $A \in \mathcal{P}(H)$.

Theorem 1.9. *Let* $A \in \mathcal{P}(H)$. *Then*

$$\sigma_p(H) \subset \sigma_p(A) \subset \sigma_p(H) \cup \sigma_r^*(H).$$ (1.5)

Proof. The first inclusion in (1.5) is obvious. Let us prove the second one. Assume that $\lambda \in \sigma_p(A)$ and x_0 is the corresponding eigenvector. Since $\mathcal{D}_H \subset \mathcal{D}_A$, we have $((H - \bar{\lambda}I)x, x_0) = (x, (A - \lambda I)x_0) = 0$ for any x from $\mathcal{D}_H$. Hence, $x_0 \perp \Delta_{H-\bar{\lambda}I}$. This means that $\bar{\lambda} \bar{\in} \rho(H)$ and $\bar{\lambda} \bar{\in} \sigma_c(H)$, i.e., $\lambda \bar{\in} \rho^*(H)$ and $\lambda \bar{\in} \sigma_c^*(H)$. But then

$$\lambda \in \sigma_p^*(H) \cup \sigma_r^*(H) = \sigma_p(H) \cup \sigma_r^*(H).$$

∎

Theorem 1.10. *Let* $A \in \mathcal{P}(H)$. *Then*

$$\sigma_c(H) \subset \sigma_c(A) \subset \sigma_c(H) \cup \sigma_r(H).$$ (1.6)

Proof. Let $\lambda \in \sigma_c(H)$. Then there exists a densely defined unbounded operator $B = (H - \lambda I)^{-1}$. Let us show that $\lambda \bar{\in} \sigma_p(A)$. Indeed, assume that $Ax_0 = \lambda x_0$ $(x_0 \neq 0)$. Since $\mathcal{D}_H \subset \mathcal{D}_{A_0}$ and $\bar{\lambda} = \lambda$, we have

$$((H - \lambda I)x, x_0) = (x, (A - \lambda I)x_0) = 0$$

for any vector x from $\mathcal{D}_H$ and, consequently, $x_0 \perp \Delta_{H-\lambda I} (= \mathcal{D}_B)$. This contradicts the assumption that the operator B is densely defined.

Thus, $\lambda \bar{\in} \sigma_p(A)$ and, hence, the operator $(A - \lambda I)^{-1}$ exists. This operator satisfies the relation

$$(A - \lambda I)^{-1}y = (H - \lambda I)^{-1}y \quad (y \in \mathcal{D}_B)$$ (1.7)

and, therefore, it is unbounded and densely defined, i.e., $\lambda \in \sigma_c(A)$. The first inclusion in (1.6) is thus proved.

To prove the second inclusion in (1.6), we note that, by virtue of the inclusions $\sigma_p(H) \subset \sigma_p(A)$ and $\rho(H) \subset \rho(A)$, it follows from $\lambda \in \sigma_c(A)$ that $\lambda \bar{\in} \sigma_p(H)$ and $\lambda \bar{\in} \rho(H)$, i.e., $\lambda \in \sigma_c(H) \cup \sigma_r(H)$.

∎

Remark. If $A \in \mathcal{P}(H)$, then, generally speaking, $\sigma_c(H) \neq \sigma_c(A)$.

Indeed, let H be an Hermitian operator from $\mathcal{H}$ and let $\lambda \in \sigma_c(H)$. Consider the

Hilbert space $\tilde{\mathcal{H}} = \mathcal{H} \oplus \mathcal{H}_1$, where $\mathcal{H}_1$ is a one-dimensional space. We define an operator A on the lineal $\mathfrak{D}_{\tilde{H}} = \mathfrak{D}_H \oplus \mathcal{H}_1$ as follows:

$$A(x + x_1) = Hx + ax_1 \quad (x \in \mathfrak{D}_H, \; x_1 \in \mathcal{H}_1),$$

where $a \neq \lambda$ is a fixed number. It is easy to see that $A \in \mathcal{P}(H)$ and

$$(A - \lambda I)^{-1}(y + y_1) = (H - \lambda I)^{-1}y + \frac{1}{a-\lambda} y_1,$$

where $y \in \Delta_{H-\lambda I}$ and $y_1 \in \mathcal{H}_1$. Hence, $\lambda \in \sigma_c(A)$. However, if we consider H as an operator from $\tilde{\mathcal{H}}$, we get $\lambda \,\bar{\in}\, \sigma_c(H)$ (because, in this case, the operator $(H - \lambda I)^{-1}$ is nondensely defined in $\tilde{\mathcal{H}}$).

Theorem 1.11. *If $A \in \mathcal{P}(H)$, where H is an Hermitian operator with finite defect numbers, then $\sigma_c(A) \subset \mathbb{R}$.*

Proof. Let $\lambda \,\bar{\in}\, \sigma_p(A)$, $\bar{\lambda} \neq \lambda$ and $y = (A - \lambda I)^{-1}x$ $(x \in \Delta_{H-\lambda I})$. Since $\mathcal{H} = \Delta_{H-\lambda I} \oplus \mathfrak{N}_\lambda$, where $\mathfrak{N}_\lambda$ is a defect space of the operator H, we have

$$x = (A - \lambda I)y = x_1 + x_2 \quad (x_1 \in \Delta_{H-\lambda I}, \; x_2 \in \mathfrak{N}_\lambda),$$

where

$$x_1 = (H - \lambda I)y_1, \quad x_2 = (A - \lambda I)y_2, \quad y_1 + y_2 = y.$$

Thus,

$$(A - \lambda I)^{-1}x = (H - \lambda I)^{-1}P_1 x + (A - \lambda I)^{-1}P_2 x,$$

where P_1 and P_2 are orthoprojectors onto $\Delta_{H-\lambda I}$ and $\mathfrak{N}_\lambda$, respectively. Since

$$\|(H - \lambda I)^{-1}\| \leq \frac{1}{|\operatorname{Im}\lambda|}$$

and the operator $(A - \lambda I)^{-1}P_2$ is finite-dimensional, the operator $(A - \lambda I)^{-1}$ is bounded, i.e., $\lambda \,\bar{\in}\, \sigma_c(A)$. ∎

Theorem 1.12. *Assume that* $A \in \mathcal{P}(H)$, *where* H *is an Hermitian operator with deficiency index* (m, n), $\lambda \in \sigma_p(A)$, *and* $\mathcal{H}_\lambda$ *is the corresponding eigensubspace of the operator* A. *Then*

$$\dim \mathcal{H}_\lambda \le \begin{cases} n & (\mathrm{Im}\,\lambda > 0), \\ m & (\mathrm{Im}\,\lambda < 0). \end{cases}$$

Proof. Let $\lambda \in \sigma_p(A)$, $\mathrm{Im}\,\lambda \ne 0$, and let $x \in \mathcal{H}_\lambda$. Then, for any vector $y \in \mathfrak{D}_H$,

$$((H - \bar{\lambda}I)y, x) = (y, (A - \lambda I)x) = 0.$$

Therefore, $x \in \mathfrak{N}_{\bar{\lambda}}$, i.e., $\mathcal{H}_\lambda \subset \mathfrak{N}_{\bar{\lambda}}$. But then, for $\mathrm{Im}\,\lambda > 0$,

$$\dim \mathcal{H}_\lambda \le \dim \mathfrak{N}_{\bar{\lambda}} = \dim \mathfrak{N}_{-i} = n.$$

Similarly, $\dim \mathcal{H}_\lambda \le m$ for $\mathrm{Im}\,\lambda < 0$.

■

1.4. K^r-Operators

A closed operator A acting in a Hilbert space $\mathcal{H}$ is called a quasi-Hermitian operator of rank r (a K^r-operator) if $A_0 = A|_{\mathfrak{S}_A}$ is an Hermitian operator with deficiency index (r, r) $(0 < r < \infty)$ and $\rho(A) \ne \varnothing$.

Examples of K^r-operators are the following: bounded operators with finite-dimensional imaginary components, differential operators generated by a self-adjoint differential expression and nonself-adjoint boundary conditions.

Denote the set of all K^r-operators acting in $\mathcal{H}$ by $\mathcal{K}^r(\mathcal{H})$ or, simply, by $\mathcal{K}^r$.

Proposition 1.13. *Let* $A \in \mathcal{K}^r$. *Then*

$$\dim \mathfrak{D}_A = r \,(\mathrm{mod}\ \mathfrak{S}_A). \tag{1.8}$$

Proof. Assume that $\lambda \in \rho(A)$ $(\mathrm{Im}\,\lambda \ne 0)$, $\mathfrak{N}_A$ is a defect space of the operator A_0, and $x_1, x_2, \ldots, x_r$ is the basis of $\mathfrak{N}_\lambda$. Then the vectors $y_k = R_\lambda x_k$ $(k = \overline{1, r})$ are linearly independent modulo $\mathfrak{S}_A$; at the same time, any $r + 1$ vectors from $\mathfrak{D}_A$ are linearly dependent modulo $\mathfrak{S}_A$.

■

Proposition 1.14. *Suppose that a closed operator* A *satisfies condition (1.8) and* $\{\lambda, \bar{\lambda}\} \subset \rho(A)$ *(Im $\lambda \neq 0$). Then* $A \in \mathcal{K}^r$.

Proof. To prove the proposition, it suffices to show that

$$\dim \mathfrak{N}_\lambda = \dim \mathfrak{N}_{\bar{\lambda}} = r,$$

where $\mathfrak{N}_z$ is the defect space of the Hermitian part A_0 of the operator A.

In $\mathfrak{D}_A$, we take r vectors $x_1, x_2, \ldots, x_r$ linearly independent modulo $\mathfrak{G}_A$. Then the vectors $y_k = (A - \lambda I) x_k$ $(k = \overline{1, r})$ are linearly independent modulo $\mathfrak{M}_\lambda = (A - \lambda I)\mathfrak{G}_A$. Consequently, $\dim \mathfrak{N}_\lambda \geq r$.

Conversely, if $y_1, y_2, \ldots, y_m$ is a basis of the subspace $\mathfrak{N}_\lambda$, then the vectors $x_k = R_\lambda y_k$ $(k = \overline{1, m})$ are linearly independent modulo $\mathfrak{G}_{\bar{A}}$. But, in this case, $m \leq r$ and, hence, $\dim \mathfrak{N}_\lambda = r$.

Similarly, one can establish that $\dim \mathfrak{N}_{\bar{\lambda}} = r$.

∎

Theorem 1.15. *If an N-operator* $A \in \mathcal{K}^r$, *then* $A^* \in \mathcal{K}^r$.

Proof. Indeed, since $(A^*)_0 = A_0$, the deficiency index of the Hermitian part $(A^*)_0$ of the operator A^* is (r, r). Further, $\rho(A^*) = \rho^*(A) \neq \varnothing$ and, therefore, $A^* \in \mathcal{K}^r$.

∎

Theorem 1.16. *If* $A \in \mathcal{K}^r$, *then*

$$\sigma(A) \backslash \mathbb{R} \subset \sigma_p(A). \tag{1.9}$$

Proof. Assume that $\lambda \in \sigma(A)$ and Im$\lambda \neq 0$. Since $\sigma_c(A) \subset \mathbb{R}$ in view of Theorem 1.11, we have $\lambda \overline{\in} \sigma_c(A)$. Let $\lambda \in \sigma_r(A)$. Then there exists a nonzero vector $x \perp \Delta_{A-\lambda I}$. Since $\mathcal{H} = \mathfrak{M}_\lambda \oplus \mathfrak{N}_\lambda$ and $\mathfrak{M}_\lambda \subset \Delta_{A-\lambda I}$, we have $x \in \mathfrak{N}_\lambda$. Consequently, the dimensionality m of the lineal $\Delta_{A-\lambda I}$ modulo $\mathfrak{M}_\lambda$ is less than r, which is impossible. Indeed, if the vectors $x_1, x_2, \ldots, x_r$ are linearly independent modulo $\mathfrak{G}_A = \mathfrak{D}_{A_0}$, then the vectors $y_k = (A - \lambda I) x_k$ $(k = \overline{1, r})$ are linearly independent modulo $\mathfrak{M}_\lambda$.

Thus, $\lambda \overline{\in} \sigma_r(A)$, i.e., $\lambda \in \sigma_p(A)$.

∎

1.5. Classification of K^r-Operators

A K^r-operator A acting in $\mathcal{H}$ is called

a K^r_I-operator if $\mathfrak{D}_{A^*} = \mathfrak{D}_A$;

a K^r_{II}-operator if $\overline{\mathfrak{G}}_A = \mathcal{H}$;

a K^r_{III}-operator if $\overline{\mathfrak{D}}_A = \mathcal{H}$, $\mathfrak{D}_{A^*} \neq \mathfrak{D}_A$, $\overline{\mathfrak{G}}_A \neq \mathcal{H}$;

a K^r_{IV}-operator if $\overline{\mathfrak{D}}_A \neq \mathcal{H}$.

We denote the corresponding operator classes by $\mathcal{K}^r_I$, $\mathcal{K}^r_{II}$, $\mathcal{K}^r_{III}$, and $\mathcal{K}^r_{IV}$, respectively.

The class $\mathcal{K}^r_I$ contains, in particular, all bounded operators with r-dimensional imaginary components.

The class $\mathcal{K}^r_{II}$ contains nonself-adjoint extensions of symmetric operators with the deficiency index (r, r) and, in particular, differential operators generated by self-adjoint differential expressions and nonself-adjoint boundary conditions.

The class $\mathcal{K}^r_{III}$ contains nonself-adjoint extensions A of Hermitian (but nonsymmetric) operators with the deficiency index (r, r) such that $\mathfrak{D}_{A^*} \neq \mathfrak{D}_A$.

The class $\mathcal{K}^r_{IV}$ contains extensions A of Hermitian operators with nondense domains of definition and the deficiency index (r, r) such that $\rho(A) \neq \varnothing$ (an example of this situation is given in the next subsection).

Proposition 1.17. *Let* $A \in \mathcal{K}^r_{II}$. *Then* $\mathfrak{G}_A = \mathfrak{D}_A \cap \mathfrak{D}_{A^*}$ *and*

$$A_0 \subset A \subset A_0^*, \qquad A_0 \subset A^* \subset A_0^*. \tag{1.10}$$

Proof. If $y \in \mathfrak{G}_A$, then, for any $x \in \mathfrak{D}_A$,

$$(Ax, y) = (x, A_0 y) \qquad (A_0 = A \mid \mathfrak{G}_A)$$

and, hence, $A^* y = A_0 y$. Thus, $A_0 \subset A^*$. Moreover, $A_0 \subset A$. Since in the considered case A and A_0 are N-operators, we have $A \subset A_0^*$ and $A^* \subset A_0^*$. These inclusions prove relations (1.10).

Let $x \in \mathfrak{D}_A \cap \mathfrak{D}_{A^*}$. Then, in view of (1.10),

$$(A^* x, y) = (A_0^* x, y) = (x, A_0 y) = (Ax, y)$$

for any $y \in \mathfrak{D}_{A_0}$ and, hence, $A^* x = Ax$. Thus,

$$\mathcal{D}_A \cap \mathcal{D}_{A^*} \subset \text{Ker}\,(A - A^*) = \mathcal{G}_A.$$

Since the inverse inclusion is obvious, we have $\mathcal{G}_A = \mathcal{D}_A \cap \mathcal{D}_{A^*}$.

∎

According to the proposition just proved, we come to the conclusion that the classes $\mathcal{K}_I^r$ and $\check{\mathcal{K}}_{II}^r$ have no common elements.

1.6. An Example of K_{IV}^r-Operator

Assume that H is a symmetric operator with the deficiency index (r, r) $(0 < r < \infty)$ acting in $\mathcal{H}_1$ and let $\mathcal{H}_2$ be an r-dimensional space. In the space $\mathcal{H} = \mathcal{H}_1 \oplus \mathcal{H}_2$, we define an operator A in the following way:

$$\mathcal{D}_A = \mathcal{D}_{H^*} = \mathcal{D}_H \dotplus \mathfrak{N}_i \dotplus \mathfrak{N}_{-i}.$$

If in this case $x \in \mathcal{D}_A$, i.e.,

$$x = x_0 + x_i + x_{-i} \quad (x_0 \in \mathcal{D}_H, \ x_\lambda \in \mathfrak{N}_\lambda),$$

then we set

$$Ax = H^* x + K x_i + L x_{-i}, \tag{1.11}$$

where linear operators K and L map, respectively, $\mathfrak{N}_i$ and $\mathfrak{N}_{-i}$ into $\mathcal{H}_2$.

Relation (1.11) implies that $H^* = PA$, where P is the orthoprojector onto $\mathcal{H}_1$ in $\mathcal{H}$.

Assume that the operator L is invertible. Let us show that $i \in \rho(A)$. For this purpose, we consider an equation

$$Ax - ix = y \quad (y \in \mathcal{H}) \tag{1.12}$$

and show that it can be uniquely solved for x.

Indeed, since

$$\mathcal{H} = \mathfrak{M}_i \oplus \mathfrak{N}_i \oplus \mathcal{H}_2 \quad (\mathfrak{M}_i = \Delta_{H-iI});$$

we have $y = y_1 + y_2 + y_3$, where $y_1 = P_1 y \in \mathfrak{M}_i$, $y_2 = P_2 y \in \mathfrak{N}_i$, and $y_3 = P_3 y \in \mathcal{H}_2$ (P_1, P_2, and P_3 are, respectively, the orthoprojectors onto the subspaces $\mathfrak{M}_i$, $\mathfrak{N}_i$, and $\mathcal{H}_2$ in $\mathcal{H}$. Therefore, in view of (1.11), we can rewrite eqn. (1.12) as follows:

$$Hx_0 - ix_i + ix_{-i} + Kx_i + Lx_{-i} - ix_0 - ix_i - ix_{-i} = y_1 + y_2 + y_3,$$

whence

$$(H - il)x_0 = y_1, \quad -2ix_i = y_2, \quad Kx_i + Lx_{-i} = y_3.$$

Consequently,

$$x_0 = (H - il)^{-1} y_1, \quad x_i = \frac{i}{2} y_2, \quad x_{-i} = L^{-1} y_3 - \frac{i}{2} L^{-1} K y_2,$$

which proves the unique solvability of eqn. (1.12). Moreover,

$$(A - il)^{-1} y = (H - il)^{-1} P_1 y + \frac{i}{2} P_2 y + L^{-1} P_3 y - \frac{i}{2} L^{-1} K P_2 y.$$

Therefore, the operator $(A - il)^{-1}$ is defined and bounded in the entire space, i.e., $i \in \rho(A)$.

Similarly, we can show that in the case where the operator K is invertible, $-i \in \rho(A)$.

In addition, one can easily check that $\mathcal{G}_A = \mathcal{D}_H$. But then the defect space of the operator $A_0 (= H)$ with $\bar{\lambda} \neq \lambda$ (in the space $\mathcal{H}$) is $\mathfrak{N}_\lambda \oplus \mathcal{H}_2$. Thus, A_0 is a Hermitian operator with the deficiency index $(2r, 2r)$, i.e., A is a K_{IV}^{2r}-operator.

Remarks. The notion of regular extensions of Hermitian operators was introduced and studied for the first time by the author in [22].

2. Auxiliary Operators and Their Application

In this section, we consider some auxiliary operators that are important for what follows.

2.1. Operator $T_{\mu\lambda}$

Assume that H is a closed Hermitian operator acting in the Hilbert space $\mathcal{H}$, $A \in \mathcal{P}(H)$, and $\mathfrak{N}_A(\lambda) = \Delta_A(\lambda) \ominus \Delta_H(\lambda)$, where $\Delta_T(\lambda) = \Delta_{T-\lambda l}$ ($T \in \{A; H\}$). Clearly,

$\mathfrak{N}_A(\lambda) \subset \mathfrak{N}_\lambda$, where $\mathfrak{N}_\lambda$ is the defect space of the operator H, and if $\lambda \in \rho(A)$, then $\mathfrak{N}_A(\lambda) = \mathfrak{N}_\lambda$.

Let $\lambda \bar{\in} \sigma_p(A)$. We introduce an operator $T_{\mu\lambda}: \Delta_A(\lambda) \to \Delta_A(\mu)$ defined by the equality

$$T_{\mu\lambda} = (A - \mu I)(A - \lambda I)^{-1}. \tag{2.1}$$

Theorem 2.1. *If $A \in \mathcal{P}(H)$ and $\lambda \bar{\in} \sigma_p(A)$, then*

$$T_{\mu\lambda}: \Delta_A(\lambda) \cap \mathfrak{N}_{\bar\mu} \to \mathfrak{N}_{\bar\lambda}. \tag{2.2}$$

In particular, if $\lambda \in \rho(A)$, then

$$T_{\mu\lambda}: \mathfrak{N}_{\bar\mu} \to \mathfrak{N}_{\bar\lambda}, \quad T_{\mu\lambda}^*: \mathfrak{N}_\mu \to \mathfrak{N}_\lambda. \tag{2.3}$$

Proof. Let $f \in \Delta_A(\lambda) \cap \mathfrak{N}_{\bar\mu}$, i.e., $f = (A - \lambda I)\varphi$ ($\varphi \in \mathfrak{D}_A$), and let

$$(f, (H - \bar\mu I)x) = 0 \quad (\forall\, x \in \mathfrak{D}_H). \tag{2.4}$$

Since $f = (A - \lambda I)\varphi$, it follows from (2.1) that $T_{\mu\lambda}f = (A - \mu I)\varphi = f + (\lambda - \mu)\varphi$. Therefore, for $x \in \mathfrak{D}_H$,

$$(T_{\mu\lambda}f, (H - \bar\lambda I)x) = (f, (H - \bar\lambda I)x) + (\lambda - \mu)(\varphi, (H - \bar\lambda I)x), \tag{2.5}$$

where, by virtue of (2.4),

$$(f, (H - \bar\lambda I)x) = (\mu - \lambda)(f, x). \tag{2.6}$$

Furthermore, since $x \in \mathfrak{D}_A$, we have

$$(\varphi, (H - \bar\lambda I)x) = (\varphi, Hx) - \lambda(\varphi, x) = (A\varphi, x) - \lambda(\varphi, x) = (f, x). \tag{2.7}$$

But then, by virtue of (2.6) and (2.7), for any $x \in \mathfrak{D}_A$, the right-hand side of relation (2.5) is equal to zero. Hence, $T_{\mu\lambda}f \in \mathfrak{N}_{\bar\lambda}$ and we obtain (2.2).

The first relation in (2.3) follows from (2.2) for $\lambda \in \rho(A)$. To prove the second relation, we note that if $f \in \mathfrak{N}_\mu$ and $x \in \mathfrak{D}_H$, then

$$(T_{\mu\lambda}^*f, (H - \lambda I)x) = (f, (H - \mu I)x) = 0$$

and, hence, $T_{\mu\lambda}^*f \in \mathfrak{N}_\lambda$. ∎

Corollary 2.2. *Assume that* $\{\mu, \lambda\} \subset \rho(A)$. *Then*

$$\dim \mathfrak{N}_\mu = \dim \mathfrak{N}_\lambda, \quad \dim \mathfrak{N}_{\bar{\mu}} = \dim \mathfrak{N}_{\bar{\lambda}}. \tag{2.8}$$

Proof. If $\{\mu, \lambda\} \subset \rho(A)$, then the operator $T_{\mu\lambda}$ is completely invertible (i.e., $T_{\mu\lambda}$ is injective and surjective) and $T_{\mu\lambda}^{-1} = T_{\lambda\mu}$. Therefore, we have

$$T_{\mu\lambda}: \mathfrak{N}_{\bar{\mu}} \to \mathfrak{N}_{\bar{\lambda}}, \quad T_{\mu\lambda}^{-1}: \mathfrak{N}_{\bar{\lambda}} \to \mathfrak{N}_{\bar{\mu}},$$

and, hence, the second equality in (2.8) is true.

Similarly, by using the fact that

$$T_{\mu\lambda}^*: \mathfrak{N}_\mu \to \mathfrak{N}_\lambda, \quad (T_{\mu\lambda}^*)^{-1} = T_{\lambda\mu}^*: \mathfrak{N}_\lambda \to \mathfrak{N}_\mu,$$

we establish the first equality in (2.8).

2.2. Values of Resolvents on Defect Spaces

Theorem 2.3. *Suppose that* $A \in \mathcal{P}(H)$, $\lambda \in \rho(A)$, *and* $\mathfrak{N}_z$ ($z \in \{\mu, \bar{\mu}, \lambda, \bar{\lambda}\}$) *are the defect spaces of the operator* H. *Then the elements* $R_\lambda g_{\bar{\mu}}$ ($g_{\bar{\mu}} \in \mathfrak{N}_{\bar{\mu}}$) *and* $R_\lambda^* g_\mu$ ($g_\mu \in \mathfrak{N}_\mu$) *with* $\mu \neq \lambda$ *can be represented in the form*

$$R_\lambda g_{\bar{\mu}} = \frac{g_{\bar{\lambda}} - g_{\bar{\mu}}}{\lambda - \mu}, \quad R_\lambda^* g_\mu = \frac{g_\lambda - g_\mu}{\bar{\lambda} - \bar{\mu}}, \tag{2.9}$$

where $g_{\bar{\lambda}}$ *and* g_λ *are elements of* $\mathfrak{N}_{\bar{\lambda}}$ *and* $\mathfrak{N}_\lambda$, *respectively.*

Proof. Since $T_{\mu\lambda} = I + (\lambda - \mu) R_\lambda$, by virtue of Theorem 2.1, we have

$$T_{\mu\lambda} g_{\bar{\mu}} = g_{\bar{\mu}} + (\lambda - \mu) R_\lambda g_{\bar{\mu}} = g_{\bar{\lambda}} \in \mathfrak{N}_{\bar{\lambda}}.$$

The first equality in (2.9) is proved.

Similarly, by using the equality $T_{\mu\lambda}^* = I + (\bar{\lambda} - \bar{\mu}) R_\lambda^*$ and Theorem 2.1, we can conclude that the second equality in (2.9) is also true. ∎

Relations (2.9) are especially suitable in the case where $\overline{\mathcal{G}_A} = \mathcal{H}$ (in particular, if the operator H is symmetric). In this case, the elements $g_{\bar{\lambda}}$ and g_λ are uniquely determined (for $\lambda \in \sigma_p(A)$) by the following conditions:

$$g_{\bar{\lambda}} - g_{\bar{\mu}} \in \mathfrak{D}_A, \quad g_\lambda - g_\mu \in \mathfrak{D}_{A^*}.$$

Indeed, if, for example, $g_{\bar{\lambda}} - g_{\bar{\mu}} \in \mathfrak{D}_A$ and $\tilde{g}_{\bar{\lambda}} - g_{\bar{\mu}} \in \mathfrak{D}_A$, then

$$x_{\bar{\lambda}} = g_{\bar{\lambda}} - \tilde{g}_{\bar{\lambda}} \in \mathfrak{D}_A \cap \mathfrak{N}_{\bar{\lambda}}.$$

Since, according to Corollary 1.6, $\mathfrak{D}_A \cap \mathfrak{N}_{\bar{\lambda}} = \{0\}$, we have $x_{\bar{\lambda}} = 0$, i.e., $\tilde{g}_{\bar{\lambda}} = g_{\bar{\lambda}}$.

Note that, in many cases (e.g., for a broad class of differential operators), the established property of a resolvent significantly simplifies the calculation of elements $R_\lambda\, g_{\bar{\mu}}$ and, hence, of characteristic functions of a K_{II}^r-operator (see Chapter 2).

2.3. Operators B_α and $\tilde{B}_\alpha$

Assume that $\rho(A) \neq \varnothing$ and $\alpha \in \rho(A)$ (Im $\alpha \neq 0$). Consider the following bounded self-adjoint operators:

$$B_\alpha = iR_\alpha - iR_\alpha^* + 2\,\mathrm{Im}\,\alpha\, R_\alpha^* R_\alpha, \tag{2.10}$$

$$\tilde{B}_\alpha = iR_\alpha - iR_\alpha^* + 2\,\mathrm{Im}\,\alpha\, R_\alpha R_\alpha^*, \tag{2.11}$$

where $R_\alpha = (A - \alpha I)^{-1}$. These operators can be represented as

$$B_\alpha = \frac{1}{2\,\mathrm{Im}\,\alpha}\,[T_\alpha^* T_\alpha - I], \quad \tilde{B}_\alpha = \frac{1}{2\,\mathrm{Im}\,\alpha}\,[T_\alpha T_\alpha^* - I], \tag{2.12}$$

where

$$T_\alpha = (A - \bar{\alpha}I)(A - \alpha I)^{-1} \quad (= T_{\bar{\alpha}\alpha}). \tag{2.13}$$

Further, it is obvious that

$$(B_\alpha(A - \alpha I)x,\ (A - \alpha I)y) = i(x,\, Ay) - i(Ax,\, y) \tag{2.14}$$

for all vectors x and y from $\mathfrak{D}_A$. By analogy, if $\{x, y\} \subset \mathfrak{D}_{A^*}$, then

$$(\tilde{B}_\alpha(A^* - \bar{\alpha}I)x,\ (A^* - \bar{\alpha}I)y) = i(A^*x,\, y) - i(x,\, A^*y). \tag{2.15}$$

Theorem 2.4. *Suppose that* $A \in \mathcal{P}(H)$ *and* $\alpha \in \rho(A)$ *(*$\operatorname{Im} \alpha \neq 0$*). Then*

$$B_\alpha : \mathcal{H} \to \mathfrak{N}_\alpha, \qquad \tilde{B}_\alpha : \mathcal{H} \to \mathfrak{N}_{\bar{\alpha}}, \tag{2.16}$$

where $\mathfrak{N}_\alpha$ *and* $\mathfrak{N}_{\bar{\alpha}}$ *are defect spaces of the operator* H.

Proof. Indeed, assume that f and g are arbitrary elements from $\mathcal{H}$ and $\mathfrak{M}_\alpha = \Delta_H(\alpha)$, respectively. Then there exist vectors $x \in \mathcal{D}_A$ and $y \in \mathcal{D}_H$ such that

$$f = (A - \alpha I)x, \qquad g = (H - \alpha I)y = (A - \alpha I)y.$$

By virtue of (2.14) and the fact that $y \in \mathcal{G}_A$, we have

$$(B_\alpha f, g) = i(x, Ay) - i(Ax, y) = 0 \qquad (\forall \, y \in \Delta_H(\alpha)).$$

Therefore, $B_\alpha f \in \mathfrak{N}_\alpha$, which proves the first relation in (2.16). Moreover, it is obvious that $B_\alpha f = 0$ for $g \in \Delta_H(\alpha)$.

To prove the second relation in (2.16), we use the second equality in (2.12) (similarly, one can use the first equality of (2.12) to prove the first relation in (2.16), but the proof presented above is simpler).

Let $g_{\bar{\alpha}} \in \mathfrak{N}_{\bar{\alpha}}$. Since $T_\alpha = T_{\bar{\alpha}\alpha}$, we have $g_\alpha = T_\alpha^* g_{\bar{\alpha}} \in \mathfrak{N}_\alpha$ by virtue of Theorem 2.1. Then it follows from Theorem 2.1 that $T_\alpha T_\alpha^* g_{\bar{\alpha}} = T_\alpha g_\alpha \in \mathfrak{N}_{\bar{\alpha}}$. Consequently, by virtue of (2.12), the defect space $\mathfrak{N}_{\bar{\alpha}}$ is invariant under the action of the operator $\tilde{B}_\alpha$.

It remains to show that the operator $\tilde{B}_\alpha$ is equal to zero on the subspace $\mathfrak{M}_{\bar{\alpha}}$.

Let $h \in \mathfrak{M}_{\bar{\alpha}}$. Then $h = (A - \bar{\alpha} I)x$, where $x \in \mathcal{D}_H$. We have established above that the operator B_α is equal to zero on the subspace $\mathfrak{M}_\alpha$. Hence, $B_\alpha (A - \alpha I)x = 0$. Since

$$B_\alpha = iR_\alpha - iR_\alpha^* T_\alpha, \qquad T_\alpha (A - \alpha I)x = A - \bar{\alpha} I, \tag{2.17}$$

we have

$$0 = B_\alpha (A - \alpha I)x = ix - iR_\alpha^* (A - \bar{\alpha} I)x = ix - iR_\alpha^* h$$

and, thus, $R_\alpha^* h = x$. But then

$$\tilde{B}_\alpha h = iR_\alpha h - ix + 2\operatorname{Im}\alpha R_\alpha x$$

$$= iR_\alpha\,[A - \bar{\alpha}I)x - (A - \alpha I)x + (\bar{\alpha} - \alpha)x] = 0.$$

∎

Theorem 2.5. *Let* $\alpha \in \rho(A)$ $(\bar{\alpha} \neq \alpha)$. *Then*

$$\mathrm{Ker}\,B_\alpha = \Delta_{A_0}(\alpha), \qquad \mathrm{Ker}\,\tilde{B}_\alpha = \Delta_{A_0}(\bar{\alpha}), \tag{2.18}$$

where the operator A_0 is the Hermitian part of the operator A.

Proof. Since $A \in \mathcal{P}(A_0)$, it follows from Theorem 2.4 that the operator B_α $(\tilde{B}_\alpha)$ is equal to zero on the subspace $\mathfrak{M}_\alpha = \Delta_{A_0}(\alpha)$ $(\mathfrak{M}_{\bar{\alpha}} = \Delta_{A_0}(\bar{\alpha}))$. Therefore,

$$\Delta_{A_0}(\alpha) \subset \mathrm{Ker}\,B_\alpha, \qquad \Delta_{A_0}(\bar{\alpha}) \subset \mathrm{Ker}\,\tilde{B}_\alpha.$$

Thus, to prove the theorem, it remains to establish the inverse inclusions.

Let $x \in \mathrm{Ker}\,B_\alpha$ and $y = R_\alpha x$. Since $B_\alpha x = 0$, equality (2.10) yields

$$y = R_\alpha^* x + (\alpha - \bar{\alpha})\,R_\alpha^* y. \tag{2.19}$$

But then, for any φ from $\mathfrak{D}_A$,

$$(y, A\varphi) = (x, R_\alpha A\varphi) + (\alpha - \bar{\alpha})(y, R_\alpha A\varphi), \tag{2.20}$$

where, obviously,

$$R_\alpha A\varphi = \varphi + \alpha R_\alpha \varphi. \tag{2.21}$$

Therefore, relations (2.19)–(2.21) yield

$$(y, A\varphi) = (x, \varphi) + \bar{\alpha}\,(x, R_\alpha \varphi) + (\alpha - \bar{\alpha})(y, \varphi) + (\alpha - \bar{\alpha})\,\bar{\alpha}\,(y, R_\alpha \varphi)$$

$$= (x, \varphi) + (\alpha - \bar{\alpha})(y, \varphi) + \bar{\alpha}\,[(R_\alpha^* x, \varphi) + (\alpha - \bar{\alpha})(R_\alpha^* y, \varphi)]$$

$$= (x, \varphi) + (\alpha - \bar{\alpha})(y, \varphi) + \bar{\alpha}\,(y, \varphi) = (x + \alpha y, \varphi). \tag{2.22}$$

Since $y = R_\alpha x$, we have $x + \alpha y = Ay$ and, thus, equality (2.22) takes the form

$$(Ay, \varphi) = (y, A\varphi) \quad (\forall\,\varphi \in \mathfrak{D}_A).$$

Consequently, $y \in \mathfrak{I}_A = \mathfrak{D}_{A_0}$. But this implies that

$$x = (A - \alpha I)y \in \Delta_{A_0}(\alpha),$$

i.e., $\operatorname{Ker} B_\alpha \subset \Delta_{A_0}(\alpha)$. The first equality in (2.18) is proved.

Assume that $x \in \operatorname{Ker} \tilde{B}_\alpha$ and $y = R_\alpha^* x$. Then equality (2.11) yields $y = R_\alpha x + (\bar{\alpha} - \alpha) R_\alpha y$. By applying the operator $A - \alpha I$ to both the sides of this equality, we get $Ay = x + \bar{\alpha} y$. Therefore, in view of equality (2.21), for any $\varphi \in \mathcal{D}_A$, we have

$$(y, A\varphi) = (x, R_\alpha A\varphi) = (x, \varphi) + \bar{\alpha}(x, R_\alpha \varphi)$$

$$= (x, \varphi) + \bar{\alpha}(y, \varphi) = (x + \bar{\alpha} y, \varphi) = (Ay, \varphi)$$

and, thus, $y \in \mathcal{G}_A$. Consequently, by virtue of the equality $Ay = x + \bar{\alpha} y$, we obtain

$$x = (A - \bar{\alpha} I)y = (A_0 - \bar{\alpha} I)y \in \Delta_{A_0}(\bar{\alpha}),$$

i.e., $\operatorname{Ker} \tilde{B}_\alpha \subset \Delta_{A_0}(\bar{\alpha})$, which proves the second equality in (2.18).

∎

It follows from Theorem 2.5 that the closures of the lineals $B_\alpha \mathcal{H}$ and $\tilde{B}_\alpha \mathcal{H}$ coincide with $\mathfrak{N}_\alpha$ and $\mathfrak{N}_{\bar{\alpha}}$, respectively:

$$\overline{B_\alpha \mathcal{H}} = \mathfrak{N}_\alpha, \qquad \overline{\tilde{B}_\alpha \mathcal{H}} = \mathfrak{N}_{\bar{\alpha}}. \tag{2.23}$$

2.4. Criterion of the Equality $\mathcal{D}_A = \mathcal{D}_{A^*}$

Theorem 2.6. *Assume that* $A \in \mathcal{P}(H)$ *and, for some nonreal* α *such that* $\{\alpha, \bar{\alpha}\} \subset \rho(A)$, *the following inclusions are true:*

$$\mathfrak{N}_\alpha \subset \mathcal{D}_A, \qquad \mathfrak{N}_{\bar{\alpha}} \subset \mathcal{D}_{A^*}, \tag{2.24}$$

where $\mathfrak{N}_\alpha$ *and* $\mathfrak{N}_{\bar{\alpha}}$ *are the defect spaces of the operator* H. *Then*

$$\mathcal{D}_A = \mathcal{D}_{A^*}. \tag{2.25}$$

Proof. If $\{\alpha, \bar{\alpha}\} \subset \rho(A)$, then in view of equality (2.17), an arbitrary vector x from $\mathcal{D}_{A^*}$ can be represented in the form

$$x = iR_\alpha^* T_\alpha y = iR_\alpha y - B_\alpha y, \tag{2.26}$$

where y is a vector from $\mathcal{H}$. Since, by virtue of the Theorem 2.4 and condition (2.24), $B_\alpha y \in \mathfrak{N}_\alpha \subset \mathfrak{D}_A$, we have $x \in \mathfrak{D}_A$ by virtue of equality (2.26). Thus, $\mathfrak{D}_{A^*} \subset \mathfrak{D}_A$.

Similarly, by using the equality $\tilde{B}_\alpha = iR_\alpha T_\alpha^* - iR_\alpha^*$ and the inclusion $\mathfrak{N}_{\bar\alpha} \subset \mathfrak{D}_{A^*}$, we can conclude that $\mathfrak{D}_A \subset \mathfrak{D}_{A^*}$.

∎

Note that the statement inverse to Theorem 2.6 is not always true. Indeed, if $\mathfrak{D}_A = \mathfrak{D}_{A^*}$ and $\{\alpha, \bar\alpha\} \subset \rho(A)$, then inclusions (2.24), generally speaking, are not true. For example, let H be an unbounded self-adjoint operator acting in $\mathcal{H}$ and let $\{-1, 1\} \subset \rho(H)$. Consider the operator $A = iH$. By definition, we have $\mathfrak{D}_A = \mathfrak{D}_{A^*}$, $\{-1, 1\} \subset \rho(A)$, and the Hermitian domain $\mathcal{G}_A = \{0\}$. Therefore, $\mathfrak{N}_{-i} = \mathfrak{N}_i = \mathcal{H}$, where $\mathfrak{N}_z$ is the defect space of A_0. Thus, in the case under consideration, $\mathfrak{D}_A = \mathfrak{D}_{A^*}$, $\mathfrak{N}_{-i} \not\subset \mathfrak{D}_A$, and $\mathfrak{N}_i \not\subset \mathfrak{D}_A$.

However, in the case of K^r-operators, the inverse statement is true and can be strengthened.

Theorem 2.7. *Let $A \in K^r(\mathcal{H})$ and let $\mathfrak{N}_\alpha$ be the defect space of the Hermitian part A_0 of the operator A. If $\mathfrak{N}_\alpha \subset \mathfrak{D}_A$ for some $\alpha \in \rho(A)$ (Im $\alpha \neq 0$), then $\mathfrak{D}_{A^*} = \mathfrak{D}_A$. Conversely, if $\mathfrak{D}_{A^*} = \mathfrak{D}_A$, then the inclusion $\mathfrak{N}_\alpha \subset \mathfrak{D}_A$ is true for any nonreal α such that $\alpha \in \rho(A)$.*

Proof. If $\mathfrak{N}_\alpha \subset \mathfrak{D}_A$ ($\alpha \in \rho(A)$, Im $\alpha \neq 0$), then, by analogy with the proof of Theorem 2.6, we establish that $\mathfrak{D}_{A^*} \subset \mathfrak{D}_A$. On the other hand, the lineals $\mathfrak{D}_A$ and $\mathfrak{D}_{A^*}$ contain exactly r ($r < \infty$) vectors which are linearly independent modulo $\mathfrak{D}_{A_0}$. Thus, obviously, $\mathfrak{D}_{A^*} = \mathfrak{D}_A$.

Conversely, let $\mathfrak{D}_A = \mathfrak{D}_{A^*}$ and let $\alpha \in \rho(A)$ (Im $\alpha \neq 0$). By virtue of (2.23), the closure of the lineal $B_\alpha \mathcal{H}$ coincides with $\mathfrak{N}_\alpha$ and, consequently, $B_\alpha \mathcal{H} = \mathfrak{N}_\alpha$ (since dim $\mathfrak{N}_\alpha < \infty$). Therefore, if $y \in \mathfrak{N}_\alpha$, then, for some $x \in \mathcal{H}$, we have

$$y = B_\alpha x = iR_\alpha x - iR_\alpha^* x + 2 \operatorname{Im} \alpha R_\alpha^* R_\alpha x \in \mathfrak{D}_A,$$

i.e., $\mathfrak{N}_\alpha \subset \mathfrak{D}_A$.

∎

2.5. Analog of the von Neumann Formulas

Theorem 2.8. *Suppose that $A \in \mathcal{P}(H)$, $\lambda \bar{\in} \sigma_p(A)$, and* $\operatorname{Im} \lambda \neq 0$. *Then any vector x from $\mathfrak{D}_A$ can be represented in the form*

$$x = x_0 + x_\lambda + \Phi x_\lambda \quad (x_0 \in \mathfrak{D}_H, \; x_\lambda \in \mathfrak{D}_\Phi), \tag{2.27}$$

where Φ is a linear operator from $\mathfrak{N}_\lambda$ to $\mathfrak{N}_{\bar{\lambda}}$ such that $-1 \bar{\in} \sigma_p(\Phi)$. Furthermore, the operator A acts according to the formula

$$A(x_0 + x_\lambda + \Phi x_\lambda) = H x_0 + \bar{\lambda} x_\lambda + \lambda \Phi x_\lambda \tag{2.28}$$

and the lineals $\mathfrak{D}_H$ and $(\Phi + I)\mathfrak{D}_\Phi$ are linearly independent.

Proof. Let $y \in \Delta_A(\lambda)$. Then $y = y_0 + y_1$, where $y_0 \in \Delta_H(\lambda)$, $y_1 \in \mathfrak{N}_\lambda$, and, hence,

$$y_0 = (H - \lambda I) x_0 = (A - \lambda I) x_0 \quad (x_0 \in \mathfrak{D}_H),$$

$$y_1 = (A - \lambda I) x_1 \quad (x_1 \in \mathfrak{D}_A).$$

But then

$$y = (A - \lambda I) x = (A - \lambda I)(x_0 + x_1) \quad (x \in \mathfrak{D}_A).$$

Since $\lambda \bar{\in} \sigma_p(A)$, we have $x = x_0 + x_1$.

Let $T_\lambda = (A - \bar{\lambda} I)(A - \lambda I)^{-1} \; (= T_{\bar{\lambda}\lambda})$. Then

$$T_\lambda y = (A - \bar{\lambda} I) x \quad (y = (A - \lambda I) x).$$

These relations yield

$$x = \frac{1}{\lambda - \bar{\lambda}} (T_\lambda y - y), \quad A x = \frac{1}{\lambda - \bar{\lambda}} (\lambda T_\lambda y - \bar{\lambda} y). \tag{2.29}$$

Since $y = y_0 + y_1$, we have

$$x = \frac{1}{\lambda - \bar{\lambda}} (T_\lambda y_0 - y_0) + \frac{1}{\lambda - \bar{\lambda}} y_1 + \frac{1}{\lambda - \bar{\lambda}} T_\lambda y_1 \tag{2.30}$$

In this case, $y_0 = (H - \lambda I) x_0 \; (x_0 \in \mathfrak{D}_H)$, $y_1 \in \mathfrak{N}_\lambda$, and, consequently,

$$\frac{1}{\lambda - \bar{\lambda}}(T_\lambda y_0 - y_0) = \frac{1}{\lambda - \bar{\lambda}}[(H - \bar{\lambda}I)x_0 - (H - \lambda I)x_0] = x_0, \tag{2.31}$$

$$\frac{1}{\bar{\lambda} - \lambda} y_1 = x_\lambda \in \mathfrak{N}_\lambda, \quad \frac{1}{\lambda - \bar{\lambda}}T_\lambda y_1 = -T_\lambda x_\lambda = \Phi x_\lambda, \tag{2.32}$$

where $\Phi = -T_\lambda|_{\mathfrak{N}_\lambda \cap \mathfrak{D}_{T_\lambda}}$. By virtue of Theorem 2.1, Φ is a linear operator acting from $\mathfrak{N}_\lambda$ to $\mathfrak{N}_{\bar{\lambda}}$.

Thus, by virtue of (2.31) and (2.32), the vector x can be represented in the form (2.27).

Taking the second equality in (2.29) into account, we can represent the vector Ax as

$$Ax = \frac{1}{\lambda - \bar{\lambda}}(\lambda T_\lambda y_0 - \bar{\lambda} y_0) + \frac{\bar{\lambda}}{\bar{\lambda} - \lambda}y_1 + \frac{\lambda}{\lambda - \bar{\lambda}}T_\lambda y_1, \tag{2.33}$$

where

$$\lambda T_\lambda y_0 - \bar{\lambda} y_0 = \lambda(H - \bar{\lambda}I)x_0 - \bar{\lambda}(H - \lambda I)x_0 = (\lambda - \bar{\lambda})Hx_0 \tag{2.34}$$

and, hence, by virtue of relations (2.32)–(2.34), the vector Ax can be represented in the form (2.28).

Let us show that the lineals $\mathfrak{D}_H$ and $(\Phi + I)\mathfrak{D}_H$ are linearly independent. Indeed, assume that $x_0 \neq 0$, $x_\lambda \neq 0$, and

$$\alpha x_0 + \beta(\Phi + I)x_\lambda = 0 \quad (x_0 \in \mathfrak{D}_H, \ x_\lambda \in \mathfrak{D}_\Phi). \tag{2.35}$$

By applying the operator A to both the sides of equality (2.35) and using equality (2.28), we obtain

$$\alpha Hx_0 + \beta(\lambda \Phi x_\lambda + \bar{\lambda} x_\lambda) = 0. \tag{2.36}$$

It follows from (2.35) and (2.36) that

$$\alpha(H - \lambda I)x_0 + \beta(\bar{\lambda} - \lambda)x_\lambda = 0.$$

Consequently, $\alpha = \beta = 0$ (since $\mathrm{Im}\,\lambda \neq 0$, $\lambda \in \sigma_\rho(H)$, $(H - \lambda I)x_0 \in \Delta_H(\lambda)$, and $x_\lambda \perp \Delta_H(\lambda)$).

Theorem 2.9. *Assume that an arbitrary vector x in $\mathfrak{D}_A$ can be represented in the form (2.27), where Φ is a linear operator from $\mathfrak{N}_\lambda$ to $\mathfrak{N}_{\bar{\lambda}}$ such that*

$-1 \,\overline{\in}\, \sigma_p(\Phi)$, *and let the lineals* $\mathfrak{D}_H$ *(H is an Hermitian operator) and* $(\Phi + I)\mathfrak{D}_\Phi$
be linearly independent. Then the operator A *defined on* $\mathfrak{D}_A = \mathfrak{D}_H \dot{+} (\Phi + I)\mathfrak{D}_\Phi$
by (2.28) is a regular extension of H, $\lambda \,\overline{\in}\, \sigma_p(A)$, *and*

$$\overline{\lambda} \in \sigma_p(A) \Leftrightarrow 0 \in \sigma_p(\Phi). \tag{2.37}$$

Proof. Let us show that the equality $Ax = 0$ follows from the condition $x_0 + x_\lambda + \Phi x_\lambda = 0$. In fact, since the vectors x_0 and $(I + \Phi)x_\lambda$ are linearly independent, we have $x_0 = x_\lambda = 0$ (the latter equality follows from the fact that $-1 \,\overline{\in}\, \sigma_p(\Phi)$). Consequently, $Ax = 0$. Thus, the operator A is well defined by equality (2.28). Furthermore, if $y \in \mathfrak{D}_H$, then $Ay = Hy$ and, for all $x \in \mathfrak{D}_A$, obviously,

$$(Ax, y) - (x, Ay) = -(x_\lambda, (H - \lambda I)y) - (\Phi x_\lambda, (H - \overline{\lambda} I)y) = 0.$$

This means that $y \in \mathcal{9}_A$ and, thus, $H \subset A_0$ and $A \in \mathcal{P}(H)$.

Assume that $Ax = \lambda x$, where $x = x_0 + x_\lambda + \Phi x_\lambda$. Then, according to equality (2.28), $Hx_0 + \overline{\lambda} x_\lambda = \lambda x_0 + \lambda x_\lambda$, i.e., $(H - \lambda I)x_0 + (\overline{\lambda} - \lambda)x_\lambda = 0$. Since the vectors $(H - \lambda I)x_0$ and x_λ are orthogonal, Im $\lambda \neq 0$, and H is an Hermitian operator, we get $x_0 = x_\lambda = 0$. Thus, $x = 0$ and, hence, $\lambda \,\overline{\in}\, \sigma_p(A)$.

By analogy, we establish that if $Ax = \overline{\lambda} x$ $(x \neq 0)$, then $(H - \overline{\lambda} I)x_0 = (\overline{\lambda} - \lambda)\Phi x_\lambda$, whence $x_0 = \Phi x_\lambda = 0$. Note that $x_\lambda \neq 0$ (since $x \neq 0$) and, therefore, $0 \in \sigma_p(\Phi)$.

Similarly, if $\Phi x_\lambda = 0$ $(x_\lambda \neq 0)$, then $Ax_\lambda = \overline{\lambda} x_\lambda$. Theorem 2.9 is proved. ∎

Finally, we note that, in the case where H is an Hermitian operator with the deficiency index (1.1), it is convenient to write equalities (2.27) and (2.28) in the following form:

$$x = x_0 + a(x_\lambda + \theta x_{\overline{\lambda}}),$$

$$Ax = Hx_0 + a(\overline{\lambda} x_\lambda + \lambda \theta x_{\overline{\lambda}}),$$

where $x_0 \in \mathfrak{D}_H$, x_λ and $x_{\overline{\lambda}}$ are nonzero fixed vectors from the defect spaces $\mathfrak{N}_\lambda$ and $\mathfrak{N}_{\overline{\lambda}}$, respectively, a is a complex number, and θ is a complex number defined by the equality $\Phi x_\lambda = \theta x_{\overline{\lambda}}$. By choosing different complex numbers θ, we obtain various regular extensions of the operator H.

Remarks. The operators $T_{\mu\lambda}$, B_α, and $\tilde{B}_\alpha$ were studied and used by Kuzhel [2–7,

22]. The condition for the equality $\mathfrak{D}_A = \mathfrak{D}_{A^*}$ was obtained by Kuzhel [7] (see also [22]).

The von Neumann formulas were established by von Neumann in [1] for symmetric (in particular, self-adjoint) extensions of symmetric densely defined operators. Later, Krasnosel'skii [1] obtained similar formulas in the case of self-adjoint extensions of Hermitian operators (generally speaking, not densely defined).

Later, these formulas were generalized to various classes of nonself-adjoint extensions of symmetric operators (see, e.g., Tsekanovskii and Shmulyan [1]). In Kuzhel and Rudenko [1, 2], the von Neumann formulas were extended to the case of regular extensions of Hermitian operators. It should be noted that any linear operator A is a régular extension of its Hermitian part A_0 and, thus, von Neumann formulas can be applied, in fact, to any linear operator A with point spectrum $\sigma_p(A) \neq \mathbb{C}$.

3. Dissipative Operators

3.1. Definition and Properties

An operator A is called *dissipative* if

$$\text{Im}\,(Ax,\,x) \geq 0 \qquad (\forall\, x \in \mathfrak{D}_A). \tag{3.1}$$

In particular, a bounded operator A (or an operator A with $\mathfrak{D}_A = \mathfrak{D}_{A^*}$) is dissipative if and only if

$$\text{Im}\,A \;=\; \frac{A - A^*}{2i} \;\geq\; 0. \tag{3.2}$$

Proposition 3.1. *Assume that A is a dissipative operator and $\text{Im}\,\lambda < 0$. Then $\lambda \,\overline{\in}\, \sigma_p(A)$ and*

$$\| (A - \lambda I)^{-1} \| \leq \frac{1}{|\text{Im}\,\lambda|}. \tag{3.3}$$

Proof. If the operator A is dissipative and $\lambda_0 \in \sigma_p(A)$, then, evidently, $\text{Im}\,\lambda_0 \geq 0$. Consequently, the operator $A - \lambda I$ is invertible for $\text{Im}\,\lambda < 0$. Assume that $x \in \Delta_{A-\lambda I}$ ($\text{Im}\,\lambda < 0$) and $y = (A - \lambda I)^{-1} x$. Then $x = Ay - \lambda y$ and, thus,

$$\text{Im}\,(x,\,y) + \text{Im}\,\lambda \|y\|^2 \;=\; \text{Im}\,(Ay,\,y) \geq 0.$$

Therefore,

$$|\text{Im}\,\lambda|\,\|y\|^2 \le \text{Im}\,(x, y) \le |(x, y)| \le \|x\|\|y\|.$$

This implies that, for any $x \in \Delta_{A-\lambda I}$,

$$\|(A - \lambda I)^{-1} x\| = \|y\| \le \frac{1}{|\text{Im}\,\lambda|}\,\|x\|.$$

■

Proposition 3.2. *If A is a dissipative operator, then the lineals $\Delta_{A-\lambda I}$ and $\mathfrak{D}_A^{\perp}$ are linearly independent for any λ such that $\text{Im}\,\lambda < 0$.*

Proof. Let $y \in \Delta_{A-\lambda I} \cap \mathfrak{D}_A^{\perp}$. Then $y = (A - \lambda I)x$, where x is an element from $\mathfrak{D}_A$. Since $y \perp \mathfrak{D}_A$, we have

$$(Ax, x) - \lambda(x, x) = (y, x) = 0.$$

But then $\text{Im}\,\lambda\,\|x\|^2 = \text{Im}\,(Ax, x) \ge 0$, which, by virtue of the equality $\text{Im}\,\lambda < 0$, is possible only for $x = 0$. Consequently, $y = 0$.

■

Corollary 3.3. *If a closed dissipative operator A has at least one regular point in the lower half-plane, then this operator is densely defined (and, thus, it is an N-operator).*

Proof. Let λ $(\text{Im}\,\lambda < 0)$ be a regular point of a closed dissipative operator A acting in the space $\mathcal{H}$. Then $\Delta_{A-\lambda I} = \mathcal{H}$ and, hence, $\mathfrak{D}_A^{\perp} = \Delta_{A-\lambda I} \cap \mathfrak{D}_A^{\perp} = \{0\}$ by virtue of Proposition 3.2.

■

3.2. The Operator Adjoint to a Dissipative Operator

If A is a bounded dissipative operator (or a dissipative operator with $\mathfrak{D}_A = \mathfrak{D}_{A^*}$), then it is easy to verify that the operator $-A^*$ is also dissipative. Let us show that this is also also true in the case $\mathfrak{D}_A \neq \mathfrak{D}_{A^*}$.

Theorem 3.4. *An N-operator A with a nonempty resolvent set is dissipative if and only if $-A^*$ is dissipative.*

Proof. Assume that A is a dissipative operator, $\alpha \in \rho(A)$, and $\text{Im}\,\lambda \neq 0$. Consider equality (2.14). For $y = x$, this equality takes the form

$$(B_\alpha h, h) = 2\,\text{Im}\,(Ax, x) \quad (h = (A - \alpha I)x), \tag{3.4}$$

where the operator B_α is defined by (2.12). Since the operator A is dissipative, it follows from (3.4) that $B_\alpha \geq 0$. For definiteness, we assume that $\text{Im}\,\alpha < 0$. Then, by virtue of the first equality in (2.12), the operator T_α defined by (2.13) is a contraction. But then the operator T_α^* is also a contraction and, hence, the operator $\tilde{B}_\alpha$ defined by the second equality in (2.12) is nonnegative.

Consider equality (2.15). For $y = x$, this equality takes the form

$$(\tilde{B}_\alpha \varphi, \varphi) = -2\,\text{Im}\,(A^* x, x) \quad (\varphi = (A^* - \bar{\alpha} I)x).$$

This implies that $-A^*$ is a dissipative operator (since $\tilde{B}_\alpha \geq 0$).

The case where $\text{Im}\,\alpha > 0$ can be considered similarly.

Conversely, if the operator $-A^*$ is dissipative, then one can easily find that the operator $-(-A^*)^* = A$ is also dissipative. ∎

Theorem 3.5. *If A is a dissipative N-operator, then all points of the half-plane* $\text{Im}\,\alpha < 0$ *belong to the resolvent set of this operator.*

Proof. Let $\text{Im}\,\alpha < 0$. Assume that $\lambda \in \sigma(A)$. Since $\lambda \bar{\in} \sigma(A)$ and the operator $(A - \lambda I)^{-1}$ is bounded (Proposition 3.1), we have $\lambda \in \sigma_r(A)$. In this case, there exists a nonzero vector $y \in \mathcal{H}$ such that $((A - \lambda I)x, y) = 0$ $(\forall x \in \mathcal{D}_A)$. Consequently, $y \in \mathcal{D}_{A^*}$ and $A^* y = \bar{\lambda} y$ or $(-A^*)y = -\bar{\lambda} y$, which is impossible because the operator $-A^*$ is dissipative and $\text{Im}\,(-\bar{\lambda}) = \text{Im}\,\lambda < 0$.

Thus, $\lambda \bar{\in} \sigma_r(A)$ and, hence, $\lambda \in \rho(A)$. ∎

Corollary 3.6. *If a closed dissipative operator A has at least one regular point in the lower half-plane, then all points of this half-plane belong to the resolvent set $\rho(A)$.*

Proof. Let λ_0 $(\text{Im}\,\lambda_0 < 0)$ be a regular point of a closed dissipative operator A. Then, by virtue of Corollary 3.3, the operator A is densely defined and, hence, it is an N-operator. Therefore, by virtue of Theorem 3.5, the half-plane $\text{Im}\,\lambda < 0$ belongs to the set $\rho(A)$. ∎

3.3. Phillips Lemma and Its Generalization

Phillips [1, Lemma 1.15] established that an arbitrary dissipative extension A of a symmetric (densely defined) operator H is a restriction of the operator H^*.

Below, we give a simple proof of this lemma and extend it to the case of dissipative extensions of Hermitian operators.

Lemma 3.7. *If A is a dissipative extension of a symmetric operator H, then* $H \subset A \subset H^*$.

Proof. Assume that $x \in \mathfrak{D}_H$, $y \in \mathfrak{D}_A$, and $\alpha \in \mathbb{R}$. Then

$$\mathrm{Im}\,(A\,(x + \alpha y),\, x + \alpha y)$$

$$= \mathrm{Im}\,(A\,x,\, x) + \alpha\,\mathrm{Im}\,[(Ax,\, y) + (Ay,\, x)] + \alpha^2\,\mathrm{Im}\,(Ay,\, y) \geq 0. \qquad (3.5)$$

Moreover, since $\mathrm{Im}\,(A\,x,\, x) = \mathrm{Im}\,(H\,x,\, x) = 0$, inequality (3.5) is possible only if

$$\mathrm{Im}\,[(Ax,\, y) + (Ay,\, x)] = 0. \qquad (3.6)$$

Consequently,

$$\mathrm{Im}\,(A\,x,\, y) = -\mathrm{Im}\,(A\,y,\, x) = \mathrm{Im}\,\overline{(Ay,\, x)} = \mathrm{Im}\,(x,\, Ay) \qquad (3.7)$$

for all $x \in \mathfrak{D}_H$ and $y \in \mathfrak{D}_A$.

By replacing in (3.6) x by ix, we obtain

$$\mathrm{Im}\,i\,[(Ax,\, y) - (Ay,\, x)] = \mathrm{Re}\,[(Ax,\, y) - (Ay,\, x)] = 0$$

and, hence,

$$\mathrm{Re}\,(A\,x,\, y) = \mathrm{Re}\,\overline{(Ay,\, x)} = \mathrm{Re}\,(x,\, Ay). \qquad (3.8)$$

Thus, by virtue of (3.7) and (3.8),

$$(Hx,\, y) = (Ax,\, y) = (x,\, Ay) \qquad (\forall\, x \in \mathfrak{D}_H,\ \forall\, y \in \mathfrak{D}_A). \qquad (3.9)$$

Consequently, $y \in \mathfrak{D}_{H^*}$ and $Ay = H^*y$, i.e., $A \subset H^*$. Thus, $H \subset A \subset H^*$. ∎

Lemma 3.8. *An arbitrary dissipative extension A of an Hermitian operator H is a regular extension of the operator H.*

Proof. By analogy with the proof of Lemma 3.7, we establish inequality (3.5) and equality (3.6). These relations, in turn, yield equalities (3.7) and (3.8). Thus, if $x \in \mathcal{D}_H$, then

$$(Ax, y) = (x, Ay) \quad (\forall \, y \in \mathcal{D}_A).$$

Therefore, $x \in \mathcal{G}_A$, where $\mathcal{G}_A$ is the Hermitian domain of an operator A. Therefore, $\mathcal{D}_H \subset \mathcal{G}_A$ and $H \subset A_0$, i.e., $A \in \mathcal{P}[H]$.

∎

3.4. An Analog of the Sz.-Nagy–Foias Triangulation Theorem

By analogy with the results of Sz.-Nagy and Foias in the case of contractions (see Appendix 1), we consider the problem of triangulation of bounded dissipative operators.

Theorem 3.9. *Let $\mathcal{H}_1$ be an invariant subspace of a bounded dissipative operator A acting in the Hilbert space $\mathcal{H}$ $(\mathcal{D}_A = \mathcal{H})$ and let $\mathcal{H}_2 = \mathcal{H} \ominus \mathcal{H}_1$. Then this operator can be represented in the form*[1]

$$A = A_1 P_1 + A_2 P_2 + Q_{A_1} L Q_{A_2} P_2, \tag{3.10}$$

where $A_1 = A|_{\mathcal{H}_1}$, $A_2 = P_2 A|_{\mathcal{H}_2}$ are dissipative operators acting in the subspaces $\mathcal{H}_1$ and $\mathcal{H}_2$, respectively, $Q_{A_k} = (\mathrm{Im}\,A_k)^{1/2}$, and L, $\|L\| \leq 2$, is a linear operator which maps the lineal $Q_{A_2} \mathcal{H}_2$ into the subspace $\mathcal{H}_1$.

Proof. Let the operator A satisfy the conditions of the theorem. Then

$$A = A_1 P_1 + A_2 P_2 + \Gamma P_2 \quad (A_1 = A|_{\mathcal{H}_1}, \, A_2 = P_2 A|_{\mathcal{H}_2})$$

where $\Gamma : \mathcal{H}_2 \to \mathcal{H}_1$ is a linear operator. In this case, if $x = x_1 + x_2$ $(x_k \in \mathcal{H}_k)$, then

$$(Ax, x) = (A_1 x_1, x_1) + (Ax_2, x_2) + (\Gamma x_2, x_1). \tag{3.11}$$

Relation (3.11) implies that A_1 and A_2 are dissipative operators. Therefore, taking

---

[1] Clearly, the operator A can be written as the operator matrix $A = \begin{pmatrix} A_1 & Q_{A_1} L Q_{A_2} \\ 0 & A_2 \end{pmatrix}$.

(3.11) into account, we get

$$\text{Im}(Ax, x) = \|Q_{A_1}x_1\|^2 + \|Q_{A_2}x_2\|^2 + \text{Im}(\Gamma x_2, x_1) \geq 0. \tag{3.12}$$

In particular, for $x_1 \in \text{Ker}\, Q_{A_1}$, inequality (3.12) takes the form

$$\text{Im}(\Gamma x_2, x_1) + \|Q_{A_2}x_2\|^2 \geq 0. \tag{3.13}$$

By substituting ax_1 or iax_1 $(a \in \mathbb{R})$ for x_1 in (3.13), we establish that this inequality is possible only if

$$\text{Im}(\Gamma x_2, x_1) = \text{Re}(\Gamma x_2, x_1) = 0.$$

Thus,

$$(\Gamma x_2, x_1) = (x_2, \Gamma^* x_1) = 0 \quad (x_1 \in \text{Ker}\, Q_{A_1}, \ x_2 \in \mathcal{H}_2)$$

and, hence,

$$\text{Ker}\, Q_{A_1} \subset \text{Ker}\, \Gamma^*. \tag{3.14}$$

Similarly, we establish that

$$\text{Ker}\, Q_{A_2} \subset \text{Ker}\, \Gamma. \tag{3.15}$$

Define the operator B acting from $\mathcal{H}_1$ into $\mathcal{H}_2$ by the equality

$$BQ_{A_1}x_1 = \Gamma^* x_1 \quad (x_1 \in \mathcal{H}_1) \tag{3.16}$$

According to (3.14), the operator B is well defined. Furthermore, $\mathcal{D}_B = \Delta_{Q_{A_1}}$, $\Delta_B = \Delta_{\Gamma^*}$. Let $y_1 = Q_{A_1}x_1$. Then, according to relations (3.12) and (3.16), we have

$$\|Q_{A_2}x_2\|^2 + \text{Im}(\Gamma x_2, y_1) + \|y_1\|^2 \geq 0. \tag{3.17}$$

In particular, for $x_2 = i\alpha B y_1$ $(\alpha \in \mathbb{R})$, the last inequality can be rewritten as follows:

$$\|Q_{A_2}By_1\|^2 \alpha^2 + \|By_1\|^2 \alpha + \|y_1\|^2 \geq 0 \quad (\forall \alpha \in \mathbb{R}).$$

Hence,

$$\|By_1\|^4 \leq 4\|Q_{A_2}By_1\|^2\|y_1\|^2 \leq 4\|Q_{A_2}\|^2\|By_1\|^2\|y_1\|^2.$$

This yields $\| By_1 \| \leq 2 \| Q_{A_2} \| \| y_1 \|$.

Thus, the operator B is bounded and defined on $\Delta_{Q_{A_1}}$. Extending this operator by continuity to $\bar{\Delta}_{Q_{A_1}}$ and defining it to be zero on the subspace $\mathrm{Ker}\, Q_{A_1}$, we obtain the bounded linear operator $B = \mathcal{H}_1 \to \mathcal{H}_2$, which is connected with the operator Γ by the equalities

$$\Gamma^* = BQ_{A_1}, \qquad \Gamma = Q_{A_1}B^*, \tag{3.18}$$

where

$$\mathrm{Ker}\, Q_{A_1} \subset \mathrm{Ker}\, B. \tag{3.19}$$

Consider the operator L defined by the equality

$$LQ_{A_2}x_2 = B^* x_2. \tag{3.20}$$

First, let us show that this operator is well defined. Indeed, assume that $Q_{A_2}x_2 = 0$. Then, by virtue of (3.15) and (3.18), we have $B^* x_2 \subset \mathrm{Ker}\, Q_{A_1}$. In view of (3.19), we get

$$\| B^* x_2 \|^2 = (BB^* x_2, x_2) = 0,$$

i.e., $B^* x_2 = 0$. Thus, the operator L is well defined.

It follows from (3.18) and (3.19) that

$$\Gamma x_2 = Q_{A_1} L Q_{A_2} x_2 \quad (x_2 \in \mathcal{H}_2), \tag{3.21}$$

which proves equality (3.20).

To complete the proof of Theorem 3.9, it remains to show that $\| L \| \leq 2$. For this purpose, we replace x_2 by λx_2 in inequality (3.12) and denote $y_k = Q_{A_k}x_k$. As a result, in view of equality (3.21), we get

$$\| y_1 \|^2 + |\lambda|^2 \| y_2 \|^2 + \mathrm{Im}\, \lambda (L y_2, y_1) \geq 0. \tag{3.22}$$

In particular, for

$$\lambda = -\frac{i}{2 \| y_2 \|^2} \overline{(L y_2, y_1)} \qquad (y_2 \neq 0)$$

inequality (3.22) can be rewritten as follows:

$$\| y_1 \|^2 - \frac{1}{4 \| y_2 \|^2} |(L y_2, y_1)|^2 \geq 0.$$

Hence,

$$|(L y_2, y_1)| \leq 2 \| y_1 \| \| y_2 \|, \qquad (3.23)$$

where y_1 is an arbitrary element from $\Delta_{Q_{A_1}}$. Since, by virtue of (3.19) and (3.20), $L y_2 \in \Delta_{B^*}$ and $\overline{\Delta}_{B^*} \subset \overline{\Delta}_{Q_{A_1}}$, we can choose the vector y_1 in the form $y_1 = L y_2$. Then, in view of (3.23), we obtain $\| L y_2 \| \leq 2 \| y_2 \|$ ($\forall \, y_2 \in \mathcal{H}_2$).

∎

Corollary 3.10. *If a bounded dissipative operator* A *induces a self-adjoint operator* A_1 *in the subspace* $\mathcal{H}_1$, *then this subspace reduces the operator* A.

Proof. Indeed, under this condition, $Q_{A_1} = 0$ and, hence, the last term in (3.10) is equal to zero.

∎

A nonself-adjoint operator A is called *simple* if it does not induce a self-adjoint operator in any nonzero subspace. By virtue of Corollary 3.10, a simple dissipative operator has no real eigenvalues.

Theorem 3.11. *Suppose that the operator* A *is defined in the space* $\mathcal{H} = \mathcal{H}_1 \oplus \mathcal{H}_2$ *by equality (3.10) in which* A_1 *and* A_2 *are bounded dissipative operators acting in the subspaces* $\mathcal{H}_1$ *and* $\mathcal{H}_2$, *respectively. Assume also that* $L: \mathcal{H}_2 \to \mathcal{H}_1$ *is a linear operator and* $\| L \| \leq 2$. *Then the operator* A *is dissipative.*

Proof. Under the conditions of the theorem, $A = A_1 P_1 + A_2 P_2 + \Gamma P_2$, where $\Gamma = Q_{A_1} L Q_{A_2}$. Then, in view of the equalities $\mathrm{Im} \, A_k = Q_{A_k}^2$, we get

$$\frac{A - A^*}{2i} = Q_{A_1}^2 P_1 + Q_{A_2}^2 P_2 + \frac{\Gamma P_2 - \Gamma^* P_1}{2i}.$$

Hence, if $x = x_1 + x_2$ ($x_k \in \mathcal{H}_k$), then

$$\left(\frac{A - A^*}{2i} x, \, x \right) = \| Q_{A_1} x_1 \|^2 + \| Q_{A_2} x_2 \|^2 + \mathrm{Im} \, (\Gamma x_2, x_1).$$

This equality can be rewritten as follows:

$$\left(\frac{A - A^*}{2i} x, \ x\right) = \|y_1\|^2 + \|y_2\|^2 + \operatorname{Im}(Ly_2, y_1), \tag{3.24}$$

where $y_k = Q_{A_k} x_k$. Since

$$-\operatorname{Im}(Ly_2, y_1) \le |(Ly_2, y_1)| \le \|L\| \|y_1\| \|y_2\| \le 2\|y_1\| \|y_2\|, \tag{3.25}$$

by virtue of (3.24) and (3.25), we have

$$\left(\frac{A - A^*}{2i} x, \ x\right) \ge (\|y_1\| - \|y_2\|)^2 \ge 0 \quad (\forall x \in \mathcal{H})$$

and, hence, A is a dissipative operator.   ■

Remarks. A different proof of Theorem 3.4 was given by Phillips in [1]. In that work, the operator L was called dissipative if $\operatorname{Re}(Lx, x) \le 0$. Under this condition, the operator $A = -iL$ is dissipative in the sense of definition (3.1).

Different proofs of Lemma 3.7 are given in Phillips [1] and V. Gorbachuk and M. Gorbachuk [1].

The results of subsection 3.4 were presented (without proof) in Kuzhel [21].

2. Characteristic Functions

1. Characteristic Matrix Functions of Regular Extensions

1.1. Δ-Basis

Let H be an Hermitian operator with deficiency index $(r, r) \: (0 < r < \infty)$, $A \in \mathcal{P}(H)$, let α (Im $\alpha \neq 0$) be some fixed regular point of the operator A, and let the corresponding operator B_α be defined by equality (2.10) (Chapter 1). Then (see Theorem 2.4, Chapter 1) $B_\alpha : \mathcal{H} \to \mathfrak{N}_\alpha$, where $\mathfrak{N}_\alpha$ is the defect subspace of the operator H.

For $n \geq r$, the operator B_α can be represented as

$$B_\alpha = \sum_{i,\, s=1}^{n} (\cdot, g_i) J_{is} g_s, \tag{1.1}$$

where $\{g_k\}$ is a collection of vectors (we do not require that they are linearly independent) whose linear span $\mathcal{L}_A$ contains the defect subspace $\mathfrak{N}_\alpha$ and the matrix $J = \| J_{ki} \|$ is Hermitian and invertible.

Note that, generally speaking, $\mathcal{L}_A \neq \mathfrak{N}_\alpha$. For example, if $A^* = A$, then $\mathfrak{N}_\alpha = \{0\}$ ($r = 0$) and $B_\alpha = 0$. But this operator can be represented in the form (1.1) with, e.g., $n = 2$; in this case, $J_{ki} = 0$ $(k \neq i)$, $J_{11} = -J_{22} = 1$, $g_1 = g_2 = g$, and g is an arbitrary fixed vector in $\mathcal{H}$. Thus, $\mathcal{L}_A = \langle g \rangle \neq \mathfrak{N}_\alpha$ in the case under consideration (the last relation holds for $g \neq 0$).

The collection of vectors $\{g_k\}_1^n$ given by equality (1.1) is called the Δ-basis of the operator A and the matrix $J = \| J_{ki} \|$ $(J^* = J, \det J \neq 0)$ is its matrix of coefficients.

Let $\omega = \| \omega_{ki} \|$ be a nondegenerate $n \times n$-matrix.

Consider vectors

$$\tilde{g}_k = \sum_{i=1}^{n} w_{ki} g_i \quad (k = \overline{1,\, n}), \tag{1.2}$$

where $\{g_i\}$ is a Δ-basis of the operator A. If, in addition, $\omega^{-1} = \| t_{ki} \|$, then

$$g_m = \sum_{k=1}^{n} t_{mk} \tilde{g}_k \qquad (m = \overline{1, n}). \tag{1.3}$$

Substituting (1.3) in (1.1), we obtain

$$B_\alpha = \sum_{k, m=1}^{n} (\cdot, \tilde{g}_k) \tilde{J}_{km} \tilde{g}_m, \qquad \tilde{J}_{km} = \sum_{i, s=1}^{n} t_{ki}^* J_{is} t_{sm}$$

where $t_{ki}^* = \bar{t}_{ik}$. Hence, if $\tilde{J} = \| \tilde{J}_{ki} \|$, then

$$\tilde{J} = \omega^{\rightarrow *} J \omega^{\rightarrow -1} \qquad (J = \omega^* \tilde{J} \omega, \quad \omega^{\rightarrow *} = (\omega^{-1})^*). \tag{1.4}$$

Thus, in view of (1.1), the collection of vectors $\{ \tilde{g}_k \}$ defined by (1.2) is a Δ-basis of the operator A. Moreover, the corresponding matrix of coefficients $\tilde{J}$ is connected with the matrix J by equalities (1.4).

Thus, by choosing a proper matrix ω, one can always guarantee the unitarity of the (Hermitian) coefficient matrix J, i.e., $J^* = J = J^{-1}$. However, in the calculation of specific characteristic functions, this is not always convenient (see Section 4).

It is clear that $\mathcal{L}_A = \mathfrak{N}_A$ for $n = r$. Consequently, for $n = r$, any basis of the defect subspace $\mathfrak{N}_A$ can be regarded as a Δ-basis of the operator A. As far as the coefficient matrix is concerned, various methods for its determination for given Δ-basis are presented in Section 3.

1.2. Characteristic Matrix Function

A matrix function $\chi_A (\lambda)$ defined by the equality

$$\chi_A^*(\alpha) J^{-1} \chi_A (\lambda) = J^{-1} + i(\bar{\alpha} - \lambda) \| (T_{\alpha\lambda} g_k, g_m) \| \qquad (\lambda \in \rho (A)) \tag{1.5}$$

is called the *characteristic matrix function* of the operator A.

In (1.5), $\{g_k\}_{k=1}^{r}$ is a Δ-basis of the operator A, $J = \| J_{ki} \|$ is the corresponding coefficient matrix, $T_{\alpha\lambda} = (A - \alpha I)(A - \lambda I)^{-1}$, and α (Im $\alpha \neq 0$) is a fixed number from $\rho (A)$.

In what follows, we also assume that $\bar{\alpha} \in \rho (A)$.

The solvability of equation (1.5) is established in the subsequent subsections of this section. Here, we only note that if the matrix M satisfies the condition $M^* J^{-1} M = J^{-1}$, then the matrix $\hat{\chi}_A(\lambda) = M \chi_A (\lambda)$ is also a solution of equation (1.5).

1.3. Auxiliary Statements

Consider a matrix

$$\tau = E + 2\operatorname{Im}\alpha JG, \tag{1.6}$$

where E is the $n \times n$ identity matrix, $G = \|(g_k, g_i)\|$, $\{g_k\}_{k=1}^n$ is a Δ-basis of the operator A, and J is the corresponding coefficient matrix. The matrix τ just defined is called the *deviation matrix* in the Δ-basis $\{g_k\}$.

Proposition 1.1. $\det \tau \neq 0$.

Proof. Assume that $x\tau = 0$, where $x = (x_1, x_2, \ldots, x_n)$. Then, taking into account the expressions for the matrices J and G and equality (1.6), we obtain

$$x_k = -2\operatorname{Im}\alpha \sum_{i,\,s=1}^{n} x_i J_{is}(g_s, g_k) = (\varphi, g_k) \quad (k = \overline{1, n}), \tag{1.7}$$

where

$$\varphi = -2\operatorname{Im}\alpha \sum_{i,\,s=1}^{n} x_i J_{is} g_s. \tag{1.8}$$

If we replace x_i in this relation by its value from (1.7), then we get

$$\varphi = -2\operatorname{Im}\alpha \sum_{i,\,s=1}^{n} (\varphi, g_i) J_{is} g_s. \tag{1.9}$$

Thus, by virtue of (1.9) and (1.1), $\varphi = -2\operatorname{Im}\alpha B_\alpha \varphi$. In view of (2.12) (see Chapter 1), the last equality can be rewritten as

$$\varphi = \varphi - T_\alpha^* T_\alpha \varphi, \tag{1.10}$$

where the operator $T_\alpha = (A - \overline{\alpha}I)(A - \alpha I)^{-1}$ is invertible. But then $\varphi = 0$ by (1.10) and, consequently, $x = 0$. Thus, $\det \tau \neq 0$. ∎

Proposition 1.2. *The matrix S defined by the equality*

$$S = E + 2\operatorname{Im}\alpha RJR, \tag{1.11}$$

where $R = \sqrt{G}$ is nonnegative and nondegenerate.

Proof. In $\mathbb{C}^n$, we consider a lineal $L = \{yR \,|\, y \in \mathbb{C}^n\}$. An arbitrary vector x from $\mathbb{C}^n$ can be represented in the form $x = h + yR$, where $h \perp L$ and, hence, $hR = 0$. Further, in view of the fact that

$$x^* = h^* + Ry^*,$$

we have

$$xSx^* = hh^* + yGy^* + 2\operatorname{Im}\alpha yGJGy^*. \tag{1.12}$$

By virtue of (1.1), we get $\| (B_\alpha g_m, g_r) \| = GJG$. In this equality, we replace B_α by its value from (2.12) (see Chapter 1). This gives

$$\tilde{G} = G + 2\operatorname{Im}\alpha GJG, \tag{1.13}$$

where $\tilde{G} = \| (T_\alpha g_m, T_\alpha g_r) \|$. But then $xSx^* = hh^* + y\tilde{G}y^* \geq 0$ by virtue of (1.12) and (1.13) and, hence, $S \geq 0$.

To prove that the matrix S is nondegenerate, note that

$$SR = R\tau \tag{1.14}$$

in view of (1.6) and (1.11).

Hence, if $xS = 0$, then $xR = 0$ by virtue of equality (1.14) and the fact that the matrix τ is invertible. Finally, taking into account (1.11), we get $xS = x = 0$. This implies that $\det S \neq 0$.

∎

Proposition 1.3. *The matrices τ^{-1} and S^{-1} exist and are defined by the equalities*

$$\tau^{-1} = E - 2\operatorname{Im}\alpha JRS^{-1}R, \tag{1.15}$$

$$S^{-1} = E - 2\operatorname{Im}\alpha R\tau^{-1}JR. \tag{1.16}$$

Proof. Let $X = \tau^{-1}$. Since $\tau = E + 2\operatorname{Im}\alpha JR^2$, we have

$$X + 2\operatorname{Im}\alpha JR^2X = E. \tag{1.17}$$

Multiplying this equality from the left by R and taking into account (1.11), we get

$SRX = R$ and, consequently, $RX = S^{-1}R$. By using the last equality and (1.17), we obtain

$$X = E - 2 \operatorname{Im} \alpha JRS^{-1}R,$$

which proves (1.15).

To find the matrix S^{-1}, we act in a similar way: If $y = S^{-1}$, then

$$Y + 2 \operatorname{Im} \alpha YRJR = E$$

by (1.11). Multiplying this equality from the right by R, we get that $YR\tau = R$ and, consequently, $YR = R\tau^{-1}$. Then we get $Y = E - 2 \operatorname{Im} \alpha R\tau^{-1} JR$, which implies (1.16).

∎

Proposition 1.4. *The matrix equation*

$$X^* J^{-1} X = J^{-1} \tau \tag{1.18}$$

is solvable and its solution is given by the equality

$$X = W[E + 2 \operatorname{Im} \alpha JR (E + \sqrt{S})^{-1}R], \tag{1.19}$$

where the matrix S is determined by equality (1.11) *and W is an arbitrary matrix satisfying the condition: $W^* J^{-1} W = J^{-1}$.*

Proof. We set $N = E + \sqrt{S}$ and $a = 2 \operatorname{Im} \alpha$. By using expression (1.19) for the matrix X, we get

$$X^* J^{-1} X = (E + aRN^{-1}RJ)(J^{-1} + aRN^{-1}R)$$

$$= J^{-1} + aRN^{-1}(2E + aRJRN^{-1})R, \tag{1.20}$$

where

$$aRJR = S - E = N(N - 2E) \tag{1.21}$$

in view of the expressions for the matrices S and N. But then

$$N^{-1}(2E + aRJRN^{-1}) = E \tag{1.22}$$

and, consequently,

$$X^* J^{-1} X = J^{-1} + aR^2 = J^{-1} \tau$$

by virtue of (1.20) and (1.6).

■

In the case where the Δ-basis $\{g_k\}$ is linearly independent (and, consequently, the matrices G and R are nondegenerate), the expression for the matrix X can be considerably simplified. In fact,

$$X = WR^{-1} [E + 2 \operatorname{Im} \alpha R J R (E + \sqrt{S})^{-1}] R$$

$$= WR^{-1} [(E + \sqrt{S}) + (S - E)] (E + \sqrt{S})^{-1} R = WR^{-1} \sqrt{S} R$$

by virtue of (1.19) and (1.11), which means that $X = W\sqrt{G}^{-1} \sqrt{S} \sqrt{G}$.

As a conclusion, we note that if the matrix X is given by equality (1.19), then it is easy to check that

$$X^{-1} = [E - 2 \operatorname{Im} \alpha J R (S + \sqrt{S})^{-1} R] W^{-1}. \qquad (1.23)$$

1.4. Correctness of the Definition of Characteristic Matrix Functions

For $\lambda = \alpha$, we can rewrite equality (1.5) in the form

$$\chi_A^*(\alpha) J^{-1} \chi_A (\alpha) = J^{-1} + 2 \operatorname{Im} \alpha G, \qquad (1.24)$$

where $J^{-1} + 2 \operatorname{Im} \alpha G = J^{-1} \tau$. Equality (1.24) holds by virtue of Proposition 1.4 with $\chi_A (\alpha) = X$, where the matrix X is given by equality (1.19). Consequently, the definition of the characteristic matrix function $\chi_A (\alpha)$ is correct, and this matrix is determined to within the left factor W satisfying the condition $W^* J^{-1} W = J^{-1}$.

Remark. The definition of the characteristic matrix function used above was introduced by the author [2, 3, 22]. A survey of the existing definitions of the characteristic function is presented in Section 9.

2. A Criterion of Unitary Equivalence

2.1. Characteristic Matrix Functions of Unitary Equivalent Operators

Assume that operators A and $\hat{A}$ are unitary equivalent, i.e., $\hat{A} = VAV^*$, where V is a unitary operator. Then $(\hat{A} - \alpha I)^{-1} = V(A - \alpha I)^{-1} V^*$ and, consequently, $\hat{B}_\alpha = VB_\alpha V^*$. Taking into account (1.1), we get

$$\hat{B}_\alpha = \sum_{i,\,s=1}^{n} (\cdot,\, \hat{g}_i)\, J_{is} \hat{g}_s \qquad (\hat{g}_k = Vg_k).$$

The system $\{\hat{g}_k\}$ is a Δ-basis of the operator $\hat{A}$ and $\hat{J} = J$ is the corresponding co-efficient matrix. By using equality (1.5), we conclude that $\chi_{\hat{A}}^*(\alpha) J^{-1} \chi_{\hat{A}}(\lambda) = \chi_A^*(\alpha) J^{-1} \chi_A(\lambda)$ and, consequently, the corresponding characteristic matrix functions $\chi_{\hat{A}}(\lambda)$ and $\chi_A(\lambda)$ can regarded as coinciding.

Thus, if the operators $\hat{A}$ and A are unitary equivalent, then some of their characteristic matrix functions coincide. However, the converse statement is not always true. For example, if $\hat{A} = A \oplus H$, where H is nonzero nonself-adjoint operator, then the characteristic matrix functions of the operators $\hat{A}$ and A coincide, although these operators are not unitary equivalent.

2.2. L-Simple Part of an Operator

Denote by $\mathcal{H}_A$ the closure of the linear span of the lineals $R_\alpha^m L_A$ $(m = 0, 1, 2, \dots)$, where L_A is the linear span of the Δ-basis $\{g_k\}_{k=1}^r$ of an operator A.

The operator $A_s = A\big|_{\mathcal{H}_A \cap \mathcal{D}_A}$ is called the *L-simple part* of the operator A. Since Δ-basis of the operator A is not uniquely determined, the *L*-simple part A_s of the operator A is also not uniquely determined in the general case.

If $A = A_s$, the operator A is called *L-simple*.

In the case where $L_A = \mathfrak{N}_\lambda$, we shall say 'the simple part of an operator' and 'a simple operator' instead of 'the *L*-simple part of an operator' and 'an *L*-simple operator', respectively.

2.3. A Criterion of Unitary Equivalence of L-Simple Parts of Operators

Assume that $\{\alpha, \bar{\alpha}\} \subset \rho(A) \cap \rho(\tilde{A})$, where A and $\tilde{A}$ are operators from $\mathcal{P}(H)$.

Theorem 2.1. *If $\chi_A(\lambda) = \chi_{\tilde{A}}(\lambda)$ and $J = \tilde{J}$ for certain Δ-bases, then the L-simple parts of the operators A and $\tilde{A}$ are unitary equivalent.*

Proof. By virtue of the condition of the theorem and equality (1.5), we have

$$\| (T_{\alpha\lambda} g_k, g_m) \| = \| (\tilde{T}_{\alpha\lambda} \tilde{g}_k, \tilde{g}_m) \|, \quad \| (g_k, g_m) \| = \| (\tilde{g}_k, \tilde{g}_m) \|, \qquad (2.1)$$

where the second equality can be obtained from the first one if we set $\lambda = \alpha$.

Since $T_{\alpha\lambda} = I + (\lambda - \alpha) R_\lambda$ and $\tilde{T}_{\alpha\lambda} = \tilde{I} + (\lambda - \alpha) \tilde{R}_\lambda$ ($\tilde{R}_\lambda = (\tilde{A} - \lambda \tilde{I})^{-1}$), it follows from (2.1) that

$$\| (R_\lambda g_k, g_m) \| = \| (\tilde{R}_\lambda \tilde{g}_k, \tilde{g}_m) \|. \qquad (2.2)$$

In a sufficiently small neighborhood of the point α, the resolvents R_α and $\tilde{R}_\alpha$ can be represented in the form

$$R_\lambda = \sum_{m=0}^{\infty} (\lambda - \alpha)^m R_\alpha^{m+1}, \qquad \tilde{R}_\lambda = \sum_{m=0}^{\infty} (\lambda - \alpha)^m \tilde{R}_\alpha^{m+1}. \qquad (2.3)$$

By substituting (2.3) in (2.2) and equating corresponding coefficients, we obtain

$$\| (R_\alpha^m g_k, g_i) \| = \| (\tilde{R}_\alpha^m \tilde{g}_k, \tilde{g}_i) \| \qquad (m \in \mathbb{N}). \qquad (2.4)$$

It follows from the expression for the operator B_α that

$$R_\alpha^* R_\alpha = \frac{1}{2 \operatorname{Im} \alpha} [B_\alpha + i R_\alpha^* - i R_\alpha].$$

Then, for any natural m and s, we get

$$(R_\alpha^m g_k, R_\alpha^s g_i) = (R_\alpha^* R_\alpha (R_\alpha^{m-1} g_k), R_\alpha^{s-1} g_i)$$

$$= \frac{1}{2 \operatorname{Im} \alpha} [(B_\alpha R_\alpha^{m-1} g_k, R_\alpha^{s-1} g_i) + i(R_\alpha^{m-1} g_k, R_\alpha^s g_i) - i(R_\alpha^m g_k, R_\alpha^{s-1} g_i)].$$

$$(2.5)$$

By applying a similar procedure to the scalar products on the right-hand side of (2.5) in which the operator R_α acts on the vectors g_k and g_i, after a finite number of steps, we arrive at an equality whose right-hand side is the sum of numbers of the form

$$(R_\alpha^p \, g_k, \, g_i), \quad (g_k, \, R_\alpha^q \, g_i), \quad (B_\alpha R_\alpha^l \, g_k, \, R_\alpha^j \, g_i), \tag{2.6}$$

where p, q, l, and j are natural numbers. In addition, we have

$$(R_\alpha^p \, g_k, \, g_i) = (\tilde{R}_\alpha^p \, \tilde{g}_k, \, \tilde{g}_i), \quad\quad (g_k, \, R_\alpha^q \, g_i) = (\tilde{g}_k, \, \tilde{R}_\alpha^q \, \tilde{g}_i) \tag{2.7}$$

by virtue of (2.4). Further, let us show that

$$(B_\alpha R_\alpha^l \, g_k, \, R_\alpha^j \, g_i) = (\tilde{B}_\alpha \, \tilde{R}_\alpha^l \, \tilde{g}_k, \, \tilde{R}_\alpha^j \, \tilde{g}_i).$$

Indeed, taking the equalities (1.1), (2.7), and $J_{ms} = \tilde{J}_{ms}$ into account, we get

$$(B_\alpha R_\alpha^l \, g_k, \, R_\alpha^j \, g_i) = \sum_{m, \, s=1}^{n} (R_\alpha^l \, g, \, g_m) J_{ms} (g_s, \, R_\alpha^j \, g_i) = (\tilde{B}_\alpha \, \tilde{R}_\alpha^l \, \tilde{g}_k, \, \tilde{R}_\alpha^j \, \tilde{g}_i).$$

This and equalities (2.5) – (2.7) imply that

$$(R_\alpha^m \, g_k, \, R_\alpha^s \, g_i) = (\tilde{R}_\alpha^m \, \tilde{g}_k, \, \tilde{R}_\alpha^s \, \tilde{g}_i) \tag{2.8}$$

for any nonnegative integers m and s.

Consider a linear operator V that acts on the vectors $R_\alpha^m \, g_k$ as follows:

$$V(R_\alpha^m \, g_k) = \tilde{R}_\alpha^m \, \tilde{g}_k. \tag{2.9}$$

This operator maps the set of vectors dense in $\mathcal{H}_A$ into the set of vectors dense in $\mathcal{H}_{\tilde{A}}$ and, moreover, by virtue of (2.8) and (2.9),

$$(V(R_\alpha^m \, g_k), \, V(R_\alpha^s \, g_i)) = (R_\alpha^m \, g_k, \, R_\alpha^s \, g_i).$$

Extending the operator V by continuity, we obtain the unitary operator $V : \mathcal{H}_A \to \mathcal{H}_{\tilde{A}}$. Furthermore,

$$VR_\alpha (R_\alpha^m \, g_k) = \tilde{R}_\alpha (\tilde{R}_\alpha^m \, \tilde{g}_k) = \tilde{R}_\alpha V(R_\alpha^m \, g_k).$$

This yields

$$VR_\alpha x = \tilde{R}_\alpha Vx \quad (x \in \mathcal{H}_A). \tag{2.10}$$

Let $y = R_\alpha x$. Recall that the subspace $\mathcal{H}_A$ is invariant under the action of the operator R_α. Therefore, $y \in \mathfrak{D}_A \cap \mathcal{H}_A$, i.e., $y \in \mathfrak{D}_{A_s}$. This implies that $x = (A_s - \alpha I)y$ and equality (2.10) can be rewritten in the following form:

$$Vy = \tilde{R}_\alpha V(A_s - \alpha I)y. \tag{2.11}$$

Moreover, $Vy \in \mathfrak{D}_{\tilde{A}}$ and $Vy \in \mathcal{H}_{\tilde{A}}$, i.e., $Vy \in \mathfrak{D}_{\tilde{A}_s}$. Acting by the operator $\tilde{A}_s - \alpha \tilde{I}$ on both the sides of (2.11) and taking the equality $(\tilde{A}_s - \alpha \tilde{I})\tilde{R}_\alpha = (\tilde{A} - \alpha \tilde{I})\tilde{R}_\alpha = \tilde{I}$ into account, we get $(\tilde{A}_s - \alpha \tilde{I})Vy = V(A_s - \alpha I)y$ and, hence, $\tilde{A}_s V = VA_s$. ∎

Remark. Theorem 2.1 (a criterion of unitary equivalence) was presented (without proof) by Kuzhel in [2, 22].

3. Determination of Characteristic Matrix Functions

3.1. General Case

Assume that $A \in \mathcal{P}(H)$, where H is an Hermitian operator with deficiency index (r, r). For simplicity, we consider the case where $n = r$. Then, according to Subsection 1.1, $L_A = \mathfrak{N}_\alpha$ and any basis of the defect subspace $\mathfrak{N}_\alpha$ may be regarded as a Δ-basis of the operator A.

Let us fix some Δ-basis $\{g_k\}_{k=1}^r$ of the operator A and let J be the corresponding coefficient matrix. By multiplying both the sides of equality (1.5) by JG, where $G = \|(g_k, g_i)\|$ is the Gramm matrix of the system $\{g_k\}$, we get

$$\omega_A(\lambda)JG = G + i(\bar{\alpha} - \lambda)\|(T_{\alpha\lambda} g_k, g_m)\| JG, \tag{3.1}$$

where $\omega_A(\lambda) = \chi_A^*(\alpha)J^{-1}\chi_A(\lambda)$. By virtue of equality (1.1), we have

$$\|(T_{\alpha\lambda} g_k, g_m)\| JG = \left\| \sum_{i,\,s=1}^r (T_{\alpha\lambda} g_k, g_i) J_{is}(g_s, g_m) \right\|$$

$$= \left\| \left(\sum_{i,\,s=1}^{r} (T_{\alpha\lambda}\, g_k,\, g_i)\, J_{is}\, g_s,\, g_m \right) \right\| = \| (B_\alpha T_{\alpha\lambda}\, g_k\, g_m) \|. \quad (3.2)$$

By using the expressions for the operators B_α and $T_{\alpha\lambda}$, we obtain

$$B_\alpha T_{\alpha\lambda} = iR_\alpha T_{\alpha\lambda} - iR_\alpha^* T_{\alpha\lambda} + 2\,\mathrm{Im}\,\alpha R_\alpha^* R_\alpha T_{\alpha\lambda}$$

$$= iR_\lambda - iR_\alpha^*(A - \alpha I)R_\lambda + i(\bar{\alpha} - \alpha)\,R_\alpha^* R_\lambda = iR_\lambda - iR_\alpha^*(A - \bar{\alpha} I)R_\lambda,$$

i.e.,

$$B_\alpha T_{\alpha\lambda} = iR_\lambda - iR_\alpha^* T_{\bar{\alpha}\lambda}. \quad (3.3)$$

Further, since $T_{\bar{\alpha}\lambda} = I + (\lambda - \bar{\alpha})R_\lambda$ and $T_{\bar{\lambda}\alpha} = I + (\alpha - \bar{\lambda})R_\lambda$, we have

$$(\lambda - \bar{\alpha})R_\lambda = T_{\bar{\alpha}\lambda} - I, \quad (\bar{\alpha} - \lambda)R_\alpha^* = T_{\bar{\lambda}\alpha}^* - I. \quad (3.4)$$

It follows from (3.3) and (3.4) that

$$i(\bar{\alpha} - \lambda)B_\alpha T_{\alpha\lambda} = T_{\bar{\lambda}\alpha}^* T_{\bar{\alpha}\lambda} - I. \quad (3.5)$$

By using equalities (3.1), (3.2), and (3.5), we obtain

$$\omega_A(\lambda)\,JG = \| (T_{\bar{\lambda}\alpha}^* T_{\bar{\alpha}\lambda} g_k\, g_m) \|. \quad (3.6)$$

Since $T_{\bar{\alpha}\lambda} \colon \mathfrak{N}_\alpha \to \mathfrak{N}_{\bar{\lambda}}$, we have

$$T_{\bar{\alpha}\lambda} g_k = \sum_{m=1}^{r} a_{km}(\lambda)\,\varphi_m, \quad (3.7)$$

where $\{\varphi_m\}$ is some basis of the subspace $\mathfrak{N}_{\bar{\lambda}}$. On the other hand, since $T_{\bar{\lambda}\alpha}^* \colon \mathfrak{N}_{\bar{\lambda}} \to \mathfrak{N}_\alpha$, we get

$$T_{\bar{\lambda}\alpha}^* \varphi_m = \sum_{s=1}^{r} b_{ms}(\lambda)\,g_s. \quad (3.8)$$

Thus, by virtue (3.7) and (3.8), we have

$$T_{\bar\lambda\alpha}^* \, T_{\bar\alpha\lambda} g_k \;=\; \sum_{s=1}^{r} c_{ks}(\lambda)\, g_s, \tag{3.9}$$

where

$$c_{ks}(\lambda) \;=\; \sum_{m=1}^{r} a_{km}(\lambda)\, b_{ms}(\lambda). \tag{3.10}$$

By inserting (3.9) in (3.6), we obtain

$$\omega_A(\lambda)\, JG \;=\; C(\lambda)\, G \quad (C(\lambda) = \| c_{ki}(\lambda) \|). \tag{3.11}$$

If, in addition, $A(\lambda) = \| a_{ki}(\lambda) \|$, $B(\lambda) = \| b_{ki}(\lambda) \|$, then $C(\lambda) = A(\lambda)B(\lambda)$ by virtue of (3.10). Taking into account the expression for the matrix $\omega_A(\lambda)$, equalities (3.11), and the fact that the matrix G is invertible, we get

$$\chi_A^*(\alpha)\, J^{-1} \chi_A(\lambda) \;=\; A(\lambda)B(\lambda)\, J^{-1}. \tag{3.12}$$

For example, in the scalar case (i.e., if $r = 1$), equality (3.12) can be rewritten as

$$\overline{\chi_A(\alpha)}\, \chi_A(\lambda) \;=\; a(\lambda)\, b(\lambda), \tag{3.13}$$

where $a(\lambda)$ and $b(\lambda)$ are defined by the conditions

$$T_{\bar\alpha\lambda} g = a(\lambda)\, \varphi, \quad T_{\bar\lambda\alpha}^* \varphi = b(\lambda)\, g \quad (g \in \mathfrak{N}_\alpha, \; \varphi \in \mathfrak{N}_{\bar\lambda}). \tag{3.14}$$

Clearly, the matrix $A(\lambda)B(\lambda)$ in (3.12) is independent of the choice of a basis $\{\varphi_m\}$ in $\mathfrak{N}_{\bar\lambda}$ (in spite of the fact that $A(\lambda)$ and $B(\lambda)$ depend on this choice).

To determine the matrix J, we set $\lambda = \alpha$ in (1.5) and (3.12). Then

$$J^{-1} + 2\,\mathrm{Im}\,\alpha G \;=\; A(\lambda)B(\lambda)\, J^{-1}.$$

Therefore,

$$J \;=\; \frac{1}{2\,\mathrm{Im}\,\alpha}\, G^{-1}[A(\lambda)B(\lambda) - E]. \tag{3.15}$$

Thus, for $n = r$, the characteristic matrix function of the operator A is determined by equality (3.12) and the elements of the matrices $A(\lambda)$ and $B(\lambda)$ are given by equalities (3.7) and (3.8), respectively, in which $\{\varphi_m\}$ is an arbitrary fixed basis of the defect subspace $\mathfrak{N}_{\bar{\lambda}}$.

3.2. The Case of K_{II}^r-Operators

Assume that A is a K_{II}^r-operator, i.e., $A \in \mathcal{P}(H)$, where H is a symmetric operator with deficiency index (r, r). In view of the first equality in (3.4), we can rewrite (3.7) as $\cdot$ ᵗ

$$(\bar{\alpha} - \lambda) R_\lambda g_k = g_k - \sum_{m=1}^{r} a_{km}(\lambda) \varphi_m.$$

This yields

$$g_k - \sum_{m=1}^{r} a_{km}(\lambda) \varphi_m \in \mathfrak{D}_A. \tag{3.16}$$

In addition, since $\{\bar{\alpha}, \lambda\} \subset \rho(A)$, the subspaces $\mathfrak{N}_\alpha$ and $\mathfrak{N}_{\bar{\lambda}}$ are linearly independent with $\mathfrak{D}_A$ (see Corollary 1.6, Chapter 1). Consequently, the elements $a_{km}(\lambda)$ of the matrix $A(\lambda)$ are uniquely determined by conditions (3.16).

Similarly, taking into account the second equality in (3.4) and equality (3.8), we establish that the elements $b_{ms}(\lambda)$ of the matrix $B(\lambda)$ are uniquely determined by the conditions

$$\varphi_m - \sum_{s=1}^{r} b_{ms}(\lambda) g_{s\cdot} \in \mathfrak{D}_{A^*}. \tag{3.17}$$

Thus, in the case of K_{II}^r-operators, elements of the matrices $A(\lambda)$ and $B(\lambda)$ are determined by conditions (3.16) and (3.17), in which $\{g_k\}$ is a Δ-basis of the operator A and $\{\varphi_m\}$ is an arbitrary fixed basis of the defect subspace $\mathfrak{N}_{\bar{\lambda}}$.

For example, if $r = 1$, equality (3.12) can be rewritten in the form (3.13) with $a(\lambda)$ and $b(\lambda)$ satisfying the following conditions:

$$g - a(\lambda)\varphi. \in \mathfrak{D}_A, \qquad \varphi - b(\lambda)g. \in \mathfrak{D}_{A^*}. \tag{3.18}$$

In (3.18), g and φ are arbitrary fixed nonzero vectors belonging to the one-dimensional defect subspaces $\mathfrak{N}_\alpha$ and $\mathfrak{N}_{\bar{\lambda}}$, respectively.

3.3. The Case of K_I^r-Operators

Assume that $\mathfrak{D}_A = \mathfrak{D}_{A^*}$ and the operator $(A - A^*)/2i$ can be represented as

$$\frac{A - A^*}{2i} = \sum_{i,\ s=1}^{n} (\cdot, h_i) J_{is} h_s, \tag{3.19}$$

where $\{h_k\}$ is a system of vectors and $J = \|J_{ki}\|$ is an Hermitian invertible matrix. It is clear that, in the case under consideration,

$$B_\alpha = 2 R_\alpha^* \frac{A - A^*}{2i} R_\alpha. \tag{3.20}$$

Therefore, by virtue of (3.19) and (3.20), the operator B_α can be represented in the form (1.1) with $g_k = \sqrt{2}\, R_\alpha^* h_k$. Then

$$\| (T_{\alpha\lambda}\, g_k,\, g_m) \| = 2\| (R_\alpha^* h_k,\, R_\lambda^* h_m) \|$$

and, hence, equality (1.5) can be rewritten as follows:

$$\chi_A^*(\alpha) J^{-1} \chi_A(\lambda) = J^{-1} + 2i (\overline{\alpha} - \lambda) \| (R_\alpha^* h_k,\, R_\lambda^* h_m) \|. \tag{3.21}$$

In the scalar case $(n = 1)$, equality (3.21) takes the form

$$\overline{\chi_A(\alpha)}\, \chi_A(\lambda) = 1 + 2i (\overline{\alpha} - \lambda)(R_\alpha^* h,\, R_\lambda^* h)\, J, \tag{3.22}$$

where h is a vector defined by the equality $(A - A^*)/2i = (\cdot, h)\, h$. In particular, if

$$R_\lambda^* h_k = \sum_{i=1}^{n} r_{ki}(\lambda)\, h_i,$$

then

$$\| (R_\alpha^* h_k,\, R_\lambda^* h_m) \| = R(\alpha) \| (h_i,\, h_j) \| R^*(\lambda), \tag{3.23}$$

where $R(\lambda) = \| r_{ki}(\lambda) \|$. By virtue of (3.21) and (3.22), this yields

$$\chi_A^*(\alpha) J^{-1} \chi_A(\lambda) = J^{-1} + 2i (\overline{\alpha} - \lambda)\, R(\alpha) \| (h_i,\, h_j) \| R^*(\lambda) \tag{3.24}$$

If, in addition, $n = 1$ and $R_\lambda^* h = r(\lambda) h$, then, according to (3.22),

$$\overline{\chi_A(\alpha)} \chi_A(\lambda) = 1 + 2i(\overline{\alpha} - \lambda) \|h\|^2 r(\alpha)\overline{r(\lambda)} J. \tag{3.25}$$

3.4. Case of Deficiency Index (1, 1)

In the previous subsections, we have studied the case of the deficiency index $(1, 1)$ as a particular case of the corresponding general approach. Here, we consider operators with deficiency index $(1, 1)$ under the condition that they are given by the Neumann formulas.

Assume that $A \in \mathcal{P}(H)$, where H is an Hermitian operator with deficiency index $(1, 1)$. If $\alpha \overline{\in} \sigma_p(A)$, then an arbitrary element x from $\mathcal{D}_A$ can be represented as

$$x = x_0 + a(x_\alpha + \theta x_{\overline{\alpha}}), \tag{3.26}$$

where $x_0 \in \mathcal{D}_H$, $a \in \mathbb{C}$, and $x_{\overline{\alpha}}$ and x_α are fixed nonzero vectors from the defect subspaces $\mathfrak{N}_\alpha$ and $\mathfrak{N}_{\overline{\alpha}}$ of the operator H, respectively. Moreover,

$$Ax = Hx_0 + a(\overline{\alpha} x_\alpha + \alpha\theta x_{\overline{\alpha}}). \tag{3.27}$$

Since $B_\alpha: \mathcal{H} \to \mathfrak{N}_\alpha$ and $B_\alpha = (\cdot, g) Jg$, then $g = kx_\alpha$. This implies that $B_\alpha x_\alpha = |k|^2 \|x_\alpha\|^2 Jx_\alpha$. On the other hand, in view of the expression for B_α, we get

$$B_\alpha x_\alpha = iR_\alpha x_\alpha - iR_\alpha^*[x_\alpha - (\overline{\alpha} - \alpha) R_\alpha x_\alpha]. \tag{3.28}$$

Assume that $y = R_\alpha x_\alpha$. Since $y \in \mathcal{D}_A$, we have $y = y_0 + b(x_\alpha + \theta x_{\overline{\alpha}})$. Consequently,

$$x_\alpha = Ay - \alpha y = (H - \alpha I) y_0 + b(\overline{\alpha} - \alpha) x_\alpha. \tag{3.29}$$

In addition, since the vectors x_α and $(H - \alpha I) y_0$ are orthogonal and $\operatorname{Im} \alpha \neq 0$, we get $y_0 = 0$ and $b = 1/(\overline{\alpha} - \alpha)$ by virtue of (3.29). But then

$$R_\alpha x_\alpha = \frac{x_\alpha + \theta x_{\overline{\alpha}}}{\overline{\alpha} - \alpha}. \tag{3.30}$$

By inserting (3.30) in (3.28), we obtain

$$B_\alpha x_\alpha = iR_\alpha x_\alpha + i\theta R_\alpha^* x_{\overline{\alpha}}. \tag{3.31}$$

By using (3.26) and (3.27), we find that any element x from $\mathcal{D}_{A^*}$ can be represented as

$$x = x_0 + c(\bar{\theta} x_\alpha + x_{\bar{\alpha}}) \qquad (c \in \mathbb{C}) \tag{3.32}$$

and, moreover,

$$A^* x = H x_0 + c(\bar{\alpha}\,\bar{\theta} x_\alpha + \alpha x_{\bar{\alpha}}). \tag{3.33}$$

The last two equalities imply that if $|\theta| = 1$, then $A^* = A$. Therefore, below, we restrict ourselves to the case where $|\theta| \neq 1$. By analogy with the derivation of equality (3.30), we obtain

$$R_\alpha^* x_{\bar{\alpha}} = \frac{x_{\bar{\alpha}} + \theta x_\alpha}{\alpha - \bar{\alpha}}. \tag{3.34}$$

By substituting (3.30) and (3.34) in (3.31), we get

$$B_\alpha x_\alpha = \frac{|\theta|^2 - 1}{2 \operatorname{Im} \alpha} x_\alpha.$$

On the other hand, $B_\alpha x_\alpha = |k|^2 \|x_\alpha\|^2 J x_\alpha$. In view of the last two equalities, one can easily find that J and k can be represented in the form

$$J = \frac{|\theta|^2 - 1}{2 \operatorname{Im} \alpha}, \qquad k = \frac{1}{\|x_\alpha\|}.$$

In the case under consideration, by virtue of (1.5), we have

$$\overline{\chi_A(\alpha)} \chi_A(\lambda) = 1 + i(\bar{\alpha} - \lambda)(T_{\alpha\lambda}\,g, g) J,$$

where $g = k x_\alpha$. Therefore, taking the expressions for J and k into account, we obtain

$$\overline{\chi_A(\alpha)} \chi_A(\lambda) = 1 + i(\bar{\alpha} - \lambda)\frac{|\theta|^2 - 1}{2\|x_\alpha\|^2 \operatorname{Im} \lambda}(T_{\alpha\lambda} x_\alpha, x_\alpha). \tag{3.35}$$

Let $\lambda \in \rho(A)$ and $R_\lambda x_\alpha = \varphi = \varphi_0 + a(x_\alpha + \theta x_{\bar{\alpha}})$. Then

$$x_\alpha = A\varphi - \lambda\varphi = (H - \lambda I)\varphi_0 + a(\bar{\alpha} - \lambda)x_\alpha + a\theta(\alpha - \lambda)x_{\bar{\alpha}}. \tag{3.36}$$

Multiplying this equality scalarly by x_λ, we obtain

$$(x_\alpha, x_\lambda) = a(\bar{\alpha} - \lambda)(x_\alpha, x_\lambda) + a\theta(\alpha - \lambda)(x_{\bar{\alpha}}, x_\lambda). \tag{3.37}$$

If, in addition, $a = 0$, i.e., $R_\lambda x_\alpha = \varphi_0$, then $T_{\alpha\lambda} x_\alpha = (A - \alpha I)\varphi_0 \in \Delta_{H-\alpha I}$, and, consequently, $(T_{\alpha\lambda} x_\alpha, x_\alpha) = 0$. Thus,

$$\chi_A(\lambda) = \overline{\chi_A^{-1}(\alpha)},$$

for $a = 0$.

Let $a \neq 0$. Assume that $(x_\alpha, x_\lambda) = 0$ in (3.37), i.e., $x_\alpha \in \Delta_{H-\lambda I}$. Then $x_\alpha = (H - \lambda I)\psi = (A - \lambda I)\psi$, where $\psi \in \mathcal{G}_A$. On the other hand, $x_\alpha = (A - \lambda I)\varphi$ by virtue of (3.36). This implies that $(A - \lambda I)(\varphi - \psi) = 0$; here, $\varphi - \psi \neq 0$ because $\varphi \in \Delta_{H-\lambda I}$. The relation obtained contradicts the fact that $\lambda \notin \sigma_p(A)$.

Thus, if $a \neq 0$, then $(x_\alpha, x_\lambda) \neq 0$. By virtue of (3.37), this yields

$$a = \frac{(x_\alpha, x_\lambda)}{(\bar{\alpha} - \lambda)(x_\alpha, x_\lambda) + \theta(\alpha - \lambda)(x_{\bar{\alpha}}, x_\lambda)}. \tag{3.38}$$

Obviously, the denominator in (3.38) is not equal to zero.

Further, it follows from $R_\lambda x_\alpha = \varphi_0 + a(x_\alpha + \theta x_{\bar{\alpha}})$ that $T_{\alpha\lambda} x_\alpha = (H - \alpha I)\varphi_0 + a(\bar{\alpha} - \alpha)x_\alpha$ and, consequently,

$$(T_{\alpha\lambda} x_\alpha, x_\alpha) = a(\bar{\alpha} - \alpha)\|x_\alpha\|^2. \tag{3.39}$$

By inserting (3.39) in (3.35), we obtain

$$\overline{\chi_A(\alpha)}\chi_A(\lambda) = 1 + (\bar{\alpha} - \lambda)(|\theta|^2 - 1)a. \tag{3.40}$$

By virtue of (3.38) and (3.40), we get

$$\overline{\chi_A(\alpha)}\chi_A(\lambda) = \theta\frac{\bar{\theta}(\bar{\alpha} - \lambda)(x_\alpha, x_\lambda) + (\alpha - \lambda)(x_{\bar{\alpha}}, x_\lambda)}{(\bar{\alpha} - \lambda)(x_\alpha, x_\lambda) + \theta(\alpha - \lambda)(x_{\bar{\alpha}}, x_\lambda)}. \tag{3.41}$$

In particular, for $\lambda = \alpha$, we have $\overline{\chi_A(\alpha)}\chi_A(\alpha) = |\theta|^2$. By setting $\chi_A(\alpha) = \bar{\theta}$ in (3.41), we get

$$\chi_A(\lambda) = \frac{\bar{\theta}(\bar{\alpha} - \lambda)(x_\alpha, x_\lambda) + (\alpha - \lambda)(x_{\bar{\alpha}}, x_\lambda)}{(\bar{\alpha} - \lambda)(x_\alpha, x_\lambda) + \theta(\alpha - \lambda)(x_{\bar{\alpha}}, x_\lambda)}. \tag{3.42}$$

Remark. The principal results of this section were first presented (without proof) in [22].

4. Examples

4.1. One-Dimensional Operator

In $\mathcal{H}$, we consider an operator

$$A = a(\cdot, g)g \quad (\|g\| = 1, \quad \bar{a} \neq a). \tag{4.1}$$

For this operator, $\sigma_p(A) = \{a\}$, $G_A = \langle g \rangle^{\perp}$, and $\mathfrak{N}_\lambda = \langle g \rangle$ for $\lambda \neq a$.

Since $T_{\bar{\alpha}\lambda}: \mathfrak{N}_\alpha \to \mathfrak{N}_{\bar{\lambda}}$, we have $T_{\bar{\alpha}\lambda}\, g = a(\lambda)\, g$. This implies that $(A - \bar{\alpha}I)g = a(\lambda)(A - \lambda I)g$, and, consequently, $ag - \bar{\alpha}g = a(\lambda)(ag - \lambda g)$ by virtue of (4.1). Thus, $a(\lambda) = (a - \bar{\alpha})/(a - \lambda)$.

Similarly, it follows from $T_{\bar{\lambda}\alpha}^*\, g = b(\lambda)\, g$ and $A^*x = \bar{a}(x, g)\, g$ that $b(\lambda) = (\bar{\alpha} - \lambda)/(\bar{a} - \bar{\alpha})$. According to (3.13), this yields

$$\overline{\chi_A(\alpha)}\,\chi_A(\lambda) = a(\lambda)b(\lambda) = \frac{a - \bar{\alpha}}{\bar{a} - \bar{\alpha}}\frac{\bar{a} - \lambda}{a - \lambda}.$$

This implies that $\chi_A(\lambda) = (\bar{\alpha} - \lambda)/(a - \lambda)$. In addition, by virtue of (3.15), we have

$$J = \frac{1}{2\,\mathrm{Im}\,\alpha}[a(\lambda)b(\lambda) - 1] = \frac{\mathrm{Im}\,a}{|a - \alpha|^2}.$$

4.2. Integration Operator

In the space $L_2(0, a)$, we consider an operator A given by

$$(Af)(x) = i\int_0^x f(t)\,dt.$$

Then

$$(A^*f)(x) = -i\int_x^a f(t)\,dt$$

and, consequently,

$$\frac{A - A^*}{2i} = (\cdot, h) h, \qquad h(x) = \frac{1}{\sqrt{2}}.$$

Let $\varphi_\lambda = R_\lambda^* h$. Then

$$h(x) = (A^* \varphi_\lambda)(x) - \bar{\lambda}\varphi_\lambda(x) = -i \int_x^a \varphi_\lambda(t)\,dt - \bar{\lambda}\varphi_\lambda(x).$$

This implies that

$$\varphi_\lambda(x) = \frac{-1}{\lambda\sqrt{2}} \exp\left\{\frac{i}{\lambda}(x - a)\right\}$$

and, hence,

$$(R_\alpha^* h, \; R_\lambda^* h) = (\varphi_\alpha, \varphi_\lambda) = \frac{1}{2i(\lambda - \bar{\alpha})}\left[1 - e^{ia/\lambda}e^{-ia/\bar{\alpha}}\right]. \qquad (4.2)$$

By substituting (4.2) in (3.22) and taking into account that $J = 1$ in the case under consideration, we obtain $\overline{\chi_A(\alpha)}\chi_A(\lambda) = e^{-ia/\bar{\alpha}} e^{ia/\lambda}$ and, therefore, $\chi_A(\lambda) = e^{ia/\lambda}$.

4.3. Differentiation Operator

In the space $L_2(0, a)$, we consider the minimal differential operator H defined on the set $\mathcal{D} \subset L_2(0, a)$ of absolutely continuous functions as follows[2]:

$$Hf = -if' \quad (f(0) = f(a) = 0). \qquad (4.3)$$

Obviously, H is a closed symmetric operator with deficiency index $(1, 1)$. The defect subspace of this operator is

$$\mathfrak{N}_\lambda = \mathrm{Ker}\,(H^* - \bar{\lambda}I) = \langle e^{i\bar{\lambda}x}\rangle.$$

Furthermore, $H^* f = -if'$ and $\mathcal{D}_{H^*} = \mathcal{D}$.

Consider the following extension A of the operator H:

$$Af = H^* f, \qquad \mathcal{D}_A = \{f \in \mathcal{D}_{H^*} \,|\, f(a) = \theta f(0)\},$$

---

[2] See, e.g., Akhiezer and Glazman [1, Subsection 55, p. 144].

where θ is a fixed complex number. The operator A^* is also an extension of the operator H and, furthermore, $\mathcal{D}_{A^*} = \{f \in \mathcal{D}_{H^*} | f(0) = \bar{\theta}f(a)\}$. Consequently, $A^* = A$ if and only if $|\theta| = 1$. Therefore, below, we can restrict ourselves to the case where $|\theta| \neq 1$.

The roots of the equation $e^{i\lambda a} = \theta$ are eigenvalues of the operator A. Therefore, $\theta \bar{\in} \{e^{i\alpha a}, e^{i\bar{\alpha}a}\}$ because, by assumption, $\{\alpha, \bar{\alpha}\} \subset \rho(A)$.

Let $g(x) = e^{i\bar{\alpha}x}$ and $\varphi(x) = e^{i\lambda x}$ ($\lambda \in \rho(A)$). Since $g \in \mathfrak{N}_\alpha$ and $\varphi \in \mathfrak{N}_{\bar{\lambda}}$, it follows from (3.18) that the function $a(\lambda)$ can be determined from the condition $\psi = g - a(\lambda)\varphi \in \mathcal{D}_A$, i.e., from the condition

$$e^{i\bar{\alpha}x} - a(\lambda)e^{i\lambda a} = \theta(1 - a(\lambda)).$$

Consequently,

$$a(\lambda) = \frac{e^{i\bar{\alpha}a} - \theta}{e^{i\lambda a} - \theta}.$$

Similarly, taking into account that $h = \varphi - b(\lambda)g \in \mathcal{D}_{A^*}$, we obtain

$$b(\lambda) = \frac{1 - \bar{\theta}e^{i\lambda a}}{1 - \bar{\theta}e^{i\bar{\alpha}a}}.$$

Thus,

$$\overline{\chi_A(\alpha)}\chi_A(\lambda) = a(\lambda)b(\lambda) = \frac{e^{i\bar{\alpha}a} - \theta}{1 - \bar{\theta}e^{i\bar{\alpha}a}} \frac{1 - \bar{\theta}e^{i\lambda a}}{e^{i\lambda a} - \theta} \tag{4.4}$$

and, therefore, we can take the function $\chi_A(\lambda)$ in the form

$$\chi_A(\lambda) = \frac{1 - \bar{\theta}e^{i\lambda a}}{e^{i\lambda a} - \theta}. \tag{4.5}$$

By using equality (3.15) and (4.4), we get

$$J = \frac{1 - |\theta|^2}{|e^{i\alpha a} - \theta|^2 e^{2a\operatorname{Im}\alpha}}.$$

For $|\theta| < 1$, we have $J > 0$ and, hence, A is a dissipative operator.

4.4. Nonself-Adjoint Sturm–Liouville Operator on a Semi-Axis. Relation between the Characteristic Function and the Weyl Function

In the space $L_2(0, \infty)$, we consider a Sturm–Liouville operator A, which is given in a standard way by the differential expression

$$l(y) = -y'' + qy \tag{4.6}$$

and the boundary condition

$$y'(0) - hy(0) = 0, \tag{4.7}$$

where $\bar{h} \neq h$ and q is a real function which realizes the case of a limit point.

For convenience, we consider the linear functional

$$F_h(y) = y'(0) - hy(0). \tag{4.8}$$

Denote by $\theta(x, \lambda)$ and $\varphi(x, \lambda)$ the solutions of the differential equation

$$l(y) = \lambda y \tag{4.9}$$

which satisfy the conditions

$$\theta(0, \lambda) = \cos\beta, \qquad \varphi(0, \lambda) = \sin\beta,$$

$$\theta'_x(0, \lambda) = \sin\beta, \qquad \varphi'_x(0, \lambda) = -\cos\beta. \tag{4.10}$$

Assume also that $m(\lambda)$ is the corresponding Weyl function. Then the function

$$g(x, \lambda) = \theta(x, \lambda) + m(\lambda)\varphi(x, \lambda) \tag{4.11}$$

belongs to the space $L_2(0, \infty)$ and is a solution of equation (4.9).

Let $g_{\bar{\lambda}}(x) = g(x, \lambda)$. Then $l(g_{\bar{\lambda}}) = \lambda g_{\bar{\lambda}}$ and, thus, $g_{\bar{\lambda}} \in \mathfrak{N}_{\bar{\lambda}}$. If, in addition, $\lambda \in \sigma_p(A)$, then $g_{\bar{\lambda}} \in \mathfrak{D}_A$ and, consequently, $F_h(g_{\bar{\lambda}}) \neq 0$.

To determine the characteristic function, we consider relation (3.18) with $g = g_\alpha$ and $\varphi = g_{\bar{\lambda}}$. Since

$$\psi = g_\alpha - a(\lambda)g_{\bar{\lambda}} \in \mathfrak{D}_A,$$

we have $F_h(\psi) = F_h(g_\alpha) - a(\lambda)F_h(g_{\bar{\lambda}}) = 0$. This implies that

$$a(\lambda) = \frac{F_h(g_\alpha)}{F_h(g_{\bar\lambda})}.$$ (4.12)

Similarly, by virtue of (3.18), we have

$$b(\lambda) = \frac{F_{\bar h}(g_{\bar\lambda})}{F_{\bar h}(g_\alpha)}.$$ (4.13)

Thus, in view of (3.13), (4.12) and (4.13),

$$\overline{\chi_A(\alpha)}\,\chi_A(\lambda) = \frac{F_h(g_\alpha)\,F_{\bar h}(g_{\bar\lambda})}{F_{\bar h}(g_\alpha)\,F_h(g_{\bar\lambda})}$$ (4.14)

Let us show that $\bar g_\lambda = g_{\bar\lambda}$. Indeed, since $l(\bar\theta) = \bar\lambda\bar\theta,$ we have $\overline{\theta(x,\ \lambda)} = \theta(x,\ \bar\lambda)$ by virtue of (4.10). Similarly,

$$\overline{\varphi(x,\ \lambda)} = \varphi(x,\ \bar\lambda).$$

By virtue of (4.11), we have

$$\overline{g_\lambda(x)} = \theta(x,\ \bar\lambda) + \overline{m(\lambda)}\,\varphi(x,\ \bar\lambda),$$

where $\bar g_\lambda \in L_2(0,\ \infty)$. On the other side, $g_{\bar\lambda} \in L_2(0,\ \infty)$, where

$$g_{\bar\lambda}(x) = \theta(x,\ \bar\lambda) + m(\bar\lambda)\,\varphi(x,\ \bar\lambda).$$

Consequently, $\overline{m(\lambda)} = m(\bar\lambda)$ and $\bar g_\lambda = g_{\bar\lambda}$.
Therefore, we have

$$F_h(g_\alpha) = F_h(\bar g_{\bar\alpha}) = \overline{F_{\bar h}(g_{\bar\alpha})}, \qquad F_{\bar h}(g_\alpha) = \overline{F_h(g_{\bar\alpha})}.$$ (4.15)

By virtue of (4.14) and (4.15), this implies that

$$\overline{\chi_A(\alpha)}\,\chi_A(\lambda) = \left(\frac{\overline{F_{\bar h}(g_{\bar\alpha})}}{F_h(g_{\bar\alpha})}\right)\frac{F_{\bar h}(g_{\bar\lambda})}{F_h(g_{\bar\lambda})}.$$

This yields

$$\chi_A(\lambda) = \frac{F_{\bar h}(g_{\bar\lambda})}{F_h(g_{\bar\lambda})}.$$ (4.16)

Note that, in view of (4.8) and (4.10), we have

$$F_h(\theta) = \sin \beta - h \cos \beta, \quad F_h(\varphi) = -\cos \beta - h \sin \beta. \tag{4.17}$$

By virtue of (4.11) and (4.17), we obtain

$$F_h(g_{\bar{\lambda}}) = (1 - hm(\lambda)) \sin \beta - (h + m(\lambda)) \cos \beta,$$

$$F_{\bar{h}}(g_{\bar{\lambda}}) = (1 - \bar{h}m(\lambda)) \sin \beta - (\bar{h} + m(\lambda)) \cos \beta.$$

Thus,

$$\chi_A(\lambda) = \frac{(1 - \bar{h}m(\lambda)) \sin \beta - (\bar{h} + m(\lambda)) \cos \beta}{(1 - hm(\lambda)) \sin \beta - (h + m(\lambda)) \cos \beta}.$$

In particular, for $\beta = 0$, we have

$$\chi_A(\lambda) = \frac{\bar{h} + m(\lambda)}{h + m(\lambda)}, \tag{4.18}$$

whence

$$m(\lambda) = \frac{h\chi_A(\lambda) - \bar{h}}{1 - \chi_A(\lambda)}.$$

4.5. Nonself-Adjoint Operators Generated by Jacobian Matrices

Consider the Jacobian matrix

$$\begin{Vmatrix} a_0 & b_0 & 0 & 0 & \cdots \\ b_0 & a_1 & b_1 & 0 & \cdots \\ 0 & b_1 & a_2 & b_2 & \cdots \\ \cdots\cdots\cdots\cdots\cdots \\ \cdots\cdots\cdots\cdots\cdots \end{Vmatrix} \quad (b_k > 0, \ \bar{a}_k = a_k, \ k = 0, 1, 2 \ldots). \tag{4.19}$$

In a separable Hilbert space $\mathcal{H}$, we define a linear operator H' as follows:

$$H' e_k = b_{k-1} e_{k-1} + a_k e_k + b_k e_{k+1} \quad (k = 0, 1, 2 \ldots; \ b_{-1} = 0),$$

where $\{e_k\}_{k=0}^{\infty}$ is an orthonormal basis in $\mathcal{H}$. The operator H' is densely defined and Hermitian; therefore, it is symmetric. Denote by H the closure of the operator H'. It is well known that the operator H is either a self-adjoint operator or a symmetric operator with deficiency index $(1, 1)$. Below, we consider the latter case. Then (see, e.g., Akhiezer [1, p. 175]) the polynomials $P_k(\lambda)$ $(k = 0, 1, 2 \ldots)$ given by the relations

$$b_{k-1} P_{k-1}(\lambda) + a_k P_k(\lambda) + b_k P_{k+1}(\lambda) = \lambda P_k(\lambda), \quad k \in \mathbb{N},$$

$$P_0(\lambda) = 1, \quad P_1(\lambda) = \frac{\lambda - a_0}{b_0}$$

satisfy the condition

$$\sum_{k=0}^{\infty} |P_k(\lambda)|^2 < \infty \quad (\operatorname{Im}\lambda \ne 0).$$

In this case, the defect subspace $\mathfrak{N}_\lambda$ of the operator H is spanned by the vector

$$x_\lambda = \sum_{k=0}^{\infty} P_k(\bar{\lambda}) e_k$$

Consider the function

$$h(\lambda, \mu) = \sum_{k=0}^{\infty} P_k(\lambda) P_k(\mu).$$

Since $\overline{P_k(\lambda)} = P_k(\bar{\lambda})$, we have $h(\lambda, \mu) = (x_{\bar{\mu}}, x_\lambda)$.

Assume that the extension A_θ of the operator H is defined by the equality

$$A_\theta x = H^* x = H x_0 + a(ix_{-i} - i\theta x_i)$$

on the vectors

$$x = x_0 + a(x_{-i} + \theta x_i) \quad (x_0 \in \mathfrak{D}_H, \ a \in \mathbb{C});$$

here, θ is a fixed complex number such that $|\theta| \ne 1$. Then

$$A_\theta^* = A_{1/\bar{\theta}} \ne A_\theta.$$

By setting $\alpha = -i$ in equality (3.42) and taking the equality $(x_{\bar{\mu}}, x_\lambda) = h(\lambda, \mu)$ into account, we obtain the following expression for the characteristic function of the operator A_θ:

$$\chi_{A_\theta}(\lambda) = \frac{\overline{\theta}(i-\lambda)h(\lambda,\ i) - (i+\lambda)h(\lambda,\ -i)}{(i-\lambda)h(\lambda;\ i) - \theta(i+\lambda)h(\lambda,\ -i)}.$$

Remark. The presentation in this section is based on the author's works [12–14, 22]. For more information about the Weyl functions, see, e.g., the monographs of Titchmarsh [1], Levitan and Sargsjan [1], and Kostjuchenko and Sargsjan [1].

5. Properties of Characteristic Matrix Functions

5.1. *J*-Nonexpandability

Let $M = J^{-1}\chi_A(\alpha)$. Then, by virtue of (1.5),

$$M^*\chi_A(\lambda) = J^{-1} + i(\overline{\alpha} - \lambda)\|(T_{\alpha\lambda}g_k,\ g_m)\|,$$

$$\chi_A^*(\mu)M = J^{-1} + i(\overline{\mu} - \alpha)\|(T_{\alpha\mu}^*g_k,\ g_m)\|.$$

Consequently,

$$M^*\chi_A(\lambda)J\chi_A^*(\mu)M = J^{-1} + i(\overline{\alpha} - \lambda)\|(T_{\alpha\lambda}g_k,\ g_m)\|$$

$$+ i(\overline{\mu} - \alpha)\|(T_{\alpha\mu}^*g_k,\ g_m)\| + (\overline{\alpha} - \lambda)(\alpha - \overline{\mu})F, \qquad (5.1)$$

where, in view of equality (1.1),

$$F = \|(T_{\alpha\lambda}g_k,\ g_m)\|\,J\,\|(T_{\alpha\mu}^*g_k,\ g_m)\|$$

$$= \left\|\sum_{i,\ s=1}^{n}(T_{\alpha\lambda}g_k,\ g_i)J_{is}(T_{\alpha\mu}^*g_s,\ g_m)\right\|$$

$$= \left\|\left(T_{\alpha\mu}^*\sum_{i,\ s=1}^{n}(T_{\alpha\lambda}g_k,\ g_i)J_{is}g_s,\ g_m\right)\right\| = \|(T_{\alpha\mu}^*B_\alpha T_{\alpha\lambda}g_k,\ g_m)\|. \qquad (5.2)$$

Furthermore, by virtue of (1.24), we have

$$J^{-1} = M^* JM + i(\alpha - \bar{\alpha}) \| (g_k, g_m) \|. \tag{5.3}$$

By inserting (5.2) and (5.3) in (5.1), we obtain

$$M^* [\chi_A(\lambda) J \chi_A^*(\mu) - J] M = \| (Ng_k, g_m) \|, \tag{5.4}$$

where

$$N = i(\alpha - \bar{\alpha})I + i(\bar{\alpha} - \lambda) T_{\alpha\lambda} + i(\bar{\mu} - \alpha) T_{\alpha\mu}^* + (\bar{\alpha} - \lambda)(\alpha - \bar{\mu}) T_{\alpha\mu}^* B_\alpha T_{\alpha\lambda}.$$
$$\tag{5.5}$$

Since $T_{\alpha\lambda} T_{\lambda\alpha} = I$ and $T_{\alpha\mu}^* T_{\mu\alpha}^* = I$, we can rewrite (5.5) in the form

$$N = T_{\alpha\mu}^* [i(\alpha - \bar{\alpha}) T_{\mu\alpha}^* T_{\lambda\alpha} + i(\bar{\alpha} - \lambda) T_{\mu\alpha}^*$$

$$+ i(\bar{\mu} - \alpha) T_{\lambda\alpha} + (\bar{\alpha} - \lambda)(\alpha - \bar{\mu}) B_\alpha] T_{\alpha\lambda}, \tag{5.6}$$

where $T_{\lambda\alpha} = (A - \lambda I)(A - \alpha I)^{-1} = I + (\alpha - \lambda) R_\alpha$ and, analogously, $T_{\mu\alpha}^* = I + (\bar{\alpha} - \bar{\mu}) R_\alpha^*$. Taking into account the last two equalities and the expression for the operator B_α, we get

$$N = T_{\alpha\mu}^* [aI + bR_\alpha + cR_\alpha^* + dR_\alpha^* R_\alpha] T_{\alpha\lambda}, \tag{5.7}$$

where

$$a = i(\alpha - \bar{\alpha}) + i(\bar{\alpha} - \lambda) + i(\bar{\mu} - \alpha) = i(\bar{\mu} - \lambda),$$

$$b = i(\alpha - \bar{\alpha})(\alpha - \lambda) + i(\bar{\mu} - \alpha)(\alpha - \lambda) + i(\bar{\alpha} - \lambda)(\alpha - \bar{\mu})$$

$$= i(\bar{\mu} - \lambda)(\alpha - \bar{\alpha}),$$

$$c = i(\alpha - \bar{\alpha})(\bar{\alpha} - \bar{\mu}) + i(\bar{\alpha} - \lambda)(\bar{\alpha} - \bar{\mu}) + i(\bar{\alpha} - \lambda)(\bar{\mu} - \alpha)$$

$$= i(\bar{\mu} - \lambda)(\bar{\alpha} - \alpha),$$

$$d = i(\alpha - \bar{\alpha})(\bar{\alpha} - \bar{\mu})(\alpha - \lambda) + i(\alpha - \bar{\alpha})(\bar{\alpha} - \lambda)(\bar{\mu} - \alpha)$$

$$= -i(\bar{\mu} - \lambda)(\alpha - \bar{\alpha})^2,$$

Substituting these expressions for $a, b, c,$ and d in (5.7), we obtain

$$N = i(\bar{\mu} - \lambda) T^*_{\alpha\mu}[I + (\alpha - \bar{\alpha})(R_\alpha - R^*_\alpha + (\bar{\alpha} - \alpha) R^*_\alpha R_\alpha)] T_{\alpha\lambda}$$

or, in other words,

$$N = i(\bar{\mu} - \lambda) T^*_{\alpha\mu}[I + i(\bar{\alpha} - \alpha) B_\alpha] T_{\alpha\lambda}.$$

Since

$$B_\alpha = \frac{1}{2 \operatorname{Im} \alpha} [T^*_{\bar{\alpha}\alpha} T_{\bar{\alpha}\alpha} - I],$$

we have

$$N = i(\bar{\mu} - \lambda) T^*_{\alpha\mu} T^*_{\bar{\alpha}\alpha} T_{\bar{\alpha}\alpha} T_{\alpha\lambda} = i(\bar{\mu} - \lambda) T^*_{\bar{\alpha}\mu} T_{\bar{\alpha}\lambda}. \tag{5.8}$$

Thus, in view of (5.4) and (5.8), we get

$$\chi_A(\lambda) J \chi_A^*(\mu) - J = i(\bar{\mu} - \lambda) M^{-*} \| T_{\bar{\alpha}\lambda} g_k, T_{\bar{\alpha}\mu} g_m \| M^{-1}.$$

This implies that

$$\chi_A(\bar{\lambda}) J \chi_A^*(\lambda) = J \tag{5.9}$$

for $\{\lambda, \bar{\lambda}\} \subset \rho(A)$. Similarly, for $\mu = \lambda (\in \rho(A))$, we have

$$\chi_A(\lambda) J \chi_A^*(\lambda) - J = 2 \operatorname{Im} \lambda M^{-*} \| T_{\bar{\alpha}\lambda} g_k, T_{\bar{\alpha}\lambda} g_m \| M^{-1} \tag{5.10}$$

and, consequently,

$$\chi_A(\lambda) J \chi_A^*(\lambda) \leq J \quad (\lambda \in \rho(A), \quad \operatorname{Im} \lambda \leq 0), \tag{5.11}$$

$$\chi_A(\lambda) J \chi_A^*(\lambda) \geq J \quad (\lambda \in \rho(A), \quad \operatorname{Im} \lambda \geq 0). \tag{5.12}$$

5.2. The Case of Dissipative K_1^r-Operators

The characteristic matrix function $\chi_A(\lambda)$ is called *minimal* if its Δ-basis is, at the same time, a basis of the defect subspace $\mathfrak{N}_\alpha$ of the Hermitian part A_0 of the operator A.

Theorem 5.1. *If A is a dissipative K_I^r-operator, then the limits*

$$V_- = \lim_{\mathrm{Im}\,\lambda \to -\infty} \chi_A(\lambda) \quad \text{and} \quad V_+ = \lim_{\mathrm{Im}\,\lambda \to +\infty} \chi_A(\lambda), \tag{5.13}$$

where $\chi_A(\lambda)$ is the minimal characteristic matrix function of the operator A exist and coincide. Moreover, the matrix $V = V_-\ (= V_+)$ is J-unitary (i.e., $VJV^* = J$).

Proof. Since $\mathfrak{D}_{A^*} = \mathfrak{D}_A$, by virtue of Theorem 2.7 (Chapter 1), we have $\mathfrak{N}_\alpha \subset \mathfrak{D}_A$ and, consequently, for any vector g from $\mathfrak{N}_\alpha$,

$$\|\lambda R_\lambda g + g\| = \|R_\lambda A g\| \le \frac{1}{|\mathrm{Im}\,\lambda|} \|Ag\| \to 0$$

as $\mathrm{Im}\,\lambda \to -0$. Thus,

$$s\text{-}\lim_{\mathrm{Im}\,\lambda \to -\infty} \lambda R_\lambda g = -g \quad (\forall\, g \in \mathfrak{N}_\alpha). \tag{5.14}$$

In view of the fact that

$$(\bar{\alpha} - \lambda)(T_{\alpha\lambda} g_k, g_m) = \frac{\bar{\alpha} - \lambda}{\lambda}(\lambda R_\lambda g_k, (A^* - \bar{\alpha}I) g_m),$$

we get

$$\lim_{\mathrm{Im}\,\lambda \to -\infty} (\bar{\alpha} - \lambda)(T_{\alpha\lambda} g_k, g_m) = (g_k, (A^* - \bar{\alpha}I) g_m) \tag{5.15}$$

by (5.14). Thus, by virtue of (1.23) and (5.15), the first limit in (5.13) exists. Moreover, by passing to the limit in (5.10) as $\mathrm{Im}\,\lambda \to -\infty$, we obtain

$$V_- J V_-^* = J \tag{5.16}$$

and, consequently, $\det V_- \ne 0$.

Passing to the limit in (5.9) as $\mathrm{Im}\,\lambda \to -\infty$, we conclude that the second limit in (5.13) exists and, moreover, the relation

$$V_+ J V_-^* = J \tag{5.17}$$

is true.

It follows from (5.16) and (5.17) that $V_+ = V_-$, and, consequently,[3] $VJV^* = J$, where $V = V_- = V_+$.

∎

Remark 5.2. It is clear that Theorem 5.1 remains true with the condition of mini-
mality of a characteristic matrix function replaced by the condition that a Δ-basis of the
operator A lies in $\mathfrak{D}_A$.

Theorem 5.1 also remains valid if the condition that the operator A is dissipative is
replaced by the following condition: $\| R_\lambda \| \to 0$ as $\operatorname{Im} \lambda \to -\infty$.

If, for some dissipative operator A, $V_+ \neq V_-$, then by virtue of Theorem 5.1, this
operator is not a K_Γ-operator (i.e., $\mathfrak{D}_{A^*} \neq \mathfrak{D}_A$). Thus, for example, the operator of differentiation introduced in Subsection 4.3 is an operator of this type. In fact, by using
equality (4.5), under the assumption that the real part of λ is constant, we conclude that
$V_- = -\bar{\theta}$ and $V_+ = -1/\theta$ and, consequently, the equality $V_+ = V_-$ holds only if
$|\theta| = 1$ (i.e., in the case where the operator A is self-adjoint).

5.3. Relationship between Characteristic Matrix Functions Defined in Different
Δ-Bases

Assume that the characteristic matrix function $\chi_A(\lambda)$ of the operator A is defined by
the equality

$$\chi_A^*(\alpha) J^{-1} \chi_A(\lambda) = J^{-1} + i(\bar{\alpha} - \lambda) \| (T_{\alpha\lambda} g_k, g_m) \|, \qquad (5.18)$$

where $\{g_k\}_{k=1}^n$ is a Δ-basis of the operator A and J is the corresponding coefficient
matrix. Consider the vectors

$$\tilde{g}_k = \sum_{i=1}^n \omega_{ki} g_i \qquad (k = \overline{1, n}), \qquad (5.19)$$

where $\omega = \| \omega_{ki} \|$ is an invertible $n \times n$-matrix. As shown in Subsection 1.1, the sys-
tem $\{\tilde{g}_k\}$ is a Δ-basis of the operator A, and the corresponding coefficient matrix is
$\tilde{J} = \omega^{-*} J \omega^{-1}$.

By analogy with (5.18), we can represent the characteristic matrix function of the
operator A in the Δ-basis $\{\tilde{g}_k\}_{k=1}^n$ as follows:

$$\tilde{\chi}_A^*(\alpha) \tilde{J}^{-1} \tilde{\chi}_A(\lambda) = \tilde{J}^{-1} + i(\bar{\alpha} - \lambda) \| (T_{\alpha\lambda} \tilde{g}_k, \tilde{g}_m) \|, \qquad (5.20)$$

---

[3] Note that if $VJV^* = J$, then $V^* J^{-1} V = J^{-1}$.

where $\tilde{J}^{-1} = \omega J^{-1} \omega^*$ and, as is easy to see,

$$\| (T_{\alpha\lambda} \, \tilde{g}_k, \, \tilde{g}_m) \| = \omega \| (T_{\alpha\lambda} g_k, \, g_m) \| \omega^*.$$

Multiplying (5.20) from the left by ω^{-1} and from the right by ω^{-*}, we obtain

$$\omega^{-1} \tilde{\chi}_A^*(\alpha) \omega J^{-1} \omega^* \tilde{\chi}_A(\lambda) \omega^{-*} = \chi_A^*(\alpha) J^{-1} \chi_A(\lambda)$$

by virtue of equalities (5.18) and (5.20). Consequently, $\tilde{\chi}_A(\lambda)$ can be chosen in the form of the following matrix function:

$$\tilde{\chi}_A(\lambda) = \omega^{-*} \chi_A(\lambda) \omega^*.$$

Thus, if the Δ-bases $\{\tilde{g}_k\}$ and $\{g_k\}$ of the operator A are connected by equality (5.19), then for each characteristic matrix function $\chi_A(\lambda)$ represented in the Δ-basis $\{g_k\}$, one can indicate a similar characteristic matrix function $\tilde{\chi}_A(\lambda)$ represented in the Δ-basis $\{\tilde{g}_k\}$.

5.4. Relationship between "Symmetric" Characteristic Matrix Functions

Assume that a characteristic matrix function $\chi_A(\lambda)$ is defined by equality (5.18). For convenience, we rewrite this equality in the form

$$W_\alpha(\lambda) = J^{-1} + i(\bar{\alpha} - \lambda) \| (T_{\alpha\lambda} g_k, \, g_m) \|, \qquad (5.21)$$

where

$$W_\alpha(\lambda) = \chi_A^*(\alpha) J^{-1} \chi_A(\lambda).$$

Since

$$B_\alpha = \frac{1}{2 \, \mathrm{Im} \, \alpha} \, [T_\alpha^* T_\alpha - I],$$

where $T_\alpha = (A - \bar{\alpha} I) R_\alpha \; (= T_{\bar{\alpha}\alpha})$, by virtue of the equality $T_\alpha T_{\bar{\alpha}} = I$, we obtain

$$T_{\bar{\alpha}}^* B_\alpha T_{\bar{\alpha}} = \frac{1}{2 \, \mathrm{Im} \, \alpha} \, [I - T_{\bar{\alpha}}^* T_{\bar{\alpha}}] = B_{\bar{\alpha}}.$$

But then

$$B_{\bar{\alpha}} = \sum_{i,\,s=1}^{n} (\cdot,\,\tilde{g}_i)\,J_{is}\,\tilde{g}_s \qquad (\tilde{g}_k = T_{\bar{\alpha}}^* g_k). \tag{5.22}$$

Thus, the system of vectors $\{\tilde{g}_k\}$, given by equalities (5.22), is a Δ-basis of the operator A. Furthermore, the corresponding coefficient matrix remains unchanged. Consequently, we can consider the characteristic matrix function $\tilde{\chi}_A(\lambda)$ defined by the equality

$$W_{\bar{\alpha}}(\lambda) = J^{-1} + i(\alpha - \lambda)\|(T_{\bar{\alpha}\lambda}\,\tilde{g}_k,\,\tilde{g}_m)\|, \tag{5.23}$$

where $W_{\bar{\alpha}}(\lambda) = \tilde{\chi}_A^*(\bar{\alpha})J^{-1}\tilde{\chi}_A(\lambda)$. The characteristic matrix function defined in this way is called *symmetric* with respect to the characteristic matrix function $\chi_A(\lambda)$.

Later (see Section 6), we shall need the relationship between $\tilde{\chi}_A(\lambda)$ and $\chi_A(\lambda)$. In particular, for a linearly independent Δ-basis $\{g_k\}_{k=1}^n$, this relationship is expressed by the relation

$$W_{\bar{\alpha}}(\lambda) = \tilde{\tau}^* W_\alpha(\lambda), \tag{5.24}$$

where $\tilde{\tau}$ is the deviation matrix in the Δ-basis $\{\tilde{g}_k\}_{k=1}^n$.

To prove equality (5.24), we note that

$$(T_{\bar{\alpha}\lambda}\,\tilde{g}_k,\,\tilde{g}_m) = (T_{\bar{\alpha}}^* g_k,\,T_{\bar{\alpha}\lambda}^* T_{\bar{\alpha}}^* g_m) = (T_{\bar{\alpha}} T_{\bar{\alpha}}^* g_k,\,T_{\bar{\alpha}\lambda}^* g_m). \tag{5.25}$$

Let us show that

$$T_{\bar{\alpha}} T_{\bar{\alpha}}^* g_k \in \mathcal{L}_A \qquad (k = \overline{1,\,n}), \tag{5.26}$$

where $\mathcal{L}_A = \langle g_1, g_2, \ldots, g_n \rangle$. Actually, assume that $y \perp \mathcal{L}_A$. Then $y \perp \mathfrak{N}_\alpha$ (since $\mathfrak{N}_\alpha \subset \mathcal{L}_A$) and, consequently, $B_\alpha y = 0$ or, in other words, $T_\alpha^* T_\alpha y = y$. But then, in view of the fact that $T_\alpha T_{\bar{\alpha}} = I$, we have

$$(T_{\bar{\alpha}} T_{\bar{\alpha}}^* g_k,\,y) = (T_{\bar{\alpha}} T_{\bar{\alpha}}^* g_k,\,T_\alpha^* T_\alpha y) = (g_k,\,y) = 0,$$

which implies (5.26).

Thus,

$$T_{\bar{\alpha}} T_{\bar{\alpha}}^* g_k = \sum_{i=1}^{n} c_{ki}\,g_i.$$

By using this equality and (5.25), we get

$$\| (T_{\bar{\alpha}\lambda}\, \tilde{g}_k,\, \tilde{g}_m) \| \;=\; c\, \| (T_{\bar{\alpha}\lambda}\, g_k,\, g_m) \| \qquad (c = \| c_{ki} \|).$$

Furthermore, since $(\alpha - \lambda)\, T_{\bar{\alpha}\lambda} = (\alpha - \bar{\alpha})\, I + (\bar{\alpha} - \lambda)\, T_{\alpha\lambda}$, we get

$$(\alpha - \lambda)\, \| (T_{\bar{\alpha}\lambda}\, \tilde{g}_k,\, \tilde{g}_m) \| \;=\; (\alpha - \bar{\alpha})\, cG \;+\; (\bar{\alpha} - \lambda)\, c\, \| (T_{\alpha\lambda}\, g_k,\, g_m) \|. \qquad (5.27)$$

This implies that, by virtue of (5.21), (5.23), and (5.27),

$$W_{\bar{\alpha}}\,(\lambda) \;=\; J^{-1} \;+\; i\,(\alpha - \bar{\alpha})\, cG \;+\; c\,\bigl(W_{\alpha}\,(\lambda) - J^{-1} \bigr),$$

i.e.,

$$W_{\bar{\alpha}}\,(\lambda) \;=\; M \;+\; c\, W_{\alpha}\,(\lambda), \qquad\qquad (5.28)$$

where $M = J^{-1} - 2\,\mathrm{Im}\,\alpha cG - cJ^{-1}$. Further, since $cG = \tilde{G}$, where $\tilde{G} = \| (\tilde{g}_k,\, \tilde{g}_m) \|$, we obtain

$$MJ \;=\; E \;-\; 2\,\mathrm{Im}\,\alpha\,\tilde{G}J \;-\; c \;=\; \tilde{\tau}^{*} \;-\; c, \qquad\qquad (5.29)$$

where $\tilde{\tau} = E - 2\,\mathrm{Im}\,\alpha J\tilde{G}$ is the deviation matrix in the Δ-basis $\{\tilde{g}_k\}$.

To study the properties of the matrix M, we scalarly multiply the equality

$$T_{\bar{\alpha}}\, T_{\bar{\alpha}}^{*}\, g_k \;=\; \sum_{i=1}^{n} c_{ki}\, g_i$$

by the vector $T_{\alpha}^{*}\, T_{\alpha}\, g_m$. As a result, in view of the equality $T_{\alpha}\, T_{\bar{\alpha}} = I$, we get

$$(g_k,\, g_m) \;=\; \sum_{i=1}^{n} c_{ki}\, (g_i,\, T_{\alpha}^{*}\, T_{\alpha}\, g_m)$$

and, consequently,

$$G \;=\; c\, \| (T_{\alpha}\, g_i,\, T_{\alpha}\, g_m) \|. \qquad\qquad (5.30)$$

On the other hand, taking the expression for B_{α} into account, we obtain

$$\| (B_{\alpha}\, g_i,\, g_m) \| \;=\; \frac{1}{2\,\mathrm{Im}\,\alpha}\, \bigl[\, \| (T_{\alpha}\, g_i,\, T_{\alpha}\, g_m) \| - G \,\bigr] \;=\; GJG.$$

By solving this equality with respect to the matrix $\| (T_\alpha g_i, T_\alpha g_m) \|$ and inserting the result obtained into (5.30), we arrive at the equality

$$G = cG + 2\,\text{Im}\,\alpha cGJG. \tag{5.31}$$

Thus,

$$MJG = 0, \tag{5.32}$$

by virtue of (5.29) and (5.31). In particular, if the Δ-basis $\{g_k\}$ is linearly independent, then $\det G \neq 0$ and, consequently, $M = 0$. But then, $c = \tilde{\tau}^*$ by virtue of (5.29), which, in view of (5.28), yields equality (5.24).

Note that, in the general case, the equality $M = 0$ is not true. Indeed, for the self-adjoint operator A introduced in Subsection 1.1, we easily find that

$$J = \begin{pmatrix} 1 & 0 \\ 0 & -1 \end{pmatrix}, \quad G = (g, g)\begin{pmatrix} 1 & 1 \\ 1 & 1 \end{pmatrix}, \quad c = E,$$

and, consequently,

$$M = 2\,\text{Im}\,\alpha\,(g, g)\begin{pmatrix} 1 & 1 \\ 1 & 1 \end{pmatrix} \neq 0.$$

Remark. The results of Subsection 5.1 were first published (without proofs) in Kuzhel's works [2, 22]. The results of the other subsections are presented here for the first time.

6. Theorem on Multiplication of Characteristic Matrix Functions

6.1. Couplings of Linear Operators

Let $\mathcal{H} = \mathcal{H}_1 \oplus \mathcal{H}_2$ and let A_1 and A_2 be operators acting in the spaces $\mathcal{H}_1$ and $\mathcal{H}_2$, respectively. In $\mathcal{H}$, consider an operator

$$S = I + KP_2 \quad (K \in \mathcal{B}[\mathcal{H}_2, \mathcal{H}_1]), \tag{6.1}$$

where P_2 is the orthoprojector in $\mathcal{H}$ onto $\mathcal{H}_2$. The operator S is invertible and $S^{-1} =$

$I - KP_2$.

The operator A defined in $\mathcal{H}$ by the equality

$$A = A_1 P_1 S^{-1} + A_2 P_2 + \Gamma P_2 \quad (\mathcal{D}_A = S[\mathcal{D}_{A_1} \oplus \mathcal{D}_{A_2}]), \tag{6.2}$$

where Γ is an operator acting from $\mathcal{H}_2$ into $\mathcal{H}_1$, is called a *coupling* of the operators A_1 and A_2 and denoted by $A = A_1 \gamma A_2$.

This definition of coupling is explained by the fact that if the operator A defined in $\mathcal{H} = \mathcal{H}_1 \oplus \mathcal{H}_2$ generates the operator A_1 acting in the subspace $\mathcal{H}_1$ and A^* generates the operator A_2^* acting in $\mathcal{H}_2$, then, generally speaking, $\mathcal{D}_A \neq \mathcal{D}_{A_1} \oplus \mathcal{D}_{A_2}$. At the same time, the operator A can be written in a form similar to (6.2).

Note that if $K \colon \mathcal{D}_{A_2} \to \mathcal{D}_{A_1}$, then it is easy to show that $\mathcal{D}_A = \mathcal{D}_{A_1} \oplus \mathcal{D}_{A_2}$ and

$$A = A_1 P_1 + A_2 P_2 + \tilde{\Gamma} P_2 \quad (\tilde{\Gamma} = \Gamma - A_1 K). \tag{6.3}$$

In what follows, we consider the only case where $\Gamma = \alpha K$, where α is a fixed point from $\rho(A)$.

Theorem 6.1. *A coupling of closed operators is a closed operator.*

Proof. Let $A = A_1 \gamma A_2$, where A_1 and A_2 are closed operators. Also let $x_n \in \mathcal{D}_A$, $x_n \to x$, and $Ax_n \to y$. Then

$$\|Ax_n - y\|^2 = \|A_1 P_1 S^{-1} x_n + \Gamma P_2 x_n - P_1 y\|^2 + \|A_2 P_2 x_n - P_2 y\|^2 \to 0$$

as $n \to \infty$ and, consequently,

$$\|A_2 P_2 x_n - P_2 y\| \to 0 \quad \text{and} \quad \|A_1 P_1 S^{-1} x_n + \Gamma P_2 x_n - P_1 y\| \to 0 \tag{6.4}$$

as $n \to \infty$. Further, since the operator A_2 is closed and $P_2 x_n \to P_2 x$ in view of the first relation from (6.4), we conclude that $P_2 x \in \mathcal{D}_{A_2}$ and $A_2 P_2 x = P_2 y$. Similarly, since

$$\Gamma P_2 x_n \to \Gamma P_2 x, \quad P_1 S^{-1} x_n \to P_1 S^{-1} x$$

as $n \to \infty$ and the operator A_1 is closed, we get $P_1 S^{-1} x \in \mathcal{D}_{A_1}$ and $A_1 P_1 S^{-1} x + \Gamma P_2 x = P_1 y$, by virtue of the second relation from (6.4). But then $S(P_1 S^{-1} x + P_2 x) = x \in \mathcal{D}_A$ and

$$Ax = A_1 P_1 S^{-1} x + A_2 P_2 x + \Gamma P_2 x = P_1 y + P_2 y = y,$$

which implies that the operator A is closed.

∎

Theorem 6.2. *A coupling of N-operators is an N-operator.*

Proof. Denote by A a coupling of N-operators A_1 and A_2. By virtue of Theorem 6.1, it suffices to prove that the operator A is densely defined.

Assume that $h \perp \mathfrak{D}_A$. Then, for any x from $\mathfrak{D}_{A_1}$, $(h, Sx) = (h, x) = 0$ and, consequently, $h \in \mathcal{H}_2$. But this implies that

$$(Sx, h) = (x, h) + (KP_2x, h) = (x, h) = 0$$

for $x \in \mathfrak{D}_{A_1} \oplus \mathfrak{D}_{A_2}$ and, hence, $h = 0$ (since $\overline{\mathfrak{D}_{A_1} \oplus \mathfrak{D}_{A_2}} = \mathcal{H}$).

∎

It is sometimes convenient to represent the operators S and $A = A_1 \gamma A_2$ as the operator matrices

$$S = \begin{pmatrix} I & K \\ 0 & I \end{pmatrix}, \quad A = \begin{pmatrix} A_1 & \Gamma \\ 0 & A_2 \end{pmatrix} S^{-1}.$$

Couplings of three or more operators, which we are used in the next chapter, are defined as a natural generalization of the notion of a coupling of two operators.

For example, if $A = (A_1 \gamma A_2) \gamma A_3$, then

$$\mathfrak{D}_A = S[\mathfrak{D}_{A_1} \oplus \mathfrak{D}_{A_2} \oplus \mathfrak{D}_{A_3}], \quad Af = S_0 S^{-1} f \quad (f \in \mathfrak{D}_A),$$

where

$$S = \begin{pmatrix} I & K_{12} & K_{13} \\ 0 & I & K_{23} \\ 0 & 0 & I \end{pmatrix}, \quad S_0 = \begin{pmatrix} A_1 & A_{12} & A_{13} \\ 0 & A_2 & A_{23} \\ 0 & 0 & A_3 \end{pmatrix}.$$

6.2. Resolvent of a Coupling

Theorem 6.3. *Let $A = A_1 \gamma A_2$ and $\lambda \in \rho(A_1) \cap \rho(A_2)$. Then $\lambda \in \rho(A)$ and the resolvents R_λ, $R_{1\lambda}$ and $R_{2\lambda}$ of operators A, A_1, and A_2, respectively, satisfy the equality*

$$R_\lambda = R_{1\lambda} P_1 + SR_{2\lambda} P_2 + \Gamma_\lambda P_2, \tag{6.5}$$

where

$$\Gamma_\lambda = R_{1\lambda}(\lambda K - \Gamma) R_{2\lambda}. \tag{6.6}$$

Proof. Let us show that $\lambda \in \sigma_p(A)$. Indeed, let $x \in \mathfrak{D}_A$ and

$$Ax = A_1 P_1 S^{-1} x + A_2 P_2 x + \Gamma_2 P_2 x = \lambda x. \tag{6.7}$$

Then $A_2 P_2 x = \lambda P_2 x$ and, consequently, $P_2 x = 0$ (since $\lambda \in \rho(A_2)$). But, then $S^{-1} x = S^{-1} P_1 x = P_1 x$, whence we get that $A_1 P_1 x = \lambda P_1 x$ by virtue of (6.7). Therefore, $x = P_1 x = 0$ and, hence, $\lambda \in \sigma_p(A)$.

Let $\varphi = (A - \lambda I)^{-1} x$. Since $\varphi \in \mathfrak{D}_A$, we have

$$\varphi = S(y_1 + y_2) = y_1 + y_2 + K y_2 \quad (y_k \in \mathfrak{D}_{A_k}).$$

But then, by virtue of (6.2), we obtain

$$x = A\varphi - \lambda\varphi = A_1 y_1 + A_2 y_2 + \Gamma y_2 - \lambda(y_1 + y_2 + K y_2)$$

whence

$$P_1 x = (A_1 - \lambda I) y_1 + \Gamma y_2 - \lambda K y_2, \quad P_2 x = (A_2 - \lambda I) y_2.$$

Consequently, $y_2 = R_{2\lambda} P_2 x$, where $R_{2\lambda} = (A_2 - \lambda I)^{-1}$ and

$$y_1 = R_{1\lambda} P_1 x + R_{1\lambda}(\lambda K - \Gamma) y_2 \quad (R_{1\lambda} = (A_1 - \lambda I)^{-1}).$$

Taking this into account, we conclude that

$$(A - \lambda I)^{-1} x = y_1 + S y_2 = R_{1\lambda} P_1 x + S R_{2\lambda} P_2 x + \Gamma_\lambda P_2 x,$$

where the operator Γ_λ is given by (6.6).

This completes the proof of equality (6.5). Moreover, the operators in the right-hand side of this equality are bounded and defined in the entire $\mathcal{H}$. Consequently, $\lambda \in \rho(A)$. ∎

In what follows, we also use another representation of the resolvent of the operator $A = A_1 \gamma A_2$. Namely,

$$R_\lambda = R_{1\lambda} P_1 + R_{2\lambda} P_2 + \Phi_\lambda P_2, \tag{6.8}$$

where the operator $\Phi_\lambda : \mathcal{H}_2 \to \mathcal{H}_1$ is given by the equality $\Phi_\lambda = KR_{2\lambda} + \Gamma_\lambda$. In fact, since $S = I + KP_2$, equality (6.8) follows directly from equality (6.5).

Note that, in view of equalities (6.6) and $\Gamma = \alpha K$, the operator Φ_λ admits the representation

$$\Phi_\lambda = T_{1\alpha\lambda} K R_{2\lambda} T_{2\alpha\lambda}, \tag{6.9}$$

where $T_{j\alpha\lambda} = (A_j - \alpha I) R_{j\lambda}$ $(j = 1, 2)$.

6.3. Δ-basis of a Coupling

Assume that, as above, $A = A_1 \gamma A_2$ and the resolvent R_λ of the operator A is given by equality (6.8). Then the operator $B_\alpha = iR_\alpha - iR_\alpha^* + 2\operatorname{Im}\alpha\, R_\alpha^* R_\alpha$ can be represented in the form

$$B_\alpha = B_{1\alpha} P_1 + B_{2\alpha} P_2 + i T_{1\alpha}^* \Phi P_2 - i \Phi^* T_{1\alpha} P_1 + 2\operatorname{Im}\alpha\, \Phi^* \Phi P_2, \tag{6.10}$$

where $T_{1\alpha} = I + (\alpha - \bar{\alpha}) R_{1\alpha}$ and $\Phi = \Phi_\alpha$.

Further, let $\{g_{1k}\}_{k=1}^n$ and $\{g_{2k}\}_{k=1}^n$ be Δ-bases of the operators A_1 and A_2, respectively. Assume that the coefficient matrices corresponding to these Δ-bases coincide, i.e., $J_1 = J_2 = J$. Without loss of generality, we can restrict ourselves to the case of $J^2 = E$.

Consider the following system of vectors in $\mathcal{H} = \mathcal{H}_1 \oplus \mathcal{H}_2$:

$$g_k = g_{1k} + h_k, \qquad h_k = \sum_{i=1}^n \omega_{ki} g_{2i}, \quad k = \overline{1,\, n}; \tag{6.11}$$

here, $\omega = \| \omega_{ki} \|$ is an invertible $n \times n$ matrix.

Let us show that the matrix ω can be chosen so that the system of vectors (6.11) is the Δ-basis of the operator $A = A_1 \gamma A_2$.

Let us fix the operator $\Phi = \Phi_\alpha$ as follows:

$$\Phi P_2 = -i \sum_{k,\, s=1}^n (\cdot, h_k) J_{ks} T_{1\bar{\alpha}}^* g_{1s}. \tag{6.12}$$

Then, by virtue of (6.10) and (6.12), we get

$$P_1 B_\alpha P_1 = B_{1\alpha} P_1 = \sum_{k,\,s=1}^{n} (\cdot, g_{1k}) J_{ks} g_{1s},$$

(6.13)

$$P_1 B_\alpha P_2 = i\, T_{1\alpha}^* \Phi P_2 = \sum_{k,\,s=1}^{n} (\cdot, h_k) J_{ks} g_{is}.$$

(6.14)

Thus, by virtue of (6.11), (6.13) and (6.14), we have

$$P_1 B_\alpha = \sum_{k,\,s=1}^{n} (\cdot, g_k) J_{ks} g_{1s}.$$

(6.15)

Taking equalities (6.10) and (6.14) into account, we obtain

$$P_2 B_\alpha P_1 = (P_1 B_\alpha P_2)^* = \sum_{k,\,s=1}^{n} (\cdot, g_{1k}) J_{ks} h_s.$$

(6.16)

Finally, by virtue of (6.10), we get

$$P_2 B_\alpha P_2 = B_{2\alpha} P_2 + 2\,\mathrm{Im}\,\alpha \Phi^* \Phi P_2.$$

(6.17)

Moreover, according to (6.12), we have

$$\Phi^* P_1 = (\Phi P_2)^* = i \sum_{m,\,r=1}^{n} (\cdot, T_{1\bar\alpha}^* g_{1m}) J_{mr} h_r.$$

(6.18)

It follows from (6.12) and (6.18) that

$$\Phi^* \Phi P_2 = \sum_{k,\,r=1}^{n} (\cdot, h_k) x_{kr} h_r,$$

(6.19)

where

$$x_{kr} = \sum_{s,\,m=1}^{n} J_{ks} (T_{1\bar\alpha}^* g_{1s}, T_{1\bar\alpha}^* g_{1m}) J_{mr}.$$

(6.20)

Note that the numbers x_{kr} are elements of the matrix

$$X = J \tilde{G}_1 J, \quad \tilde{G}_1 = \| (T_{1\bar\alpha}^* g_{1s}, T_{1\bar\alpha}^* g_{1m}) \|.$$

(6.21)

By virtue of (6.19), the second term in (6.17) can be represented as

$$2 \operatorname{Im} \alpha \, \Phi^* \, \Phi \, P_2 \;=\; \sum_{k,\, r=1}^{n} (\cdot, h_k) J_{kr} h_r \;-\; \sum_{k,\, r=1}^{n} (\cdot, h_k)(J_{kr} - 2 \operatorname{Im} \alpha x_{kr}) h_r, \qquad (6.22)$$

where $J_{kr} - 2 \operatorname{Im} \alpha x_{kr}$ are the elements of the matrix $J - 2 \operatorname{Im} \alpha X$. Consider the matrix

$$\tilde{\tau}_1 \;=\; E - 2 \operatorname{Im} \alpha J \tilde{G}_1. \qquad (6.23)$$

By virtue of (6.21) and (6.23), we have

$$J - 2 \operatorname{Im} \alpha X \;=\; \tilde{\tau}_1 J. \qquad (6.24)$$

According to Subsection 5.4, $\tilde{\tau}_1$ is the deviation matrix in the Δ-basis $\{\tilde{g}_{1k}\}$, where $\tilde{g}_{1k} = T^*_{1\bar{\alpha}} g_{1k}$ $(k = \overline{1, n})$. By virtue of Proposition 1.1, this implies that $\det \tilde{\tau}_1 \neq 0$. Since $J^{-1} = J$, it follows from Proposition 1.4 that the matrix equation

$$Y_1^* J Y_1 \;=\; J \tilde{\tau}_1 \qquad (6.25)$$

is solvable. Therefore, since the matrix $\tilde{\tau}_1$ is nondegenerate, the matrix Y_1 in (6.25) is also nondegenerate.
    Consider the matrix

$$\omega \;=\; J Y_1^{-1}. \qquad (6.26)$$

By virtue of (6.25) and (6.26), we have

$$\omega^* \tilde{\tau}_1 J \omega \;=\; Y_1^{-*} J \tilde{\tau}_1 Y_1^{-1} \;=\; J. \qquad (6.27)$$

By using equalities (6.11), we can rewrite the second sum in (6.22) as

$$\sum_{k,\, r=1}^{n} (\cdot, h_k)(J_{kr} - 2 \operatorname{Im} \alpha x_{kr}) h_r \;=\; \sum_{m,\, s=1}^{n} (\cdot, g_m) \varepsilon_{ms} g_{2s}, \qquad (6.28)$$

where

$$\varepsilon_{ms} \;=\; \sum_{k,\, r=1}^{n} \omega^*_{mk}(J_{kr} - 2 \operatorname{Im} \alpha x_{kr}) \omega_{rs}.$$

Thus, ε_{ms} is an element of the matrix

$$\omega^* (J - 2 \operatorname{Im} \alpha X) \omega \;=\; \omega^* \tilde{\tau}_1 J \omega \;=\; J,$$

i.e., $\varepsilon_{ms} = J_{ms}$. But then, by virtue of (6.28), equality (6.22) takes the form

$$2 \operatorname{Im} \alpha \Phi^* \Phi P_2 = \sum_{k, \, r=1}^{n} (\cdot, h_k) J_{kr} h_r - \sum_{m, \, s=1}^{n} (\cdot, g_{2m}) J_{ms} g_{2s},$$

or, in other words,

$$2 \operatorname{Im} \alpha \Phi^* \Phi P_2 = \sum_{k, \, r=1}^{n} (\cdot, h_k) J_{kr} h_r - B_{2\alpha} P_2. \tag{6.29}$$

By inserting (6.29) in (6.17), we obtain

$$P_2 B_\alpha P_2 = \sum_{k, \, r=1}^{n} (\cdot, h_k) J_{kr} h_r. \tag{6.30}$$

By virtue of (6.11), (6.16) and (6.30), we get

$$P_2 B_\alpha = \sum_{k, \, s=1}^{n} (\cdot, g_k) J_{ks} h_s. \tag{6.31}$$

Equalities (6.15) and (6.31) yield the following expression for the operator B_α:

$$B_\alpha = \sum_{k, \, s=1}^{n} (\cdot, g_k) J_{ks} g_s.$$

Thus, if the matrix ω is defined by equality (6.26), then the system of vectors $\{g_k\}_{k=1}$, defined by equalities (6.11) is the Δ-basis of the operator $A = A_1 \gamma A_2$.

6.4. Multiplication Theorem

The following theorem is true:

Theorem 6.4. *Let a coupling A of the operators A_1 and A_2 be given by equality (6.2) and let its Δ-basis $\{g_k\}$ be defined by (6.11). Assume also that the Δ-basis $\{g_{1k}\}$ of the operator A_1 is linearly independent, and the matrix $\omega = \chi_{A_1}^*(\alpha)$ in (6.11). Then the characteristic matrix functions of the operators A_1, A_2, and A satisfy the equality*

$$\chi_A(\lambda) = \chi_{A_2}(\lambda)\chi_{A_1}(\lambda). \tag{6.32}$$

Proof. If the operator $A = A_1 \gamma A_2$ is given by equality (6.2) with $\Gamma = \alpha K$, then its resolvent R_λ can be represented in the form (6.8). Furthermore, by virtue of (6.2), we have

$$T_{\alpha\lambda} = A_1 P_1 S^{-1} R_\lambda + A_2 P_2 R_\lambda + \Gamma P_2 R_\lambda - \alpha R_\lambda. \tag{6.33}$$

Note that, by virtue of the equality $\Phi_\lambda = KR_{2\lambda} + \Gamma_\lambda$, we have

$$P_1 S^{-1} R_\lambda = P_1(I - KP_2)R_\lambda = R_{1\lambda} P_1 + \Gamma_\lambda P_2. \tag{6.34}$$

By substituting (6.34) in (6.33) and taking into account equality (6.9) and the expressions for the operators Γ, Γ_λ, and Φ_λ, we obtain

$$T_{\alpha\lambda} = T_{1\alpha\lambda} P_1 + T_{2\alpha\lambda} P_2 + (\lambda - \alpha)\Phi_\lambda P_2.$$

In view of (6.11), we get

$$\| (T_{\alpha\lambda} g_k, g_m) \|$$

$$= \| (T_{1\alpha\lambda} g_{1k}, g_{1m}) \| + \| (T_{2\alpha\lambda} h_k, h_m) \| + (\lambda - \alpha) \| (\Phi_\lambda h_k, g_{1m}) \|. \tag{6.35}$$

By virtue of (6.9), we have $\Phi_\lambda = T_{1\alpha\lambda} \Phi T_{2\alpha\lambda}$, where $\Phi = \Phi_\alpha$. By using equality (6.12), we obtain

$$\| (\Phi_\lambda h_k, g_{1m}) \| = \| (\Phi T_{2\alpha\lambda} h_k, T_{1\alpha\lambda}^* g_{1m}) \|$$

$$= -i \| (T_{2\alpha\lambda} h_k, h_m) \| J \| (T_{1\bar\alpha}^* g_{1k}, T_{1\alpha\lambda}^* g_{1m}) \|. \tag{6.36}$$

In view of (6.35) and (6.36), this yields

$$\chi_A^*(\alpha) J \chi_A(\lambda) = J + i(\bar\alpha - \lambda) \| (T_{\alpha\lambda} g_k, g_m) \|$$

$$= J + i(\bar\alpha - \lambda) \| (T_{1\alpha\lambda} g_{1k}, g_{1m}) \| + i(\bar\alpha - \lambda) \| (T_{2\alpha\lambda} h_k, h_m) \|$$

$$- (\bar\alpha - \lambda)(\alpha - \lambda) \| (T_{2\alpha\lambda} h_k, h_m) \| J \| (T_{1\bar\alpha}^* g_{1k}, T_{1\alpha\lambda}^* g_{1m}) \|$$

$$= \chi_{A_1}^*(\alpha) J \chi_{A_1}(\lambda) + i(\bar\alpha - \lambda) \| (T_{2\alpha\lambda} h_k, h_m) \| J [J$$

$$+ i(\alpha - \lambda) \| (T^*_{1\bar{\alpha}} g_{1k}, T^*_{1\alpha\lambda} g_{1m}) \|]. \tag{6.37}$$

Furthermore, since $T^*_{1\alpha\lambda} = T^*_{1\bar{\alpha}\lambda} T^*_{1\bar{\alpha}}$, we have

$$\| (T^*_{1\bar{\alpha}} g_{1k}, T^*_{1\alpha\lambda} g_{1m}) \| = \| (T^*_{1\bar{\alpha}\lambda} \tilde{g}_{1k}, \tilde{g}_{1m}) \|,$$

where $\tilde{g}_{1k} = T^*_{1\bar{\alpha}} g_{1k}$. Therefore, the expression in square brackets in (6.37) is equal to $\tilde{\chi}^*_{A_1}(\bar{\alpha}) J \tilde{\chi}_{A_1}(\lambda)$, where $\tilde{\chi}_{A_1}(\lambda)$ is symmetric with respect to the characteristic matrix function $\chi_{A_1}(\lambda)$. Consequently, in view of equality (5.24), we get

$$\tilde{\chi}^*_{A_1}(\bar{\alpha}) J \tilde{\chi}_{A_1}(\lambda) = \tilde{\tau}^*_1 \chi^*_{A_1}(\lambda) J \chi_{A_1}(\lambda). \tag{6.38}$$

Thus,

$$\chi^*_A(\alpha) J \chi_A(\lambda)$$

$$= \chi^*_{A_1}(\alpha) J \chi_{A_1}(\lambda) + i(\bar{\alpha} - \lambda) \| (T_{2\alpha\lambda} h_k, h_m) \| J \tilde{\tau}^*_1 \chi^*_{A_1}(\alpha) J \chi_{A_1}(\lambda). \tag{6.39}$$

Note that, in view of equality (6.11), we have

$$\| (T_{2\alpha} h_k, h_m) \| = \omega \| (T_{2\alpha\lambda} g_{2k}, g_{2m}) \| \omega^*. \tag{6.40}$$

Furthermore, for $\lambda = \alpha$, equality (6.38) can be rewritten as follows:

$$J = \tilde{\tau}^*_1 \chi^*_{A_1}(\alpha) J \chi_{A_1}(\lambda).$$

This implies

$$J \tilde{\tau}^*_1 \chi^*_{A_1}(\alpha) J = \chi^{-1}_{A_1}(\alpha). \tag{6.41}$$

By inserting (6.40) and (6.41) in (6.39) and taking the equality $\omega = \chi^*_{A_1}(\alpha)$ into account, we obtain

$$\chi^*_A(\alpha) J \chi_A(\lambda) = \chi^*_{A_1}(\alpha) [J + i(\bar{\alpha} - \lambda) \| (T_{2\alpha\lambda} g_{2k}, g_{2m}) \|] \chi_{A_1}(\lambda)$$

$$= \chi^*_{A_1}(\alpha) \chi^*_{A_2}(\alpha) J \chi_{A_2}(\lambda) \chi_{A_1}(\lambda).$$

This completes the proof of equality (6.32).

Remark. The results presented in this section were obtained by the author in [2, 3] under certain additional restrictions.

7. Factorization of Characteristic Matrix Functions

In this section, we present without proof some known concepts and facts necessary for what follows.

7.1. Blaschke Product

Let $\{z_k\}_{k=1}^{N}$ $(N \le \infty)$ be an arbitrary sequence of nonzero complex numbers from the unit circle $D = \{z \in \mathbb{C} \mid |z| < 1\}$ such that, for $N = \infty$, the series

$$\sum_{k=1}^{\infty} (1 - |z_k|)$$

is convergent.
Then the function

$$B(z) = z^m \prod_{k=1}^{N} \frac{z_k - z}{1 - \bar{z}_k z} \frac{|z_k|}{z_k}, \tag{7.1}$$

where m is a fixed positive integer, is analytic in the circle D. The numbers $z_1, z_2, \ldots$ are zeros of the function $B(z)$. Note that $|B(z)| \le 1$ for $z \in D$ and $|B(e^{i\theta})| = 1$ almost everywhere on $[0, 2\pi]$.
The function $B(z)$ thus defined is called the *Blaschke product*. For details, see, e.g., Hoffman [1] or Garnett [1].

7.2. Riesz–Herglotz Formula

Assume that f is a nonzero analytic function in a circle D which maps this circle into itself ($|f(z)| < 1$ for $z \in D$). Then f can be represented in the form

$$f(z) = B(x) \exp[-h(z)],$$

where h is an analytic function in the circle D with a nonnegative real part. The function h can be uniquely represented in the form

$$h(z) = ic + \int_0^{2\pi} \frac{e^{i\theta} + z}{e^{i\theta} - z} \, d\sigma(\theta),$$

where c is a real constant and the function σ is nondecreasing and nonnegative on $[0, 2\pi]$ and has a bounded variation.

Thus, an arbitrary nonzero analytic function in a circle D which maps this circle into itself can be represented (to within a unitary factor) in the form

$$f(z) = z^m \prod_{k=1}^{N} \frac{z_k - z}{1 - \bar{z}_k z} \frac{|z_k|}{z_k} \exp\left[\int_0^{2\pi} \frac{z + e^{i\theta}}{z - e^{i\theta}} \, d\sigma(\theta)\right], \tag{7.2}$$

where $m \in \mathbf{N}$, $N \le \infty$, $\{z_k\}$ are nontrivial zeros of the function f, and σ is a nondecreasing nonnegative function of bounded variation.

Relation (7.2) is called the *Riesz–Herglotz formula* (Nevanlinna generalized this formula to the case of analytic functions with a bounded characteristic; this is why this relation is often called the *Riesz–Nevanlinna formula*).

For more detailed information about the factorization of functions, see the monographs of Hoffman [1], Garnett [1], and Koosis [1].

7.3. Potapov Theorem

An $n \times n$ matrix ω is called *J-nonexpanding* if

$$\omega J \omega^* \le J \quad (J^* = J^{-1} = J). \tag{7.3}$$

In the case where relation (7.3) turns into an equality, the matrix ω is called *J-unitary*.

If $\omega J \omega^* = J$, then $\det \omega \neq 0$. Therefore, multiplying the last equality from the left by $\omega^* J$ and from the right by $\omega^{-*} J$, we obtain $\omega^* J \omega = J$. Thus, if the matrix ω is J-unitary, then the matrix ω^* is also J-unitary.

It is more difficult to prove that the matrix ω^* is J-nonexpanding if ω is J-nonexpanding (see, e.g., Potapov [1]).

If a matrix $\omega(z)$ is nonexpanding for any $z \in \mathcal{9}$, then it is called a matrix function nonexpanding in the domain $\mathcal{9}$.

If $n = 1$ (the scalar case) and $J = 1$, the fact that $\omega(z)$ is J-nonexpanding means that $|\omega(z)| \le 1$ for $z \in \mathcal{9}$. Thus, an arbitrary analytic function f that maps a unit circle into itself can be regarded as a function nonexpanding in the domain D. For such functions, as indicated above, the Riesz–Herglotz formula is true. Potapov [1] obtained an analogue of this formula in the case of J-nonexpanding matrix functions.

Below, we consider the case where

$$J = \begin{pmatrix} E_p & 0 \\ 0 & -E_q \end{pmatrix} \quad (q \geq 0, \; p + q = n);$$

here, E_m is an identity $m \times m$-matrix.

Theorem 7.1. (The fundamental theorem of Potapov [1]). *Suppose that a matrix function $\omega(z)$ is analytic and J-nonexpanding in the unit disk D and its determinant is not identically equal to zero. Then $\omega(z)$ can be represented as*

$$\omega(z) = \mathbf{B}_\infty(z)\mathbf{B}_0(z) \overset{l}{\underset{0}{\hat{\int}}} \exp\left[\frac{z + e^{i\theta(t)}}{z - e^{i\theta(t)}} dE(t) \right]. \tag{7.4}$$

Here,

$$\mathbf{B}_\infty(z) = \overset{\frown}{\prod_k} u_k \begin{pmatrix} E_{p_k} & 0 \\ 0 & \dfrac{1 - \bar{\mu}_k z}{\mu_k - z} \dfrac{\mu_k}{|\mu_k|} E_{q_k} \end{pmatrix} u_k^{-1} \tag{7.5}$$

is the product of primary factors determined by the poles μ_k $(|\mu_k| < 1)$ of the matrix function $\omega(z)$, $q_k \leq q$, $p_k + q_k = m$, and u_k are J-unitary matrices;

$$\mathbf{B}_0(z) = \overset{\frown}{\prod_k} v_k \begin{pmatrix} \dfrac{z_k - z}{1 - \bar{z}_k z} \dfrac{|z_k|}{z_k} E_{p'_k} & 0 \\ 0 & E_{q'_k} \end{pmatrix} v_k^{-1} \tag{7.6}$$

is the product of primary factors determined by the roots $z_k \in D$ of the determinant of the matrix function $\omega_\infty(z) = \mathbf{B}_\infty^{-1}(z)\omega(z)$, $p'_k \leq p$, $p'_k + q'_k = n$, which is holomorphic in the disk D, and v_k are J-unitary matrices;

$$\overset{l}{\underset{0}{\hat{\int}}} \exp\left[\frac{z + e^{i\theta(t)}}{z - e^{i\theta(t)}} dE(t) \right] = \lim_{\max \Delta t_k \to 0} e_0(z) e_1(z) \dots e_{n-1}(z), \tag{7.7}$$

where

$$e_k(z) = \exp\left[\frac{z + e^{i\theta(\tau_k)}}{z - e^{i\theta(\tau_k)}} \Delta E(t_k)\right]$$

and

$$0 = t_0 \leq \tau_0 \leq t_1 \leq \tau_1 \leq t_2 \leq \ldots \leq t_{n-1} \leq \tau_{n-1} \leq t_n = l,$$

is the multiplicative Stieltjes integral, in which θ (t) $(0 \leq \theta(t) \leq 2\pi)$ is a non-decreasing function, $E(t)J$ is a nondecreasing family of Hermitian matrices, and the variable t satisfies the condition $\operatorname{tr} E(t)J = t$.

The arrows in (7.4)–(7.7) indicate the direction in which corresponding factors are ordered. It should be mentioned that if

$$W(z) = \int_0^l H(z, t)\, dt,$$

then

$$W^*(z) = \int_0^l H^*(z, t)\, dt$$

and

$$W^{-1}(z) = \int_0^l H^{-1}(z, t)\, dt$$

(under the condition that the matrix $H^{-1}(z, t)$ exists). The same is true for the products $B_\infty(z)$ and $B_0(z)$.

It follows directly from (7.4)–(7.7) that

$$\omega^{-1}(z) = \int_0^l \exp\left[-\frac{z + e^{i\theta(t)}}{z - e^{i\theta(t)}} dE(t)\right] B_0^{-1}(z)\, B_\infty^{-1}(z), \qquad (7.8)$$

where

$$B_0^{-1}(z) = \prod_k v_k \left(\begin{array}{cc} \dfrac{1 - \bar{z}_k z}{z_k - z} \dfrac{z_k}{|z_k|} E_{p_k'} & 0 \\ 0 & E_{q_k'} \end{array}\right) v_k^{-1},$$

$$\mathbf{B}_{\infty}^{-1}(z) = \prod_{k} u_k \begin{pmatrix} E_{p_k} & 0 \\ 0 & \dfrac{\mu_k - z}{1 - \overline{\mu}_k z} \dfrac{|\mu_k|}{\mu_k} E_{q_k} \end{pmatrix} u_k^{-1}.$$

The generalization of the Potapov theorem to the infinite-dimensional case (an operator analogue of the Riesz–Herglotz formula) was obtained by Ginzburg [1].

7.4. J-Nonexpanding Matrix Functions in the Upper Half-Plane

Let $W(\lambda)$, $\mathrm{Im}\,\lambda > 0$, be an analytic J-nonexpanding matrix function whose determinant is not identically equal to zero. Consider the mapping $\lambda \to (\lambda - i)(\lambda + i)^{-1}$ of the upper half-plane into the unit circle D. Assume that $z = (\lambda - i)(\lambda + i)^{-1}$. Then $\lambda = (1 + z)(1 - z)^{-1} i$ and the matrix function $\omega(z) = W((1 + z)(1 - z)^{-1} i)$ is analytic and J-nonexpanding in the circle D. By virtue of Theorem 7.1, this implies that $\omega(z)$ can be represented in the form (7.4). Let us substitute $(\lambda - i)(\lambda + i)^{-1}$ for z in this equality and find the expression for the corresponding factors.
    We have

$$\frac{1 - \overline{\mu}_k z}{\mu_k - z} \frac{\mu_k}{|\mu_k|} = \frac{\lambda - \overline{\delta}_k}{\lambda - \delta_k} \gamma_k,$$

where

$$\delta_k = \frac{1 + \mu_k}{1 - \mu_k} i$$

and

$$\gamma_k = \frac{\delta_k - i}{\overline{\delta}_k - i} \frac{|\overline{\delta}_k - i|}{|\delta_k - i|} \qquad (|\gamma_k| = 1).$$

Therefore, denoting $B_{\infty}(\lambda) := \mathbf{B}_{\infty}((\lambda - i)(\lambda + i)^{-1})$, we get

$$B_{\infty}(\lambda) = \prod_{k} u_k \begin{pmatrix} E_{p_k} & 0 \\ 0 & \dfrac{\lambda - \overline{\delta}_k}{\lambda - \delta_k} \gamma_k E_{q_k} \end{pmatrix} u_k^{-1}. \qquad (7.9)$$

Similarly, for $B_0(\lambda) := \mathbf{B}_0((\lambda - i)(\lambda + i)^{-1})$, we obtain

$$B_0(\lambda) = \prod_k v_k \begin{pmatrix} \dfrac{\lambda - \lambda_k}{\lambda - \bar{\lambda}_k} \beta_k E_{p_k'} & 0 \\ 0 & E_{q_k'} \end{pmatrix} v_k^{-1}, \qquad (7.10)$$

where

$$\lambda_k = \frac{1 + z_k}{1 - z_k} i, \quad \beta_k = \frac{\bar{\lambda}_k - i \, |\lambda_k - i|}{\lambda_k - i \, |\bar{\lambda}_k - i|} \quad (|\beta_k| = 1).$$

Further, we denote $I(\lambda) := \mathfrak{I}((\lambda - i)(\lambda + i)^{-1})$, where

$$\mathfrak{I}(z) = \int_0^l \exp\left[\frac{z + e^{i\theta(t)}}{z - e^{i\theta(t)}} dE(t) \right]. \qquad (7.11)$$

Let $a = \sup\{t \,|\, \theta(t) = 0\}$ and $b = \inf\{t \,|\, \theta(t) = 2\pi\}$. Then, in view of obvious properties of multiplicative integrals, we get

$$\mathfrak{I}(z) = \int_0^a \exp\left[\frac{z + 1}{z - 1} dE(t) \right] \int_a^b \exp\left[\frac{z + e^{i\theta(t)}}{z - e^{i\theta(t)}} dE(t) \right] \int_b^l \exp\left[\frac{z + 1}{z - 1} dE(t) \right]. \qquad (7.12)$$

By substituting $(\lambda - i)(\lambda + i)^{-1}$ for z in (7.12), we obtain

$$I(\lambda) = \int_0^{l_1} \exp\left[i\lambda dE_1(t) \right] \int_0^{l_2} \exp\left[i \frac{1 + \lambda\alpha(t)}{\alpha(t) - \lambda} dE_2(t) \right] \int_0^{l_3} \exp\left[i\lambda dE_3(t) \right], \qquad (7.13)$$

where $l_1 = a$, $l_2 = b - a$, $l_3 = l - b$, $E_1(t) = E(t)$, $E_2(t) = E(t - a)$, $E_3(t) = E(t - b)$, and

$$\alpha(t) = i \frac{1 + e^{i\theta(t-a)}}{1 - e^{i\theta(t-a)}} \qquad (7.14)$$

is a nondecreasing number function.

Thus, an arbitrary analytic J-nonexpanding matrix function $W(\lambda)$ ($\det W(\lambda) \neq 0$) defined in the upper half-plane can be represented as

$$W(\lambda) = B_\infty(\lambda) B_0(\lambda) I(\lambda), \qquad (7.15)$$

where the matrices $B_\infty(\lambda)$, $B_0(\lambda)$, and $I(\lambda)$ are given by equalities (7.9), (7.10), and

(7.13), respectively.

In our further discussion, we use the following theorem:

Theorem 7.2. (Potapov [1]). *An entire matrix function $W(\lambda)$ J-nonexpanding in the upper half-plane and J-unitary on the real axis can be represented as*

$$W(\lambda) = W(0) \int_0^{\hat{l}} \exp[i\lambda H(t)dt], \qquad (7.16)$$

where $H(t)$ is a J-summable Hermitian nonnegative matrix which satisfies the condition tr $H(t)J = 1$.

7.5. Multiplicative Integrals with Variable Upper Limits

Let $B(t)$ be an absolutely continuous matrix function, i.e., for any $\varepsilon > 0$, there exists $\delta > 0$ such that the inequality

$$\sum_{k=1}^{n} (\beta_k - \alpha_k) < \delta \quad (a \le \alpha_1 < \beta_1 \le \alpha_2 < \beta_2 \le \dots \le \alpha_n < \beta_n \le b)$$

yields the inequality

$$\sum_{k=1}^{n} \| B(\beta_k) - B(\alpha_k) \| < \varepsilon$$

By analogy with the scalar case, it is easy to show that the derivative $dB(t)/dt = H(t)$ of $B(t)$ exists almost everywhere and is summable; moreover, $B(t)$ can be represented in the form

$$B(t) = \int_a^t H(t)dx + c. \qquad (7.17)$$

Furthermore, $B(t)$ is a matrix function with bounded variation and

$$\operatorname*{Var}_{[a,\,b]} B(t) = \int_a^b \| H(x) \| dx \qquad (7.18)$$

Conversely, if $H(t)$ is a summable matrix function satisfying the condition

$$\int_a^b \| H(t) \| \, dt < \infty, \qquad\qquad (7.19)$$

then the matrix function $B(t)$ given by equality (7.17) is absolutely continuous and its variation is given by relation (7.18).

Under the conditions given above, the multiplicative integral

$$\int_a^b \exp[H(t)dt] := \int_a^b \exp[dB(t)]$$

is called a multiplicative Lebesgue integral.

Theorem 7.3. (Potapov [1]). *Assume that $H(t)$ is a summable matrix function which satisfies condition (7.19). Then the multiplicative Lebesgue integral with variable upper limit*

$$\omega(x) = \int_a^x \exp[H(t)dt] \qquad\qquad (7.20)$$

is differentiable almost everywhere and its derivative satisfies the relation

$$\frac{d\omega(x)}{dx} = \omega(x)H(x). \qquad\qquad (7.21)$$

Thus, the multiplicative integral (7.20) is a solution of the matrix differential equation (7.21) with the initial condition $\omega(a) = E$.

7.6. Factorization of $\chi_A(\lambda)$

By virtue of (5.12), $\chi_A(\lambda)J\chi_A^*(\lambda) \geq J$ for $\text{Im}\,\lambda > 0$ and, consequently, the matrix function $W(\lambda) \geq \chi_A^{-1}(\lambda)$ is J-nonexpanding in the upper half-plane. But this implies that $W(\lambda)$ can be represented in the form (7.5), whence, by virtue of the expression for $W(\lambda)$, we obtain

$$\chi_A(\lambda) = W_3(\lambda)W_2(\lambda)W_1(\lambda), \qquad\qquad (7.22)$$

where

$$W_1(\lambda) = B_\infty^{-1}(\lambda) = \prod_k^\frown u_k \begin{pmatrix} E_{p_k} & 0 \\ 0 & \dfrac{\lambda - \delta_k}{\lambda - \bar\delta_k} \bar\gamma_k E_{q_k} \end{pmatrix} u_k^{-1}, \tag{7.23}$$

$$W_2(\lambda) = B_0^{-1}(\lambda) = \prod_k^\frown v_k \begin{pmatrix} \dfrac{\lambda - \bar\lambda_k}{\lambda - \lambda_k} \bar\beta_k E_{p_k'} & 0 \\ 0 & E_{q_k'} \end{pmatrix} v_k^{-1}, \tag{7.24}$$

$$W_3(\lambda) = I^{-1}(\lambda)$$

$$= \int_0^{l_3} \exp\left[-i\lambda dE_3(t)\right] \int_0^{l_2} \exp\left[i\,\frac{1+\lambda\alpha(t)}{\lambda-\alpha(t)}\,dE_2(t)\right] \int_0^{l_1} \exp\left[-i\lambda dE_1(t)\right]$$

in view of (7.9), (7.10), and (7.13).
We set

$$L(k) = \begin{pmatrix} 0 & 0 \\ 0 & \delta_k E_{q_k} \end{pmatrix}, \quad \gamma(k) = \begin{pmatrix} E_{p_k} & 0 \\ 0 & \bar\gamma_k E_{q_k} \end{pmatrix}. \tag{7.26}$$

Then

$$W_1(\lambda) = \prod_k^\frown u_k[L(k) - \lambda E][L^*(k) - \lambda E]^{-1}\gamma(k) u_k^{-1} \tag{7.27}$$

by virtue of (7.23) and (7.26).
Similarly, we have

$$W_2(\lambda) = \prod_k^\frown v_k[M^*(k) - \lambda E][M(k) - \lambda E]^{-1}\beta(k) v_k^{-1}, \tag{7.28}$$

where

$$M(k) = \begin{pmatrix} \lambda_k E_{p_k'} & 0 \\ 0 & 0 \end{pmatrix}, \quad \beta(k) = \begin{pmatrix} \bar\beta_k E_{p_k'} & 0 \\ 0 & E_{q_k'} \end{pmatrix}.$$

Thus, the characteristic matrix function $\chi_A(\lambda)$ of the operator A can be represented in the form (7.22), where the corresponding matrix factors are given by (7.25), (7.27), and (7.28).

In particular, if the operator A is dissipative ($J = E$ and, consequently, $q_k = 0$), then the factor $W_1(\lambda)$ in equality (7.22) is absent. Thus, in the case of a dissipative operator, its characteristic matrix function $\chi_A(\lambda)$ can be represented in the form

$$\chi_A(\lambda) = W_3(\lambda) W_2(\lambda). \tag{7.29}$$

7.7. Singular Points

The following statements are true:

Theorem 7.4. (Potapov [1]). *If the matrix function*

$$\omega(z) = \int_0^l \exp\left[\frac{z + e^{i\theta(t)}}{z - e^{i\theta(t)}} dE(t)\right]$$

is holomorphic and J-unitary on an open arc $\alpha_0 < \theta(t) < \beta_0$ *of the unit circle, then the loading* $E(t)$ *remains constant for* $t \in (a_0, b_0)$.

For the case of the upper half-plane, we have the following assertion:

Theorem 7.5. *If the matrix function*

$$W(\lambda) = \int_0^l \exp\left[i\frac{1 + \lambda\alpha(t)}{\alpha(t) - \lambda} dE(t)\right] \tag{7.30}$$

is holomorphic and J-unitary on an open interval $a < \alpha(t) < b$ *of the real axis, then the loading* $E(t)$ *remains constant in the interval* $t \in (a_0, b_0)$.

Theorem 7.6. *The set of singular points of the matrix function (7.30) coincides with the set* $\mathfrak{M}$ *of the left and right limit values of the functions* $\alpha(\cdot)$.

Proof. Let $\lambda_0 \bar{\in} \mathfrak{M}$. By using the same reasoning as in the Potapov's work [1, p. 227], we conclude that the multiplicative integral

$$\int_0^l \exp\left[i\frac{1 + \lambda_0\alpha(t)}{\alpha(t) - \lambda_0} dE(t)\right] = W(\lambda_0)$$

exists and the matrix function $W(\lambda)$ is holomorphic at the point λ_0.

Let $\lambda_0 \in \mathfrak{M}$. Assume that $W(\lambda)$ is holomorphic at the point λ_0. Consider the equality

$$W(\lambda) J W^*(\bar{\lambda}) = J \quad (\text{Im } \lambda \neq 0), \tag{7.31}$$

which can be justified by differentiating the expression $W(\lambda) J W^*(\bar{\lambda})$ with respect to the upper limit l. Note that, by virtue of (5.9), equality (7.31) is also true for characteristic matrix functions.

It follows from (7.31) and the assumption about the point λ_0 that the matrix function $W(\lambda)$ is holomorphic and J-unitary on some interval (a, b) containing the point λ_0. We set

$$a_0 = \inf \{t \mid \alpha(t) \subset (a, b)\}, \qquad b_0 = \sup \{t \mid \alpha(t) \in (a, b)\}.$$

By virtue of Theorem 7.5, the loading $E(t)$ remains constant on the interval (a_0, b_0). Therefore,

$$W(\lambda) = \overset{a_0}{\underset{0}{\int}} \exp[H(t)dE(t)] \overset{l}{\underset{b_0}{\int}} \exp[H(t)dE(t)]$$

$$= \overset{a_0}{\underset{0}{\int}} \exp[H_1(t)dE_1(t)] \overset{l_1}{\underset{a_0}{\int}} \exp[H_1(t)dE_1(t)] = \overset{l_1}{\underset{0}{\int}} \exp[H_1(t)dE_1(t)],$$

$$\tag{7.32}$$

where

$$l_1 = l - b_0 + a_0, \qquad H(t) = i\frac{1 + \lambda\alpha(t)}{\alpha(t) - \lambda},$$

and $H_1(t)$ and $E_1(t)$ coincide on the interval $[0, a_0]$ with $H(t)$ and $E(t)$, respectively. On the other hand, we have $H_1(t) = H(t + b_0 - a_0)$ and $E_1(t) = E(t + b_0 - a_0)$ on the interval $(a_0, l_1]$. Furthermore, if $t \leq a_0$, then $\alpha(t) \leq a$. For $t > a_0$, we have $t + b_0 - a_0 > b_0$ and, hence, $\alpha_1(t) = \alpha(t + b_0 - a_0) \geq b$. Thus, $\mathfrak{M} \cap (a, b) = \varnothing$, which contradicts the assumption that $\lambda_0 \in \mathfrak{M} \cap (a, b)$.

∎

In the general case where the characteristic matrix function of the operator A is given by equality (7.22), the set of real singular points may consist not only of points of the set $\mathfrak{M}$ but also of limit points of the sequences $\{\bar{\delta}_k\}$ and $\{\lambda_k\}$. Note that these limit points lie on the real axis; indeed, the equality

$$1 - |\lambda_k|^2 = \frac{4 \operatorname{Im} \lambda_k}{|\lambda_k + i|}$$

implies that $\operatorname{Im} \lambda_k \to 0$ as

$$\left|\frac{\lambda_k - i}{\lambda_k + i}\right| \to 1.$$

8. Characteristic Operator Functions

In this section, we introduce the notion of characteristic operator functions and study their main properties.

8.1. Definition of Characteristic Operator Functions

Let H be an Hermitian operator with equal (finite or infinite) defect numbers acting in the Hilbert space $\mathcal{H}$, $A \in \mathcal{P}(H)$, and $\{\alpha, \bar{\alpha}\} \subset \rho(A)$ ($\operatorname{Im} \alpha \neq 0$). We represent the bounded self-adjoint operator B_α given by equality (2.10) in Chapter 1 as

$$B_\alpha = QJQ^* \qquad (J^* = J, \quad 0 \in \rho(J)). \tag{8.1}$$

The operators Q and J are determined not uniquely. Moreover, Q can be regarded as an operator acting from an auxiliary space $\mathcal{E}_A$ (which is called *external*) into the space $\mathcal{H}$, and J should be a bounded operator acting in $\mathcal{E}_A$. In particular, we can choose $\mathcal{E}_A$ as a subspace of the space $\mathcal{H}$, Q as a self-adjoint operator, and J as a self-adjoint unitary operator ($J^* = J = J^{-1}$).

Since $Q : \mathcal{E}_A \to \mathcal{H}$, we have $Q^* : \mathcal{H} \to \mathcal{E}_A$ and, thus, $\bar{\Delta}_{Q^*} \subset \mathcal{E}_A$. In particular, $\mathcal{E}_A$ and $\bar{\Delta}_{Q^*}$ may coincide.

The operator function $\theta_A(\lambda)$ defined by the equality

$$\theta_A(\lambda) J^{-1} \theta_A^*(\alpha) = J^{-1} + i(\bar{\alpha} - \lambda) Q^* T_{\alpha\lambda} Q \qquad (\lambda \in \rho(A)) \tag{8.2}$$

is called the *characteristic operator function* of the operator A.

To prove that the definition of the characteristic operator functions is correct, we consider a "deviation operator"

$$\tau = J^{-1} + 2\operatorname{Im}\alpha Q^*Q \quad (\tau \in \mathcal{B}[\mathcal{E}_A]) \tag{8.3}$$

and show that this operator is invertible. Actually, assume that $\tau x = 0$, for some x from $\mathcal{E}_A$. Then, by virtue of (8.3), we get

$$x = -2\operatorname{Im}\alpha JQ^*Qx. \tag{8.4}$$

Applying the operator Q to both parts of this equality and taking into account equalities (8.1) and

$$2\operatorname{Im}\alpha B_\alpha = T_\alpha^*T_\alpha - I, \tag{8.5}$$

we obtain $Qx = -2\operatorname{Im}\alpha B_\alpha Qx = Qx - T_\alpha^*T_\alpha Qx$, which implies that $Qx = 0$. But then $x = 0$, by virtue of (8.4).

Thus, $\operatorname{Ker}\tau = \{0\}$. Moreover, it can be easily verified that (by virtue of equalities (8.5) and $T_\alpha T_{\bar\alpha} = I$)

$$\tau^{-1} = J - 2\operatorname{Im}\alpha JQ^* T_{\bar\alpha} T_{\bar\alpha}^* QJ. \tag{8.6}$$

By analogy with Section 1, one can show that the operator

$$X = [I + 2\operatorname{Im}\alpha Q^* T^{-1} QJ] W, \tag{8.7}$$

where $T = I + \sqrt{T_\alpha^*T_\alpha}$ and W is an arbitrary operator satisfying the condition $WJ^{-1}W^* = J^{-1}$, is a solution of the operator equation

$$XJ^{-1}X^* = \tau \tag{8.8}$$

Furthermore,

$$X^{-1} = W^{-1}[I - 2\operatorname{Im}\alpha Q^* (T_\alpha^*T_\alpha + \sqrt{T_\alpha^*T_\alpha})^{-1} QJ].$$

For $\lambda = \alpha$, equality (8.2) can be rewritten in the form $\theta_A(\alpha) J^{-1} \theta_A^*(\alpha) = \tau$. Therefore, by virtue of the arguments presented above, the definition of the characteristic operator function $\theta_A(\lambda)$ is correct. Moreover, we can take $\theta_A(\lambda) = X$, where the invertible operator X is given by (8.7).

8.2. Criterion of Unitary Equivalence

Denote by $\mathcal{H}_A$ the closure of the linear span of lineals $R_\alpha^m Q \, \mathcal{E}_A$ ($m = 0, 1, 2, \ldots$). The operator $A_s = A \mid \mathcal{H}_A \cap \mathcal{D}_A$ is called the $\mathcal{L}$-simple part of the operator A.

Theorem 8.1. *If the characteristic operator functions* $\theta_A(\lambda)$ *and* $\theta_{\tilde{A}}(\lambda)$ *coincide and, in addition,* $J = \tilde{J}$, *then the $\mathcal{L}$-simple parts of the operators* A *and* $\tilde{A}$ *are unitary equivalent.*

Proof. Assume that $\theta_A(\lambda)$ is given by equality (8.2) and $\theta_{\tilde{A}}(\lambda)$ is given by a similar equality

$$\theta_{\tilde{A}}(\lambda)\tilde{J}^{-1}\theta_{\tilde{A}}^*(\alpha) = \tilde{J}^{-1} + i(\bar{\alpha} - \lambda)\tilde{Q}^* \tilde{T}_{\alpha\lambda} \tilde{Q}, \tag{8.9}$$

where $\tilde{T}_{\alpha\lambda} = (\tilde{A} - \alpha I)(\tilde{A} - \lambda I)^{-1}$. Taking into account the conditions of the theorem, we obtain $Q^* T_{\alpha\lambda} Q = \tilde{Q}^* \tilde{T}_{\alpha\lambda} \tilde{Q}$. In particular, for $\lambda = \alpha$, we have $Q^* Q = \tilde{Q}^* \tilde{Q}$. Thus, in view of the fact that $T_{\alpha\lambda} = I + (\lambda - \alpha) R_\lambda$ and $\tilde{T}_{\alpha\lambda} = \tilde{I} + (\lambda - \alpha)\tilde{R}_\lambda$, we get $Q^* R_\lambda Q = \tilde{Q}^* \tilde{R}_\lambda \tilde{Q}$. Substituting the expressions for R_λ and $\tilde{R}_\lambda$ from (2.3) into this equality, we obtain

$$Q^* R_\alpha^m Q = \tilde{Q}^* \tilde{R}_\alpha^m \tilde{Q} \quad (m \in \mathbb{N}).$$

Hence,

$$(R_\alpha^m Qf, \, Q\varphi) = (\tilde{R}_\alpha^m \tilde{Q}f, \, \tilde{Q}\varphi)$$

for any f and φ from $\mathcal{E}_A$. Setting in this relation $x = Qf$, $y = Q\varphi$, $\tilde{x} = \tilde{Q}f$, and $\tilde{y} = \tilde{Q}\varphi$, we obtain the following equalities $(R_\alpha^m x, y) = (\tilde{R}_\alpha^m \tilde{x}, \tilde{y})$.

By analogy with the proof of equalities (2.8), we conclude that

$$(R_\alpha^m x, \, R_\alpha^s y) = (\tilde{R}_\alpha^m \tilde{x}, \, \tilde{R}_\alpha^s \tilde{y})$$

for any nonnegative integers m and s.

To complete the proof of the theorem, it remains to consider the operator V given by the equality $V R_\alpha^m x = \tilde{R}_\alpha^m \tilde{x}$ and repeat the corresponding reasoning from Section 2.

8.3. J-Nonexpandability

Let $\theta = J^{-1} \theta_A^*(\alpha)$. Then, by virtue of (8.2),

$$\theta_A(\lambda)\theta = J^{-1} + i(\bar{\alpha} - \lambda)Q^* T_{\alpha\lambda} Q \qquad (\lambda \in \rho(A)),$$

$$\theta^* \theta_A^*(\mu) = J^{-1} - i(\alpha - \bar{\mu})Q^* T_{\bar{\alpha}\mu}^* Q \qquad (\mu \in \rho(A))$$

and, consequently,

$$\theta^* \theta_A^*(\mu) J\theta_A(\lambda)\theta = J^{-1} + i(\bar{\alpha} - \lambda)Q^* T_{\alpha\lambda} Q - i(\alpha - \bar{\mu})Q^* T_{\bar{\alpha}\mu}^* Q$$

$$+ (\alpha - \bar{\mu})(\bar{\alpha} - \lambda)Q^* T_{\bar{\alpha}\mu}^* QJQ^* T_{\alpha\lambda} Q. \qquad (8.10)$$

By setting $\lambda = \alpha$ in (8.2), we get

$$J^{-1} = \theta^* J\theta - i(\bar{\alpha} - \alpha)Q^* Q. \qquad (8.11)$$

Substituting (8.11) in (8.10) and taking equality (8.1) into account, we obtain

$$\theta^* [\theta_A^*(\mu) J\theta_A(\lambda) - J]\theta = Q^* NQ, \qquad (8.12)$$

where the operator N is given by equality (5.5) and, consequently, can be represented in the form (5.8). Substituting (5.8) in (8.12), we obtain

$$\theta^* [\theta_A^*(\mu) J\theta_A(\lambda) - J]\theta = i(\bar{\mu} - \lambda)Q^* T_{\bar{\alpha}\mu}^* T_{\bar{\alpha}\lambda} Q,$$

which implies, by analogy with Section 5, that

$$\theta_A^*(\lambda) J\theta_A(\bar{\lambda}) = J \qquad (\{\lambda, \bar{\lambda}\} \in \rho(A)), \qquad (8.13)$$

$$\left.\begin{array}{ll} \theta_A^*(\lambda) J\theta_A(\lambda) \leq J & (\text{Im } \lambda \leq 0, \quad \lambda \in \rho(A)), \\[2mm] \theta_A^*(\lambda) J\theta_A(\lambda) \geq J & (\text{Im } \lambda \geq 0, \quad \lambda \in \rho(A)). \end{array}\right\} \qquad (8.14)$$

Thus, in the lower half-plane, the characteristic operator function $\theta_A(\lambda)$ is a J-non-expanding operator function.

8.4. Relation Between "Symmetric" Characteristic Operator Functions

Let us rewrite (8.2) as

$$\theta_\alpha(\lambda) = J^{-1} + i(\bar\alpha - \lambda) Q^* T_{\alpha\lambda} Q, \qquad (8.15)$$

where $\theta_\alpha(\lambda) = \theta_A(\lambda) J^{-1} \theta_A^*(\alpha)$. Since $B_{\bar\alpha} = T_{\bar\alpha}^* B_\alpha T_{\bar\alpha}$, where $T_{\bar\alpha} = (A - \alpha I) R_{\bar\alpha}$ (see Subsection 5.4), we have

$$B_{\bar\alpha} = \hat Q J \hat Q^* \qquad (\hat Q = T_{\bar\alpha}^* Q) \qquad (8.16)$$

by virtue of (8.1).

Consequently, we can consider the characteristic operator function $\hat\theta_A(\lambda)$ defined by the equality

$$\theta_{\bar\alpha}(\lambda) = J^{-1} + i(\alpha - \lambda) \hat Q^* T_{\bar\alpha\lambda} \hat Q, \qquad (8.17)$$

where $\theta_{\bar\alpha}(\lambda) = \hat\theta_A(\lambda) J^{-1} \hat\theta_A^*(\bar\alpha)$. The characteristic operator function $\hat\theta_A(\lambda)$ defined in this way is called *symmetric* with respect to the characteristic operator function $\theta_A(\lambda)$.

Let us show that if the image of the operator $\hat Q^*$ is a dense lineal in $\mathcal{E}_A$ ($\bar\Delta_{\hat Q^*} = \mathcal{E}_A$), then

$$\theta_{\bar\alpha}(\lambda) = \theta_\alpha(\lambda) J \hat\tau, \qquad (8.18)$$

where $\hat\tau = J^{-1} + 2\operatorname{Im}\bar\alpha\, \hat Q^* \hat Q$ is the corresponding deviation operator.

Since $(\alpha - \lambda) T_{\bar\alpha\lambda} = (\alpha - \bar\alpha) I + (\bar\alpha - \lambda) T_{\alpha\lambda}$, we have

$$i(\alpha - \lambda) \hat Q^* T_{\bar\alpha\lambda} \hat Q = -2\operatorname{Im}\alpha \hat Q^* \hat Q + i(\bar\alpha - \lambda) Q^* T_{\bar\alpha} T_{\alpha\lambda} T_{\bar\alpha}^* Q$$

$$= \hat\tau - J^{-1} + i(\bar\alpha - \lambda) Q^* T_{\alpha\lambda} T_{\bar\alpha} T_{\bar\alpha}^* Q \qquad (8.19)$$

and, consequently, by virtue of (8.17) and (8.19),

$$\theta_{\bar\alpha}(\lambda) = \hat\tau + i(\bar\alpha - \lambda) Q^* T_{\alpha\lambda} Y, \qquad (8.20)$$

where $Y = T_{\bar\alpha} T_{\bar\alpha}^* Q$. By applying the operator $T_{\bar\alpha} T_{\bar\alpha}^*$ to both parts of the equality

$$T_\alpha^* T_\alpha - I = 2\operatorname{Im}\alpha\, Q J Q^* \qquad (8.21)$$

and taking into account the equality $T_{\bar{\alpha}} = T_{\alpha}^{-1}$, we obtain

$$I - T_{\bar{\alpha}} T_{\bar{\alpha}}^* = 2 \operatorname{Im} \alpha Y J Q^*. \tag{8.22}$$

We multiply the equality obtained from the right by the operator $T_{\alpha}^* T_{\alpha}$. Then, by virtue of (8.21) and (8.22), we find

$$2 \operatorname{Im} \alpha Q J Q^* = 2 \operatorname{Im} \alpha Y J Q^* T_{\alpha}^* T_{\alpha}. \tag{8.23}$$

If we delete the factor $2 \operatorname{Im} \alpha$ in (8.23) and substitute for $T_{\alpha}^* T_{\alpha}$ its value from (8.21), then we get

$$Q J Q^* = Y J Q^* + 2 \operatorname{Im} \alpha Y J Q^* Q J Q^*$$

$$= Y J [J^{-1} + 2 \operatorname{Im} \alpha Q^* Q] J Q^* = Y J \tau J Q^*. \tag{8.24}$$

Consequently, by multiplying both parts of equality (8.20) from the right by $J \tau J Q^*$ and using equality (8.24), we obtain

$$\theta_{\bar{\alpha}}(\lambda) J \tau J Q^* = \hat{\tau} J \tau J Q^* + i(\bar{\alpha} - \lambda) Q^* T_{\alpha\lambda} Q J Q^* = [\hat{\tau} J \tau + \theta_{\alpha}(\lambda) - J^{-1}] J Q^*. \tag{8.25}$$

Let us show that

$$\hat{\tau} J \tau = J^{-1}. \tag{8.26}$$

In fact, by virtue of the equalities $\hat{Q} = T_{\bar{\alpha}}^* Q$ and (8.6), we have

$$\tau^{-1} = J - 2 \operatorname{Im} \alpha J \hat{Q}^* \hat{Q} J = J \hat{\tau} J,$$

which proves equality (8.26). But then, by virtue of (8.25) and (8.26), we get

$$\theta_{\bar{\alpha}}(\lambda) J \tau J Q^* = \theta_{\alpha}(\lambda) J Q^*. \tag{8.27}$$

Furthermore, if $\overline{\Delta}_{Q^*} = \mathcal{E}_A$, then it follows from (8.26) and (8.27) that equality (8.18) holds.

8.5. Regular Couplings

Let $A = A_1 \gamma A_2$ be a coupling of operators A_1 and A_2 in the sense of Subsection 6.1. Without loss of generality, we can assume that the external spaces of the operators A_1 and A_2 coincide ($\mathcal{E}_{A_1} = \mathcal{E}_{A_2}$). Furthemore, we assume that $J_1 = J_2$ and $J: J_1 = (= J_2)$. Thus, by virtue of equality (6.10), we obtain

$$B_\alpha = Q_1 J Q_1^* P_1 + Q_2 J Q_2^* P_2 + i T_{1\alpha} \Phi P_2 - i \Phi^* T_{1\alpha} P_1 + 2 \operatorname{Im} \alpha \Phi^* \Phi P_2, \quad (8.28)$$

where $\Phi = K R_{2\alpha}$. We define the operator Φ (and, thus, the operator K) by the equality

$$\Phi = -i T_{1\bar{\alpha}}^* Q_1 J \hat{\theta}_{A_1}(\alpha) Q_2^*, \quad (8.29)$$

where the symmetric characteristic operator function $\hat{\theta}_{A_1}(\lambda)$ is defined by the equality

$$\hat{\theta}_{A_1}(\lambda) J^{-1} \hat{\theta}_{A_1}^*(\bar{\alpha}) = J^{-1} + i(\alpha - \lambda) \hat{Q}_1^* T_{1\bar{\alpha}\lambda} \hat{Q}_1. \quad (8.30)$$

It follows from (8.29) that

$$T_{1\alpha}^* \Phi = -i Q_1 J \hat{\theta}_{A_1}(\alpha) Q_2^*, \quad (8.31)$$

$$\Phi^* T_{1\alpha} = i Q_2 \hat{\theta}_{A_1}^*(\alpha) J Q_1^*. \quad (8.32)$$

Moreover, since $T_{1\bar{\alpha}}^* Q_1 = \hat{Q}_1$, we have

$$\Phi^* \Phi = Q_2 \hat{\theta}_{A_1}^*(\alpha) J \hat{Q}_1^* \hat{Q}_1 J \hat{\theta}_{A_1}(\alpha) Q_2^*.$$

Consequently, in view of the equality $2 \operatorname{Im} \alpha \, \hat{Q}_1^* \hat{Q}_1 = J^{-1} - \hat{\tau}_1$, we find that

$$2 \operatorname{Im} \alpha \, \Phi^* \Phi = Q_2 \hat{\theta}_{A_1}^*(\alpha) J \hat{\theta}_{A_1}(\alpha) Q_2^* - Q_2 \hat{\theta}_{A_1}^*(\alpha) J \hat{\tau}_1 J \hat{\theta}_{A_1}(\alpha) Q_2^*. \quad (8.33)$$

Multiplying the equality $\hat{\theta}_{A_1}(\alpha) J^{-1} \hat{\theta}_{A_1}^*(\bar{\alpha}) = J^{-1}$ from the left by $\hat{\theta}_{A_1}^*(\bar{\alpha}) J$ and from the right by $\hat{\theta}_{A_1}^{-*}(\bar{\alpha}) J$, we obtain the equality $\hat{\theta}_{A_1}^*(\bar{\alpha}) J \hat{\theta}_{A_1}(\alpha) = J$, which implies that $\hat{\theta}_{A_1}^*(\bar{\alpha}) J \hat{\theta}_{A_1}(\alpha) = J$. Moreover, $\hat{\theta}_{A_1}(\bar{\alpha}) J^{-1} \hat{\theta}_{A_1}^*(\bar{\alpha}) = \hat{\tau}_1$ by virtue of (8.30). The last three equalities imply that

$$\hat{\theta}^*_{A_1}(\alpha) \, J \, \hat{\tau}_1 \, J \, \hat{\theta}_{A_1}(\alpha) \; = \; \hat{\theta}^*_{A_1}(\alpha) \, J \, \hat{\theta}^*_{A_1}(\bar{\alpha}) \, J^{-1} \, \hat{\theta}^*_{A_1}(\bar{\alpha}) \, J \, \hat{\theta}_{A_1}(\alpha) \; = \; J. \qquad (8.34)$$

Substituting (8.34) in (8.38), we obtain

$$2 \operatorname{Im} \alpha \, \Phi^* \Phi \; = \; Q_2 \, \hat{\theta}^*_{A_1}(\alpha) \, J \, \hat{\theta}_{A_1}(\alpha) \, Q^*_2 \; - \; Q_2 J Q^*_2, \qquad (8.35)$$

whence, by virtue of (8.28), (8.31), (8.32), and (8.35), we get $B_\alpha = QJQ^*$, where

$$Q \; = \; Q_1 + Q_2 \, \hat{\theta}^*_{A_1}(\alpha) \qquad (Q^* \; = \; Q^*_1 \, P_1 + \hat{\theta}_{A_1}(\alpha) \, Q^*_2 \, P_2). \qquad (8.36)$$

The coupling $A = A_1 \gamma A_2$, which satisfies the conditions given above (i.e., $\mathcal{E}_{A_1} = \mathcal{E}_{A_2}$, $J_1 = J_2$, and the operator Φ is defined by equality (8.29)), is called a *regular coupling* of the operators A_1 and A_2. Thus, if A is a regular coupling of the operators A_1 and A_2, then $B_\alpha = QJQ^*$, where the operator Q is given by equality (8.36).

8.6. Multiplication Theorem

Theorem 8.2. *Assume that A is a regular coupling of the operators A_1 and A_2 and that $\overline{\Delta}_{Q^*_1}$ is the external space $\mathcal{E}$. Then the corresponding characteristic operator functions of this operator satisfy the equality*[4]

$$\theta_A(\lambda) \; = \; \theta_{A_1}(\lambda) \, \theta_{A_2}(\lambda). \qquad (8.37)$$

Proof. Assume that the characteristic operator function $\theta_A(\lambda)$ is defined by equality (8.2). As shown in Subsection 6.4, the operator $T_{\alpha\lambda}$ can be represented as

$$T_{\alpha\lambda} \; = \; T_{1\alpha\lambda} P_1 + T_{2\alpha\lambda} P_2 + (\lambda - \alpha) \, \Phi_\lambda P_2,$$

where $\Phi_\lambda = T_{1\alpha\lambda} \Phi T_2$. By virtue of equalities (8.36), we have

$$J^{-1} + i(\bar{\alpha} - \lambda) \, Q^* T_{\alpha\lambda} Q \; = \; J^{-1} + i(\bar{\alpha} - \lambda) \, Q^*_1 T_{1\alpha\lambda} Q_1$$

$$+ \; i(\bar{\alpha} - \lambda) \, [(\lambda - \alpha) \, Q^*_1 \Phi_\lambda + \hat{\theta}_{A_1}(\alpha) \, Q^*_2 T_{2\alpha\lambda}] \, Q_2 \, \hat{\theta}_{A_1}(\alpha), \qquad (8.38)$$

where

---

[4] Since characteristic operator functions are defined not uniquely, it is clear that here we consider characteristic operator functions satisfying equality (8.37).

$$\Phi_\lambda = -i T_{1\alpha\lambda} T^*_{1\bar\alpha} Q_1 J \hat\theta_{A_1}(\alpha) Q^*_2 T_{2\alpha\lambda} \tag{8.39}$$

according to (8.29). Thus, in view of (8.2), (8.38), and (8.39), we obtain

$$\theta_A(\lambda) J^{-1} \theta^*_A(\alpha) = \theta_{A_1}(\lambda) J^{-1} \hat\theta^*_{A_1}(\alpha)$$

$$+ i(\bar\alpha - \lambda) XJ\hat\theta_{A_1}(\alpha) Q^*_2 T_{2\alpha\lambda} Q_2 \hat\theta^*_{A_1}(\alpha), \tag{8.40}$$

where $X = i(\alpha - \lambda) Q^*_1 T_{1\alpha\lambda} T^*_{1\bar\alpha} Q_1 + J^{-1}$. Furthermore, by using the equalities

$$T^*_{1\bar\alpha} Q_1 = \hat Q_1, \quad Q^*_1 T_{1\alpha\lambda} = Q^*_1 T_{1\bar\alpha} T_{1\alpha} T_{1\alpha\lambda} = \hat Q^*_1 T_{1\bar\alpha\lambda}$$

and (8.18), we get

$$X = \hat\theta_{A_1}(\lambda) J^{-1} \hat\theta^*_{A_1}(\bar\alpha) = \theta_{A_1}(\lambda) J^{-1} \theta^*_{A_1}(\alpha) J\hat\tau_1. \tag{8.41}$$

Substituting (8.41) in (8.40), we obtain

$$\theta_A(\lambda) J^{-1} \theta^*_A(\alpha) = \theta_{A_1}(\lambda) J^{-1} \theta^*_{A_1}(\alpha) JS, \tag{8.42}$$

where

$$S = J^{-1} + i(\bar\alpha - \lambda) \hat\tau_1 J\hat\theta_{A_1}(\alpha) Q^*_2 T_{2\alpha\lambda} Q_2 \hat\theta^*_{A_1}(\alpha).$$

Multiplying both parts of equality (8.34) by the operator $\hat\theta_{A_1}(\bar\alpha) J^{-1}$ and using the equality $\hat\theta_{A_1}(\bar\alpha) J^{-1}\hat\theta^*_{A_1}(\alpha) = J^{-1}$, we conclude that $\hat\tau_1 J\hat\theta_{A_1}(\alpha) = \hat\theta_{A_1}(\bar\alpha)$. Taking the last two equalities into account, we find that

$$S = \hat\theta_{A_1}(\bar\alpha)[J^{-1} + i(\bar\alpha - \lambda) Q^*_2 T_{2\alpha\lambda} Q_2] \hat\theta^*_{A_1}(\alpha),$$

i.e., $S = \hat\theta_{A_1}(\bar\alpha) \theta_{A_2}(\lambda) J^{-1} \theta^*_{A_2}(\alpha)\hat\theta^*_{A_1}(\alpha)$. If we substitute the value of S in (8.42), then we get

$$\theta_A(\lambda) J^{-1} \theta^*_A(\alpha) = \theta_{A_1}(\lambda) W \theta_{A_2}(\lambda) J^{-1} Y, \tag{8.43}$$

where

$$W = J^{-1} \theta^*_{A_1}(\alpha) J\hat\theta_{A_1}(\bar\alpha), \quad Y = \theta^*_{A_2}(\alpha) \hat\theta_{A_1}(\alpha). \tag{8.44}$$

Furthermore, since $\hat{\theta}_{A_1}(\overline{\alpha}) J^{-1} \hat{\theta}_{A_1}^*(\overline{\alpha}) = \hat{\tau}_1$, we have

$$W J^{-1} W^* = J^{-1} \theta_{A_1}^*(\alpha) J \hat{\tau}_1 J \theta_{A_1}(\alpha) J^{-1}, \tag{8.45}$$

where

$$J \hat{\tau}_1 J = \tau_1^{-1} = \theta_{A_1}^{-*}(\alpha) J \theta_{A_1}^{-1}(\alpha) \tag{8.46}$$

by virtue of (8.26). Substituting (8.46) in (8.45), we obtain $W J^{-1} W^* = J^{-1}$. In view of the fact that the characteristic operator function is determined to within an operator W satisfying the last equality, we can take the characteristic operator function $\tilde{\theta}_{A_1}(\lambda) = \theta_{A_1}(\lambda) W$ instead of $\theta_{A_1}(\lambda)$.

Let us rewrite the operator Y from (8.44) in the form $Y = \theta_{A_2}^*(\alpha) W^* \theta_{A_1}^*(\alpha) R$, where $R = \theta_{A_1}^{-*}(\alpha) W^{-*} \theta_{A_1}^*(\alpha)$. To find R, we multiply the operator W in (8.44) by $\theta_{A_1}(\alpha)$. Then, by virtue of (8.26),

$$\theta_{A_1}(\alpha) W = \tau_1 J \hat{\theta}_{A_1}(\overline{\alpha}) = J^{-1} \hat{\tau}_1^{-1} \theta_{A_1}(\overline{\alpha}),$$

where $\hat{\tau}_1^{-1} = \hat{\theta}_{A_1}^{-*}(\overline{\alpha}) J \hat{\theta}_{A_1}^{-1}(\overline{\alpha})$. Thus, $\theta_{A_1}(\alpha) W = J^{-1} \hat{\theta}_{A_1}^{-*}(\overline{\alpha}) J = \hat{\theta}_{A_1}(\alpha)$. But this means that $R^{-*} = \theta_{A_1}(\alpha) W \theta_{A_1}^{-1}(\alpha) = I$, i.e., $R = I$.

Thus, we obtain

$$\theta_A(\lambda) J^{-1} \theta_A^*(\alpha) = \theta_{A_1}(\lambda) W \theta_{A_2}(\lambda) J^{-1} \theta_{A_2}^*(\alpha) W^* \theta_{A_1}^*(\alpha),$$

whence we get that $\theta_A(\lambda) = \theta_{A_1}(\lambda) W \theta_{A_2}(\lambda)$, where $W J^{-1} W^* = J^{-1}$. Furthermore, according to the facts presented above, we can omit the operator W in the expression for $\theta_A(\lambda)$.

Remark. The definition of characteristic operator functions considered above is due to Kuzhel [22]. The principal results of this section are published here for the first time.

9. Characteristic Functions. Brief Survey

In this section, we present a brief survey of the development and applications of the concept of the characteristic function for various classes of operators.

9.1. Nonunitary (Quasiunitary) Operators

9.1.1. The concept of the characteristic function was first introduced by Livsic [1] in the study of nonunitary extensions of an isometric operator V with deficiency index $(1, 1)$; in other words, $\mathcal{H} \ominus \mathfrak{D}_V$ and $\mathcal{H} \ominus \Delta_V$ are the one-dimensional subspaces of $\mathcal{H}$ ($\mathcal{H}$ is the Hilbert space in which the operator V acts).

According to Livsic [1], the operator T_α acting in $\mathcal{H}$ is called a *quasiunitary extension* of the operator V if $V \subset T_\alpha$ and $T_\alpha g = ag'$, where a is a fixed complex number called a contraction coefficient and g and g' are fixed unit vectors from the subspaces $\mathfrak{N}_0 = \mathcal{H} \ominus \mathfrak{D}_V$ and $\mathfrak{N}'_0 = \mathcal{H} \ominus \Delta_V$, respectively.

The operator $T_\alpha(z) = (T_\alpha - zI)(I - \bar{z}T_\alpha)^{-1}$ is a quasiunitary extension of the isometric operator $V(z) = (V - zI)(I - \bar{z}V)^{-1}$ ($|z| \neq 1$) and, in this case, the vectors g_z from $\mathfrak{N}_z = \mathcal{H} \ominus \mathfrak{D}_{V(z)}$ and g'_z from $\mathfrak{N}'_z = \mathcal{H} \ominus \Delta_{V(z)}$ are determined by the equalities $g_z = (I - zT_a^*)^{-1}g$ and $g'_z = (I - \bar{z}T_\alpha)^{-1}g'$.

The function $\omega(z, a)$ is called a characteristic function of the quasiunitary operator T_α if it satisfies the equality

$$T_\alpha(z)\, g_z = \omega(z, a)\, \bar{g}_z \qquad (\bar{g}_z = (I - \bar{z}T_\alpha)^{-1}\bar{g}).$$

In this case, $\bar{f} = Kf$, where K is a "mirror" operator defined by the following equalities:

(1) $K(\alpha f + \beta g) = \bar{\alpha}Kf + \bar{\beta}Kg$;

(2) $(Kf, Kf) = (f, f)$;

(3) $K^2 f = f$.

It is established that $\omega(z, a)$ is a regular function which maps the unit disk D onto itself and can be represented as follows:

$$\omega(z, a) = \frac{a + \omega(z, 0)}{1 + \bar{a}\omega(z, 0)}, \qquad \omega(z, 0) = -z\,\frac{(g_z, \bar{g})}{(g_z, g)}.$$

Furthermore, in the same paper, Livsic established the criterion of unitary equivalence of the simple parts of the operators under consideration and described the spectrum of quasiunitary operators is described in terms of the characteristic functions.

9.1.2. Later Livsic generalized these results for the case of quasiunitary extensions of isometric operators with deficiency index (m, m), where $m < \infty$ (see [2]). In this case, the characteristic functions are defined as an $m \times m$ matrix

$$\omega(T, z) = B^{-1}(z) A(z),$$

where

$$B(z) = g(z) - z g'(z) \tau^*, \quad A(z) = g(z) \tau - z g'(z),$$

$$g(z) = \| (g_k(z), g_s) \|, \quad g'(z) = \| (g_k(z), g'_s) \|,$$

$\{g_k(z)\}_{k=1}^m$ is an arbitrary fixed basis of the defect subspace $\mathfrak{N}_z$, while $\{g_k\}_{k=1}^m$ and $\{g'_k\}_{k=1}^m$ are orthonormal bases in the subspaces $\mathfrak{N}_0$ and $\mathfrak{N}'_0$, respectively. The matrix $\tau = \| \tau_{ki} \|$ is determined by the equalities

$$T g_k = \sum_{j=1}^m \tau_{kj} g'_j \quad (k = \overline{1, m}). \tag{9.1}$$

9.1.3. In the paper by Livsic and Potapov [1], the characteristic functions considered in the previous subsection were "normalized". This approach enabled the authors to introduce the concept of a normalized characteristic function under the assumption that $\det (E - \tau^* \tau) \neq 0$, where the elements of the matrix τ are determined by equalities (9.1).

The normalized characteristic function $\omega_T(z)$ of a quasiunitary operator T is defined by the equality

$$\omega_T(z) = u^* |E - \tau \tau^*|^{1/2} (I - \omega(z) \tau^*)^{-1} (\tau - \omega(z)) |E - \tau^* \tau|^{1/2} v^*,$$

where u and v are unitary matrices, $\omega(z) = -z g^{-1}(z) g'(z)$ is the characteristic function of an isometric operator V ($V \subset T$), and the matrices $g(z)$ and $g'(z)$ have the same sense as in Subsection 9.1.2. Livsic and Potapov [1] also clarified the conditions, under which a given matrix function $\omega(z)$ is a characteristic function of some quasiunitary operator T, established the criterion of unitary equivalence of the simple parts of quasiunitary operators in terms of the characteristic functions, and formulate the multiplication theorem for the case where

$$T = \begin{pmatrix} T_1 & \Gamma \\ 0 & T_2 \end{pmatrix}$$

and the quasiunitary operators T_1 and T_2 have the same rank m (i.e., the maximal isometric operators, whose extensions are given by the indicated operators, are isometric operators with deficiency index (m, m)).

9.1.4. Developing the ideas of Livsic and Potapov [1], Shmulyan introduced in [1] the concept of the normalized characteristic function $\omega(T, z)$ of a nonunitary extension T of an isometric operator V whose defect numbers are assumed to be neither finite nor equal.

According to Shmulyan's definition, for any f from $\mathfrak{N}_0 = \mathcal{H} \ominus \mathfrak{D}_V$, we have

$$\omega(T, z)f = |I - TT^*|^{-1/2}(T - zI)(I - zT^*)^{-1}|I - T^*T|^{1/2}f. \tag{9.2}$$

Since $T|I - T^*T|^{1/2} = |I - TT^*|^{1/2}T$ and

$$(T - zI)(I - zT^*)^{-1} = T - z(I - TT^*)(I - zT^*)^{-1},$$

we can also write

$$\omega(T, z)f = Tf - zJ'|I - TT^*|^{1/2}(I - zT^*)^{-1}|I - T^*T|^{1/2}f, \tag{9.3}$$

for $f \in \mathfrak{N}_0$, where $J' = \text{sign}\,(I - TT^*)$. In the same paper, the multiplication theorem is formulated under certain conditions, and some general properties of the characteristic functions are discussed.

In [1], Polyatskii succeeded in constructing the (triangular) models for a broad class of nonunitary operators by using the definition of characteristic functions in the form (9.2) and the results of Livsic [3] concerning the construction of model operators for the bounded nonself-adjoint operators.

9.1.5. Sz.-Nagy and Foias [1, 2] introduced the concept of characteristic function for a contraction T in a completely different way, namely, by using the unitary dilations of the operator T. Thus, if T is contracting, then we consider the operators $D_T = (I - T^*T)^{1/2}$ and $D_{T^*} = (I - TT^*)^{1/2}$ and the subspaces $\mathfrak{N}_T = \overline{D_T \mathcal{H}}$ and $\mathfrak{N}_{T^*} = \overline{D_{T^*} \mathcal{H}}$. In this case, the characteristic function $\theta_T(z)$ of the operator T is defined by the equality

$$\theta_T(z) = [-T + zD_{T^*}(I - zT^*)^{-1}D_T]|_{\mathfrak{N}_T}. \tag{9.4}$$

Comparing (9.4) with (9.3), we conclude that $\theta_T(z)$ is a special case of the characteristic function $\omega(T, z)$ in Shmulyan's sense (more exactly, it only differs from $\omega(T, z)$ by the sign). A more detailed exposition of the properties of the characteristic function $\theta_T(z)$ and its applications can be found in the work of Sz.-Nagy and Foias [4].

9.1.6. In the paper of Kuzhel [2], the characteristic matrix function of an operator T is defined in a somewhat different way. Namely, the author considered a bounded invertible operator T for which $\dim(I - T^*T)\,\mathcal{H} = r$ $(r < \infty)$. A collection $\{g_k\}_{k=1}^s$ is called an α-*basis* of the operator T if the operator $I - T^*T$ can be represented as follows:

$$I - T^*T = \sum_{k, i=1}^s (\cdot, g_k)\, J_{ki}\, g_i,$$

where $J = \| J_{ki} \|$ is a unitary Hermitian matrix. Under some additional conditions imposed on the α-basis, the characteristic matrix function $W(\lambda)$ of the operator T is given by the equality

$$W(z)\, W(0) = E - \| ((I - zT^*)^{-1} g_k, g_i) \|\, J,$$

where $W(0)$ is a nonnegative Hermitian matrix. This definition of a characteristic matrix function enables one to justify the multiplication theorem in a more general form than in the work of Livsic and Potapov [1]. The characteristic matrix function $W(z)$ (more exactly, $\omega(z) = W^*(\bar{z})$) was also studied by Sahnovich [2]; in particular, he established the conditions under which the operator T is similar to a unitary operator.

9.1.7. In [17], by analogy with (9.4), Kuzhel defined the characteristic operator function $\theta_T(z)$ of an arbitrary bounded nonunitary operator T by the equality

$$\theta_T(z) = TJ_T - zQ_{T^*}(I - zT^*)^{-1}Q_T, \tag{9.5}$$

where $Q_T = |I - T^*T|^{1/2}$ and $J_T = \operatorname{sign}(I - T^*T)$.

The characteristic function thus defined satisfies, in particular, the following conditions:

$$\theta_T^*(z) = \theta_{T^*}(\bar{z}), \qquad \theta_T(\bar{z})\, J_T\, \theta_T^*(1/z) = J_{T^*}, \tag{9.6}$$

$$\theta_T^*(0)\, J_{T^*}\,\theta_T(z) = J_T - Q_{T^*}(I - zT^*)^{-1}Q_T, \tag{9.7}$$

$$\theta_T(z)\, J_T\, \theta_T^*(0) = J_{T^*} - Q_{T^*}(I - zT^*)^{-1}Q_T. \tag{9.8}$$

By analogy with (9.4), we can consider the restriction of $\theta_T(z)$ to $\mathfrak{N}_T$:

$$\theta_T(z) = [TJ_T - zQ_{T^*}(I - zT^*)^{-1}Q_T]|_{\mathfrak{N}_T}.$$ (9.9)

In this case, if T is a contraction, then $J_T|_{\mathfrak{N}_T} = I$ and the characteristic operator function $\theta_T(z)$ determined by equality (9.9) differs from the characteristic operator function determined by (9.4) only by the sign.

Also note that since $TJ_Tx = J_{T^*}Tx$ for $x \in \mathfrak{N}_T$, we can apply the operator J_{T^*} to the sides of equality (9.3), compare the equality obtained with (9.9), and conclude that the characteristic operator functions $\omega(T, z)$ and $\theta_T(z)$ are connected by the equality $\theta_T(z) = J_{T^*}\omega(T, z)$.

The characteristic operator function $\theta_T(z)$ defined by equality (9.5) (or (9.9)) has been used in the investigation of various properties of operators by Kuzhel [1, 2] (the theorem on factorization of characteristic operator functions), by Davis and Foias [2] (the construction of a J-unitary dilation, the investigation of it, and the proof of the Sahnovich theorem [2] on the similarity, under certain conditions, between the operator T and a unitary operator), by Clark [2] (the construction of an abstract model for a bounded operator T which is not contracting), by McEnnis [1–3] (the construction of models and the J-unitary dilations), by Makarov [1] (the problem of stability of the essential spectrum for nonunitary operators), etc.

9.1.8. In the case of a bounded operator T, invertible in the restricted sense $(0 \in \rho(T))$, Brodskii, Gohberg, and Krein [1] defined the characteristic operator function $\theta_{\mathfrak{J}}(z)$ of the knot $\mathfrak{J} = (\mathcal{H}, \mathcal{E}; T, R, J)$ by the equality

$$\theta_{\mathfrak{J}}(z) = J(K^*)^{-1}(J - R^*(I - zT^*)^{-1}R),$$ (9.10)

where J is a self-adjoint unitary operator acting in the space $\mathcal{E}$, the operator R belongs to $\mathcal{B}[\mathcal{E}, \mathcal{H}]$ and satisfies the condition $I - T^*T = RJR^*$, and K is a bounded invertible operator acting in the space $\mathcal{E}$ and solving the operator equation $I - R^*R = K^*JK$ (the solvability of this equation can be established by different methods).

After certain transformations, the characteristic operator function (9.10) can be rewritten in the form

$$\theta_{\mathfrak{J}}^*(0)J\theta_{\mathfrak{J}}(z) = J - R^*(I - zT^*)^{-1}R.$$

Using the definition of characteristic operator function (9.10), V. Brodskii proved in [1] that the product of operator knots is associated with the product of the corresponding characteristic operator functions (the multiplication theorem).

Finally, we note that the characteristic functions of nonunitary operators are widely applied not only in the study of nonunitary operators but also in scattering theory, in prediction theory, and so on. For more details, see the works by Adamjan and Arov [1–4] (see also Appendix 3).

9.2. Bounded Nonself-Adjoint Operators

The characteristic function of a bounded operator A with a completely continuous imaginary component $\operatorname{Im} A$ was introduced in the paper of Livsic [3] by the equality

$$W(\lambda) = I + i \operatorname{sign}\left[\frac{A-A^*}{i}\right]\sqrt{\left|\frac{A-A^*}{i}\right|}(A^*-\lambda I)^{-1}\cdot\sqrt{\left|\frac{A-A^*}{i}\right|}$$

In this paper, the properties of characteristic functions are investigated, the multiplication theorem is proved, and the triangular model is constructed.

In [2. 3], by using Livsic's results, Sahnovich obtained the similarity conditions for a broad class of nonself-adjoint operators and studied the problem of reducing these operators to "diagonal" form.

A more general definition of the characteristic function of a bounded operator A was suggested by M. Brodskii [1], namely,

$$W(\lambda) = I - 2iK^*(A - \lambda I)^{-1}KJ, \tag{9.11}$$

where J is a self-adjoint unitary operator acting in a certain auxiliary ("exterior") space $\mathcal{E}$ and K is a bounded operator acting from $\mathcal{E}$ into $\mathcal{H}$ and satisfying the condition

$$\frac{A-A^*}{2i} = KJK^*.$$

This definition of the characteristic function enables one to prove the multiplication theorem under weaker and much simpler restrictions; one can also investigate many general problems in the theory of nonself-adjoint operators (the triangular representations of the Volttera operators, the one-cell property of the Volttera and dissipative operators, etc).

Rutkas [1] and Karpenko [1] generalized the definition of the characteristic operator functions (9.11) for the case of unbounded operators with $\mathfrak{D}_A = \mathfrak{D}_{A^*}$. A series of results due to M. Brodskii was generalized for this class of operators by Karpenko [1].

9.3. Unbounded Nonself-Adjoint Operators

9.3.1. The characteristic function of an unbounded operator B was first defined by Livsic [1] under the condition that B is a nonself-adjoint extension of a symmetric (densely defined) operator A with deficiency index $(1, 1)$. It was shown in this paper that, under this condition, any element u from $\mathcal{D}_B$ can be uniquely represented in the form

$$u = \varphi_\lambda + u_\lambda + \omega_B(\lambda) K u_\lambda,$$

for any fixed nonreal λ; here, K is the "mirror" operator mentioned above.

The function $\omega_B(\lambda)$ is called a characteristic function of the operator B. In the same paper, the properties of the characteristic function thus defined are examined. In particular, it is shown that this function is connected with the characteristic function $\omega_T(z)$ of the corresponding quasiunitary operator $T = (B - iI)(B + iI)^{-1}$ by the equality $\omega_B(\lambda) = e^{i\alpha} \omega_T((\lambda - i)(\lambda + i)^{-1})$, where α is a real constant.

9.3.2. In the papers by Kuzhel [1–3], the characteristic function was defined for the quasi-Hermitian extensions of an Hermitian operator H (which is not necessarily densely defined) with finite and equal defect numbers (the K^r-operators). If A is a K^r-operator and $-i \in \rho(A)$, then the self-adjoint operator $B = iR_{-i} - iR^*_{-i} - 2R^*_{-i}R_{-i}$ can be represented in the form

$$B = \sum_{k, \, i=1}^{s} (\cdot, g_k) J_{ki} g_i \quad (J = \|J_{ki}\|; \quad J^* = J = J^{-1}).$$

Then the characteristic matrix function $W_A(\lambda)$ of the operator A is defined by the equality

$$W_A(\lambda) W_A(i) = I + i(\lambda + i) \| ((A^* - iI)(A^* - \lambda I)^{-1} g_k, g_m) \| J, \qquad (9.12)$$

where $W_A(i)$ is a nonnegative Hermitian matrix. In the same papers, the general properties of such characteristic functions are studied, the triangular model for the K^r-operators is constructed, and the spectrum of these operators is investigated.

9.3.3. At the same time, Shtraus [5, 6] introduced the concept of the characteristic function $X(\lambda)$ for an arbitrary closed densely defined operator A with a nonempty set of regular points $\rho(A)$. We now reproduce his scheme.

Consider a factor space $\mathcal{L}_A = \mathcal{D}_A / G_A$, where

$$G_A = \{x \in \mathcal{D}_A \, | \, (Ax, y) = (x, Ay) \quad (\forall \, y \in \mathcal{D}_A)\}.$$

In this space, we introduce an indefinite scalar product

$$[\tilde{x}, \tilde{y}] = \frac{1}{i}[(Ax, y) - (x, Ay)] \quad (\{\tilde{x}, \tilde{y}\} \subset L_A),$$

where $x \in \tilde{x}$ and $y \in \tilde{y}$. Then we define a "limiting" operator Γ which maps $\mathfrak{D}_A$ onto a linear space L isomorphic to the space L_A by the equality

$$[\Gamma x, \Gamma y] = \frac{1}{i}[(Ax, y) - (x, Ay)].$$

Similarly, an operator Γ' is defined by the equality

$$[\Gamma'\varphi, \Gamma'\psi] = \frac{1}{i}[(\varphi, A^*\psi) - (A^*\varphi, \psi)].$$

One should also consider the operator $S_\lambda = (A^* - \lambda I)^{-1}(A - \lambda I)$. Finally, the characteristic function $X(\lambda)$ of the operator A is defined by the equality

$$X(\lambda)\Gamma x = \Gamma'S_\lambda x \quad (x \in \mathfrak{D}_A). \tag{9.13}$$

Shtraus [5, 6] also established the criterion of unitary equivalence of the simple parts of nonself-adjoint operators in terms of characteristic functions, formulated the multiplication theorem for characteristic functions, and described the relationship with the Livsic characteristic function.

In the papers of Kochubei [1], Derkach and Malamud [1], and Tsekanovskii [1], the characteristic function introduced by Shtraus and its generalizations were applied to studying various classes of extensions of symmetric (densely defined) operators.

In the works of S. Kuzhel [3] and Malamud [2], the results obtained by Kochubei [1] (the characteristic function, the description of the spectrum of extensions, etc.), were generalized to the case of nonself-adjoint extensions of nondensely defined Hermitian operators.

9.3.4. On the basis of Shtraus' ideas, Polyakov [1] introduced the concept of the characteristic function of a bounded nonunitary operator T as follows: A linear space L with a nondegenerate scalar product $[\cdot, \cdot]$ is called the boundary space of the operator T if there exists a linear operator $Z: \mathcal{H} \to L$ such that

$$((I - T^*T))h, h = [Zh, Zh] \quad (\forall h \in \mathcal{H}).$$

The operator Z is called limiting for T. The characteristic function $\theta_T(\lambda)$ of the operator T is defined by the equality

$$\theta_T(\lambda)Zf = Z'S_\lambda f \quad (f \in \mathcal{H}),$$

where Z' is the limiting operator of T^* and

$$S_\lambda = (I - \lambda T^*)^{-1}(T - \lambda I), \quad \frac{1}{\lambda} \in \rho(T).$$

9.3.5. In the survey by Tsekanovskii and Shmulyan [1], Tsekanovskii's definition of the characteristic function is presented in terms of the rigged Hilbert spaces (for a detailed description of these spaces, see, e.g., the monograph by Berezanskii, Us, and Sheftel [1]). This definition coincides formally with the definition of the characteristic function in the case of bounded operators (see equality (9.11)).

9.3.6. Rutkas [1] introduced and studied the concept of the characteristic function of an operator bundle $\lambda A + B$. He used this function when investigating various radio-physical systems.

9.3.7. By using the Kuzhel's results [2, 3, 17], Do Kong Han [1] introduced the concept of the characteristic function $\chi(\lambda)$ for an unbounded closed densely defined operator A as follows:

$$\chi^*(i) J' \chi(\lambda) = J + i(\lambda + i) Q^*(A^* - iI)(A^* - \lambda I)^{-1} Q,$$

where the operators Q, J, and J' are determined by the conditions [5]

$$B = QJQ^* \quad (B = iR_{-i} - iR_{-i}^* - 2R_{-i}^* R_{-i}), \tag{9.14}$$

$$J\tau^{-1} = CJ'C^* \quad (\tau = I - 2Q^* QJ). \tag{9.15}$$

In the same paper, Do Kong Han established a connection between this function and the characteristic function defined by (9.13).

9.3.8. By analogy with the previous definition, Gubreev [1] defined the characteristic operator function $W(\lambda)$ of a closed densely defined operator A by the equality

$$W(\lambda) J W(-i) = J - i(\lambda - i) Q^*(A + iI)(A - \lambda I)^{-1} Q,$$

where the operators Q and J are given by equality (9.14). He also proved the multiplication theorem and, moreover, solved the inverse problem (the construction of an operator knot for a given operator function).

9.3.9. In the previous subsections, we presented results obtained under assumption that the operator A is densely defined. Kuzhel [22] generalized the definition of the

---

[5] According to the results of Subsection 8.1, without loss of generality, we can assume that $J' = J$.

characteristic operator function (see equality (8.9)) to the case where the operator A is nondensely defined and the condition $J^* = J^{-1}$ is not necessarily satisfied.

9.4. Hermitian Operators

9.4.1. The first definition of the characteristic functions of Hermitian and isometric operators belongs to Livsic [1, 2]. For example, in the case of an isometric operator V with deficiency index (m, m) $(m < \infty)$, the matrix function

$$W_V(z) = B^{-1}(z)A(z)$$

with matrices $B(z)$ and $A(z)$ defined as in Subsection 9.1.2 (for $\tau = 0$) is called the characteristic function of this operator. The characteristic function of an Hermitian operator A is defined by

$$W_A(\lambda) = \theta W_V\left(\frac{\lambda - i}{\lambda + i}\right),$$

where $V = (A - iI)(A + iI)^{-1}$ and $|\theta| = 1$.

In [1, 2], Livsic established the criterion of the unitary equivalence of simple isometric (Hermitian) operators in the form of unitary equivalence of the corresponding characteristic functions and proved the multiplication theorem for characteristic functions for some special couplings of isometric operators.

The following particular result should be noted, since it gave rise to the investigation of model operators: .

Theorem (Livsic [1]). *In order that a simple Hermitian operator A with deficiency index* $(1, 1)$ *be unitary equivalent to the operator of differentiation D acting in the space $L_2(0, a)$ and defined (as usual) by the conditions*

$$Df = \frac{1}{i}f'(x) \quad (f(0) = f(a) = 0),$$

it is neccesary and sufficient that the collection of quasi-Hermitian extensions of the operator A contain an extension without spectrum.

9.4.2. Later, the theory of characteristic functions of Hermitian operators was significantly developed by Shtraus [1–4, 7–10].

The final version of the definition of the characteristic function $C(\lambda)$ of an Hermitian operator A can be found in the paper of Shtraus [9] in the following form:

For a nonreal λ (Im $\lambda > 0$), we consider the extension A_λ of an Hermitian operator

A defined on the lineal $\mathfrak{D}_{A_\lambda} = \mathfrak{D}_A \dotplus \mathfrak{N}_{\bar{\lambda}}$ by the equality

$$A_\lambda(x_0 + x_{\bar{\lambda}}) = Ax_0 + \lambda x_{\bar{\lambda}} \quad (x_0 \in \mathfrak{D}_A, \quad x_{\bar{\lambda}} \in \mathfrak{N}_{\bar{\lambda}}),$$

where $\mathfrak{N}_{\bar{\lambda}}$ is the defect subspace of the operator A.

Then the characteristic function $C(\lambda)$ of the operator A is determined by the equality

$$C(\lambda) = (A_\lambda - \lambda_0 I)(A_\lambda - \bar{\lambda} I)^{-1}\Big|_{\bar{\mathfrak{N}}_{\lambda_0}},$$

where λ_0 is a fixed number from the upper half-plane.

Since the operator A_λ is dissipative, i.e.,

$$\operatorname{Im}(A_\lambda x, x) \geq 0 \quad (\forall x \in \mathfrak{D}_{A_\lambda})$$

we have $\mu = \bar{\lambda}_0 \in \sigma_p(A_\lambda)$ and, hence, for this operator, the following analogue of the von Neumann formulas (see Chapter 1, Section 2) is valid: An arbitrary element $h = x_0 + x_{\bar{\lambda}}$ in $\mathfrak{D}_{A_\lambda}$ can be represented in the form

$$h = h_0 + h_\mu + \Phi_\lambda h_\mu \quad (h_0 \in \mathfrak{D}_A, \quad h_\mu \in \mathfrak{N}_\mu),$$

where $\Phi: \mathfrak{N}_\mu \to \mathfrak{N}_{\bar{\mu}}$ is a linear operator. In this case,

$$A_\lambda h = Ah_0 + \bar{\mu} h_\mu + \mu \Phi h_\mu.$$

A fairly simple calculation shows that the characteristic function $C(\lambda)$ of the operator A differs from the operator Φ_λ only by the sign,

$$C(\lambda) = -\Phi_\lambda. \tag{9.16}$$

An interesting application of the characteristic function $C(\lambda)$ was suggested in the paper of Kochubei [2], where Phillips' results [2] concerning the $\mathfrak{U}$-invariance of the Friedrihs extension of a $\mathfrak{U}$-invariant semibounded operator were generalized to the case of $\mathfrak{U}$-invariant self-adjoint extensions of an arbitrary $\mathfrak{U}$-invariant symmetric operator.

Philips also constructed an example of an $\mathfrak{U}$-invariant symmetric operator with deficiency index $(1, 1)$ which does not have $\mathfrak{U}$-invariant self-adjoint extensions. As shown by Kochubei, for the characteristic function of these operators, we have $C(\lambda) = 0$ ($\operatorname{Im}\lambda > 0$). In this case, one can show, by using equality (9.16), that the point spectrum of the operator A_λ covers the entire upper half-plane.

9.5. Linear Operators in the Space with an Indefinite Metric

9.5.1. *Spaces with an Indefinite Metric.* Let $\mathcal{H}$ be a Hilbert space, let $(\cdot, \cdot)$ be a scalar product in $\mathcal{H}$, and let G be a bounded continuosly invertible self-adjoint operator acting in $\mathcal{H}$. We define a new "scalar product" in $\mathcal{H}$ by the equality $[x, y] = (Gx, y)$ and regard this construction as the G-metric generated in $\mathcal{H}$ by the operator G. The space $\mathcal{H}$ equipped with this G-metric is denoted by Π. Instead of G, we often consider an operator J, which is not only self-adjoint but also unitary ($J^* = J = J^{-1}$). If the operator G is not of fixed sign (i.e., the bilinear form (Gx, x) may take values with different signs), then Π is called a space with an indefinite metric.

Additional information concerning spaces with an indefinite metric can be found, e.g., in the monograph of Asisov and Iokhvidov [1].

9.5.2. *Quasiunitary Operators.* A bounded continuously invertible operator T acting in the space Π is called a quasiunitary operator of rank r (or a K_r-operator) if

$$\dim [(I - T^+T) \Pi] = r \quad (0 \le r < \infty),$$

where $T^+ = G^{-1}T^*G$ is an operator adjoint to the operator T in the G-metric.

The K_r-operators were studied in detail in the papers of Kuzhel [4–11, 15, 16, 18].

The operator $I - T^+T$ can be represented in the form

$$I - T^+T = \sum_{k,\, i=1}^{s} [\cdot, g_k] J_{ki} g_i,$$

where $s \ge r$, $\{g_k\}_1^s$ is a collection of vectors from Π, and $J = \| J_{ki} \|$ is a unitary Hermitian matrix.

The characteristic function $\chi_T(z)$ of the operator T is defined by the equality

$$\chi_T(z) J^{(0)} \chi_T^*(0) = J - \| [(I - zT^+)^{-1} g_k, g_i] \|,$$

where $J^{(0)}$ is a unitary Hermitian matrix. These characteristic functions possess properties which are typical for the case of Hilbert spaces (the isomorphism criterion, the multiplication theorem, etc.).

Unlike the case of Hilbert spaces, the characteristic function of a K_r-operator acting in a space with an indefinite metric can be a constant matrix ($\chi_T(z) \equiv C$). In this case, as shown by the author [15], the operator T has a nontrivial invariant subspace.

Kuzhel [4–11, 15, 16, 18] constructed model operators for K_r-operators acting in the space Π_κ (the Pontryagin space) and established a criterion of completeness for the finite-dimensional invariant subspaces of a nonexpanding operator. These results generalize the corresponding results due to Polyatskii [1] obtained earlier in the case of Hilbert spaces.

In particular, the criterion of completeness can be formulated as follows: A collection of finite-dimensional invariant subspaces of a simple nonexpanding K_r-operator T, which correspond to the eigenvalues z_k ($|z_k| < 1$) and s_k ($|s_k| > 1$) of this operator, is complete in the space Π_κ if and only if the inequality

$$\det \chi_T(0) \leq \prod_{k=1}^{K} |s_k| \prod_{k=1}^{N} |z_k| \quad (N \leq \infty)$$

turns into the equality.

In this case, the operator T is called simple if it does not induce a unitary operator (in the G-metric) in any nontrivial invariant subspace; it is called a nonexpanding operator if

$$[Tx, Tx] \leq [x, x] \quad (\forall\, x \in \Pi).$$

9.5.3. *Bounded Nonself-Adjoint Operators.* Assume that A is a bounded operator such that the operator $(A - A^+)/i$ is finite-dimensional. Then this operator can be represented in the form

$$\frac{A - A^+}{i} = \sum_{k,i=1}^{s} [\cdot, g_k] J_{ki}\, g_i,$$

where, as before, $J = \| J_{ki} \|$ is a unitary Hermitian matrix. The characteristic function of the operator A is defined by the equality

$$\chi_A(\lambda) = I + i \| [(A^+ - \lambda I)^{-1} g_k,\, g_i] \| J.$$

In this case, one can also establish the isomorphism criterion, the multiplication theorem, the criterion of completeness for dissipative operators, etc. These results were obtained by the author in [10, 11, 18] as the generalization of the corresponding Livsic' results [1] to the case of spaces with an indefinite metric.

9.5.4. *Quasi-Hermitian Operators.* A closed operator A densely defined in Π is called a quasi-Hermitian operator of rank r (or a K^r-operator) if the following conditions are satisfied:

(i) $i \in \rho(A) \cap \rho(A^+)$;

(ii) $\dim \mathfrak{D}_A = r \,(\mathrm{mod}\, \mathcal{G}_A)$, where

$$\mathcal{G}_A = \{f \in \mathfrak{D}_A \,|\, [Af, g] = [f, Ag] \; (\forall \, g \in \mathfrak{D}_A)\}$$

is the Hermitian $(G$-Hermitian$)$ domain of A.

By analogy with Section 3, we consider an auxiliary operator

$$B_\alpha = iR_\alpha - iR_\alpha^+ + 2\,\mathrm{Im}\,\alpha R_\alpha^+ R_\alpha \quad (\alpha \in \rho(A)),$$

which is assumed to be finite-dimensional. For $\alpha = -i$, the operator B_{-i} can be represented in the form

$$B_{-i} = \sum_{k,\,j=1}^{s} [\cdot, g_k] J_{k_j} g_i,$$

where $J^* = J = J^{-1}$ $(J = \|J_{ki}\|)$. The characteristic function of the operator A is defined by the equality

$$\chi_A(\lambda)\,J^{(0)}\chi_A^*(i) = J + i(\lambda + i)\| [(A^+ - \lambda I)(A^+ - \lambda I)^{-1} g_k,\, g_j]\|,$$

where $J^{(0)}$ is a unitary Hermitian matrix. In this case, together with the other results, one can also establish the isomorphism criterion, the multiplication theorem, and the criterion of completeness for the collection of finite-dimensional invariant subspaces of a simple dissipative K^r-operator A acting in the Pontryagin space Π_κ (an operator A is called dissipative or G-dissipative if $\mathrm{Im}\,[Ax, x] \geq 0$ $(\forall \, x \in \mathfrak{D}_A)$). The corresponding results were obtained by the author in [17, 18] as a generalization of similar results obtained in [1, 2].

Finally, we note that all the concepts and results considered in this subsection can be generalized to the case of $r = \infty$ and represented not in matrix form but in operator form. At the same time, one should take into account that the investigation of spaces with an indefinite metric is connected with serious difficulties. The problems mainly arise from the fact that not all subspaces of the Hilbert space are projection complete (a subspace Π_1 of the space Π is called projection complete if the decomposition $\Pi = \Pi_1 \,[+]\, \Pi_2$ holds, where $\Pi_2 = \Pi\,[-]\,\Pi_1$ and $[+]$ and $[-]$ are the signs of the orthogonal sum and the orthogonal complement in the G-metric, respectively). Various criteria of projection completeness are presented in the monograph of Azizov and Iokhvidov [1] and in the paper of Kuzhel [24]).

3. Models of Nonself-Adjoint Operators

1. Preliminary Remarks

As shown in Section 7 of Chapter 2, the characteristic matrix function $\chi_A(\lambda)$ of an arbitrary regular extension A $(\rho(A) \neq 0)$ of an Hermitian operator H with finite and equal defect numbers can be represented in the form

$$\chi_A(\lambda) = W_3(\lambda) W_2(\lambda) W_1(\lambda), \tag{1.1}$$

where the matrix factors $W_k(\lambda)$ $(k = 1, 2, 3)$ are defined by equalities (7.25), (7.27), and (7.28).

In the next sections, we consider the following special cases:

1.
$$\chi_A(\lambda) = \int_0^l \exp\left[i \frac{1 + \lambda \alpha(t)}{\lambda - \alpha(t)} dE(t)\right]; \tag{1.2}$$

2.
$$\chi_A(\lambda) = \int_0^l \exp[-i\lambda dE(t)]; \tag{1.3}$$

3.
$$\chi_A(\lambda) = \prod_k u_k [L(k) - \lambda E][L^*(k) - \lambda E]^{-1} \gamma(k) u_k^{-1}; \tag{1.4}$$

Finally, we investigate the general case where $\chi_A(\lambda)$ is representable in the form (1.1).

First, we introduce terminology the exact meaning of which will be clarified in what follows.

We say that

(a) the spectrum of an operator A is real if its characteristic matrix function $\chi_A(\lambda)$ can be represented in the form (1.2);

111

(b) the spectrum of an operator A is concentrated at infinity if $\chi_A(\lambda)$ admits representation (1.3);

(c) the spectrum of an operator A is discrete if $\chi_A(\lambda)$ can be represented in the form (1.4).

2. K^r-Operators with Real Spectrum

2.1. Model

Assume that the characteristic matrix function $\chi_A(\lambda)$ of the K^r-operator A can be represented in the form (1.2), where $\alpha(t)$ is a nondecreasing real function, $E(t)J$ is a monotonically increasing family of Hermitian matrices, and $\mathrm{tr}\, E(t)J = t$. By virtue of Theorem 7.6 (Chapter 2), we conclude that, in the case under consideration, the set $\mathfrak{M}$ of the left and right extreme values of the function $\alpha(\cdot)$ consists of spectral points of the operator A . Indeed, according to this assertion, every point λ from $\mathfrak{M}$ is a singular point of the operator $\chi_A(\lambda)$ and, hence, cannot be a regular point of the operator A .

Note that the spectrum of the operator A may contain points not from the set $\mathfrak{M}$. Indeed, if the operator A is not an L-simple operator, then the function $\alpha(\cdot)$ does not describe the spectrum of operator $\tilde{A} = A|_{\mathcal{H}\ominus\mathcal{H}_A}$.

The conditions imposed on the matrix $E(t)$ imply that $\alpha E(t)J = P^2(t)dt$, where $P(\cdot)$ is an Hermitian matrix function satisfying the condition $\mathrm{tr}\, P^2(t) = 1$. Thus, equality (1.2) can be rewritten in the form

$$\chi_A(\lambda) = \int_0^l \exp\left[i\,\frac{1+\lambda\alpha(t)}{\lambda-\alpha(t)}\,P^2(t)J\,dt\right]. \qquad (2.1)$$

We set

$$\beta(x) = [\alpha(x)+i]\,p(x), \qquad (2.2)$$

$$\omega(x) = \int_{x_0}^x \exp\left[i\alpha(t)JP^2(t)dt\right], \qquad (2.3)$$

where x_0 is a fixed number from the interval $[0, l]$ (if the function $\alpha(\cdot)$ is not summable in a neighborhood of zero, then $x_0 > 0$ because the integral in (2.3) does not exist if $x_0 = 0$ and $\mathrm{tr}\, [JP^2(t)] \neq 0$).

Let us show that the matrix $\omega(x)$ is J-unitary for $x \in (0, l)$, i.e.,

$$\omega^*(x)J\omega(x) = J. \tag{2.4}$$

Indeed, the matrix function $H(t) = i\alpha(t)JP^2(t)$ is summable on the interval $[l_1, l_2]$ $(0 < l_1 < l_2 < l)$, and

$$\int_{l_1}^{l_2} \| H(t) \| \, dt < \infty.$$

Therefore, by virtue of Theorem 7.3 (Chapter 2),

$$\frac{d}{dx} \omega(x) = H(x)\omega(x), \tag{2.5}$$

whence

$$\frac{d}{dx} \omega^*(x) = \omega^*(x)H^*(x).$$

Since $H^*(x)J + JH(x) = 0$, we have

$$\frac{d}{dx} [\omega^*(x)J\omega(x)] = \omega^*(x)[H^*(x)J + JH(x)]\omega(x) = 0.$$

Thus, $\omega^*(x)J\omega(x) = c$, where c is a constant matrix. By setting $x = x_0$ in the last equality, we obtain $c = J$, which proves equality (2.4).

Consider a Hilbert space $\vec{\mathcal{H}}$ whose elements are vector functions $f(\cdot)$ of the form

$$f(\cdot) = (f_1(\cdot), f_2(\cdot), \dots, f_n(\cdot)) \qquad (f_k(\cdot) \in L_2(0, l)).$$

The scalar product in $\vec{\mathcal{H}}$ is defined by the formula

$$(f, g) = \int_0^l f(x)g^*(x)dx \left(= \sum_{k=1}^n (f_k, g_k) \right).$$

(For simplicity, we use the same notation for the scalar products in $\mathcal{H}$ and $\vec{\mathcal{H}}$.)

In the space $\tilde{\mathcal{H}}$, we now consider the operator $\bar{A}$

$$(\bar{A}f)(x) = \alpha(x)f(x) + i\int_x^l f(t)\sigma(t)J\sigma^*(x)dt, \tag{2.6}$$

where $\sigma(x) = \beta(x)\omega(x)$. The operator $\bar{A}$ defined in this way is called a (triangular) model of the operator A.

2.2. Resolvent of the Model

Let $\mathrm{Im}\,\lambda \neq 0$ and $f \in \Delta_{\bar{A}-\lambda I}$. Then, for some g_λ from $\mathcal{D}_{\bar{A}}$, we have $f = (\bar{A} - \lambda I)g_\lambda$ or, in view of equality (2.6),

$$f(x) = (\alpha(x) - \lambda)g_\lambda(x) + y(x)\beta^*(x), \tag{2.7}$$

where

$$y(x) = i\int_x^l g_\lambda(t)\sigma(t)J\omega^*(x)dt. \tag{2.8}$$

It is easy to verify that

$$\frac{d}{dx}y(x) = -ig_\lambda(x)\sigma(x)J\omega^*(x) + i\int_x^l g_\lambda(t)\sigma(t)J\frac{d}{dx}\omega^*(x)dt,$$

where $\sigma(x)J\omega^*(x) = \beta(x)\omega(x)J\omega^*(x) = \beta(x)J$. Hence, taking equalities (2.5) and (2.8) into account, we get

$$i\int_x^l g_\lambda(t)\sigma(t)J\frac{d}{dx}\omega^*(x)dt = y(x)H^*(x).$$

Consequently,

$$\frac{d}{dx}y(x) = y(x)H^*(x) - ig_\lambda(x)\beta(x)J. \tag{2.9}$$

By virtue of (2.7), we have

$$g_\lambda(x) = \frac{f(x)}{\alpha(x) - \lambda} - \frac{y(x)\beta^*(x)}{\alpha(x) - \lambda}. \tag{2.10}$$

By inserting (2.10) in (2.9), we obtain

$$\frac{d}{dx}y(x) = -y(x)M_\lambda(x) + f(x)Q_\lambda(x)J, \tag{2.11}$$

where

$$Q_\lambda(x) = -i\frac{\beta(x)}{\alpha(x) - \lambda}.$$

In view of (2.2) and (2.5), we get

$$M_\lambda(x) = -H^*(x) - i\frac{\beta^*(x)\beta(x)}{\alpha(x) - \lambda}J = i\frac{1 + \lambda\alpha(x)}{\lambda - \alpha(x)}P^2(x)J. \tag{2.12}$$

Furthermore, it follows from (2.8) that $y(l) = 0$.

    Consider the solution

$$W(x, \lambda) = \int_0^x \exp[M_\lambda(t)dt] \tag{2.13}$$

of the equation

$$\frac{d}{dx}W(x, \lambda) = M_\lambda(x)W(x, \lambda). \tag{2.14}$$

By analogy with the proof of equality (2.4), we get

$$W^*(x, \lambda)JW(x, \bar\lambda) = J. \tag{2.15}$$

By using equalities (2.13)–(2.15), we establish that the vector function

$$y(x) = -\int_x^l f(t)Q_\lambda(t)JW(t, \lambda)J dt\, W^*(x, \bar\lambda)J \tag{2.16}$$

is a solution of Eq. (2.11). It is also easy to see that Eq. (2.7) is solvable for any f from $\mathcal{H}$. Therefore, for $\mathrm{Im}\,\lambda \neq 0$, the operator $\bar A - \lambda I$ is invertible and $g_\lambda = (\bar A - \lambda I)^{-1}f$.

Hence, by virtue of equalities (2.10) and (2.16) and the expression for the matrix $Q_\lambda(t)$, we get

$$(\bar{R}_\lambda f)(x) = \frac{f(x)}{\alpha(x) - \lambda} - i \int_x^l f(t) S(t, \lambda) J S^*(x, \bar{\lambda}) dt, \qquad (2.17)$$

where $\bar{R}_\lambda = (\bar{A} - \lambda I)^{-1}$ and

$$S(x, \lambda) = \frac{\beta(x) J W(x, \lambda)}{\alpha(x) - \lambda}. \qquad (2.18).$$

In what follows, we shall also use the equality

$$\frac{d}{dx} [W^*(x, \lambda) J W(x, \mu)] = i(\bar{\lambda} - \mu) S^*(x, \lambda) S(x, \mu). \qquad (2.19)$$

To prove (2.19), we note that (2.14) yields

$$\frac{d}{dx} [W^*(x, \lambda) J W(x, \mu)] = W^*(x, \lambda) [M_\lambda^*(x) J + J M_\mu(x)] W(x, \mu), \qquad (2.20)$$

where, by virtue of (2.12),

$$M_\lambda^*(x) J + J M_\mu(x) = i \frac{(\bar{\lambda} - \mu)(\alpha^2(x) + 1)}{(\alpha - \bar{\lambda})(\alpha - \mu)} J P^2(x) J. \qquad (2.21)$$

By inserting (2.21) in (2.20) and taking equalities (2.2) and (2.18) into account, we obtain (2.19).

Let us prove that operator $\bar{R}_\lambda$ is bounded (for $\operatorname{Im} \lambda \neq 0$). Let operators L and K be defined by the equalities

$$(Lf)(x) = \frac{f(x)}{\alpha(x) - \lambda}, \qquad (Kf)(x) = \int_x^l f(t) \Gamma(t, x; \lambda) dt, \qquad (2.22)$$

where

$$\Gamma(t, x; \lambda) = S(t, \lambda) J S^*(x, \bar{\lambda}) \qquad (2.23)$$

Thus, $\bar{R}_\lambda = L - iK$. Since the operator L is obviously bounded, we consider the operator K. According to (2.22) and (2.23), we have

$$\| Kf \|^2 = \int_0^l \varphi(x) A(x, \bar{\lambda}) \varphi^*(x) dx, \tag{2.24}$$

where

$$\varphi(x) = \int_x^l f(t) S(t, \lambda) J dt, \quad A(x, \lambda) = S^*(x, \lambda) S(x, \lambda). \tag{2.25}$$

For fixed x, we consider an operator S_λ defined by the equality $(S_\lambda \varphi)(x) = \varphi(x) S^*(x, \bar{\lambda})$ in the Euclidean space $\mathcal{E}_n$. Then[6]

$$\varphi(x) A(x, \bar{\lambda}) \varphi^*(x) = \| S_\lambda \varphi \|_{\mathcal{E}}^2 \le \mathrm{tr}\, [A(x, \bar{\lambda})] \, \| \varphi(x) \|_{\mathcal{E}}^2$$

and, hence,

$$\| Kf \|^2 = \int_0^l \mathrm{tr}\, [A(x, \bar{\lambda})] \, \| \varphi(x) \|_{\mathcal{E}}^2 \, dx. \tag{2.26}$$

Let

$$\psi(t) = f(t) S(t, \lambda) J = (\psi_1(t), \psi_2(t), \dots \psi_n(t)). \tag{2.27}$$

Then, according to (2.25) and (2.27), we get

$$\varphi(x) = (\varphi_1(x), \varphi_2(x), \dots \varphi_n(x)), \quad \varphi_k(x) = \int_x^l \psi_k(t) dt.$$

Since

$$| \varphi_k(x) | \le \int_0^l | \psi_k(t) | dt \le \int_0^l \| \psi(t) \|_{\mathcal{E}} dt,$$

we have

$$\| \varphi(x) \|_{\mathcal{E}} \le \sqrt{n} \int_0^l \| \psi(t) \|_{\mathcal{E}} dt, \tag{2.28}$$

where

---

[6] If $y = xA$, where A is a square matrix, then, obviously, $\| y \|_{\mathcal{E}}^2 \le \mathrm{tr}\, (AA^*) \, \| x \|_{\mathcal{E}}^2$.

$$\| \psi(t) \|_{\mathcal{E}}^2 \le \mathrm{tr} [S(t,\lambda) S^*(t,\lambda)] \, \| f(t) \|_{\mathcal{E}}^2.$$

Taking the identity $\mathrm{tr}(AA^*) = \mathrm{tr}(A^*A)$ into account, we obtain

$$\| \psi(t) \|_{\mathcal{E}}^2 \le \mathrm{tr}[A(t,\lambda)] \, \| f(t) \|_{\mathcal{E}}^2. \tag{2.29}$$

Thus, by virtue of inequalities (2.28) and (2.29), we have

$$\| \varphi(x) \|_{\mathcal{E}} \le \sqrt{n} \int_0^l \mathrm{tr}^{1/2}[A(t,\lambda)] \, \| f(t) \|_{\mathcal{E}} dt.$$

By using the Cauchy–Bunyakovsky inequality, we get

$$\| \varphi(x) \|_{\mathcal{E}}^2 \le n \int_0^l \mathrm{tr}[A(t,\lambda)] \, dt \int_0^l \| f(t) \|_{\mathcal{E}}^2 \, dt, \tag{2.30}$$

where

$$\int_0^l \| f(t) \|_{\mathcal{E}}^2 \, dt = \| f \|^2$$

and

$$\int_0^l \mathrm{tr}[A(t,\lambda)] \, dt = \mathrm{tr} \int_0^l A(t,\lambda) \, dt. \tag{2.31}$$

It follows from (2.19) and (2.25) that

$$\int_0^l A(t,\lambda) \, dt = \frac{1}{2\,\mathrm{Im}\,\lambda} [W^*(t,\lambda) J W(t,\lambda)] \Big|_0^l. \tag{2.32}$$

Furthermore, by virtue of relations (2.1), (2.12), and (2.13), we have $W(l,\lambda) = \chi_A(\lambda)$ and $W(0,\lambda) = E$. Therefore, equality (2.32) takes the form

$$\int_0^l A(t,\lambda) \, dt = \frac{1}{2\,\mathrm{Im}\,\lambda} [\chi_A^*(\lambda) J \chi_A(\lambda) - J]. \tag{2.33}$$

But then, by virtue of (2.30), (2.31), and (2.33), we have

$$\| \varphi(x) \|_{\mathcal{E}}^2 \leq n M_\lambda \| f \|^2, \quad M_\lambda = \text{tr} \frac{\chi_A^*(\lambda) J \chi_A(A) - J}{2 \text{ Im } \lambda}. \tag{2.34}$$

By inserting (2.34) in (2.26), we get

$$\| Kf \|^2 \leq n M_\lambda \| f \|^2 \text{ tr} \int_0^l A(x, \bar{\lambda}) \, dx. \tag{2.35}$$

Therefore, in view of (2.33) and (2.34), we can conclude that $\| Kf \|^2 \leq n M_\lambda M_{\bar\lambda} \| f \|^2$, where M_λ and $M_{\bar\lambda}$ are finite numbers and, hence, the operator K is bounded.

Thus, for $\text{Im } \lambda \neq 0$, the operator $\vec{R}_\lambda = L - iK$ is bounded and defined on the entire space $\vec{\mathcal{H}}$. Therefore, the spectrum of operator $\vec{A}$ lies on the real axis.

2.3. Operator $\vec{B}$

Consider the operator

$$\vec{B} = i \vec{R}_{-i} - i \vec{R}_{-i}^* - 2 \vec{R}_{-i}^* \vec{R}_{-i}. \tag{2.36}$$

In view of obvious equalities $\vec{R}_{-i} = L - iK$ and $iL - iL^* - 2L^*L = 0$, where the operators L and K are defined (for $\lambda = -i$) by equalities (2.22), we have

$$\vec{B} = K + K^* - 2K^*K + 2i(L^*K - K^*L). \tag{2.37}$$

Taking equalities (2.22) and (2.23) into account, we get

$$(K^*\varphi)(x) = \int_0^x \varphi(t) \Gamma(t, x; \bar{\lambda}) \, dt. \tag{2.38}$$

But then $K^*Kf = K^*\varphi$, where

$$\varphi(t) = (Kf)(t) = \int_t^l f(\xi) \Gamma(\xi, t; \lambda) \, d\xi.$$

Hence,

$$(K^*Kf)(x) = \int_0^x \left[\int_t^l f(\xi) \Gamma(\xi, t; \lambda) \, d\xi \right] \Gamma(t, x; \bar{\lambda}) \, dt. \tag{2.39}$$

By using equalities (2.23) and (2.19), we obtain

$$\Gamma(\xi, t; \lambda)\,\Gamma(t, x; \bar{\lambda}) \;=\; \frac{1}{2\,\mathrm{Im}\,\lambda}\,S(\xi, \lambda)\,J\,\frac{d}{dt}\,[W^*(t, \bar{\lambda})\,J\,W(t, \bar{\lambda})]\,JS^*(x, \lambda).$$

Thus,

$$2\,\mathrm{Im}\,\bar{\lambda}\;(K^*Kf)(x) \;=\; \int_0^x\left(\int_t^l f(\xi)S(\xi, \lambda)\,J\,d\xi\right) d\,[W^*(t, \bar{\lambda})\,J\,W(t, \bar{\lambda})]\,JS^*(x, \lambda).$$

Integration by parts gives

$$2\,\mathrm{Im}\,\bar{\lambda}\;(K^*Kf)(x) \;=\; \Big[h(t)\Big|_0^x - g(x)\Big]\,JS^*(x, \lambda), \tag{2.40}$$

where

$$h(t) \;=\; \int_t^l f(\xi)\,S(\xi, \lambda)\,J\,d\xi\,W^*(t, \bar{\lambda})\,J\,W(t, \bar{\lambda}), \tag{2.41}$$

$$g(x) \;=\; -\int_0^x f(t)S(t, \lambda)\,J\,W^*(t, \bar{\lambda})\,J\,W(t, \bar{\lambda})\,dt. \tag{2.42}$$

Equality (2.15) implies that $W(x, \bar{\lambda})\,J\,W^*(x, \lambda) = J$. Therefore, in view of (2.18), we obtain

$$W^*(x, \bar{\lambda})\,J\,W(x, \bar{\lambda})\,JS^*(x, \lambda) \;=\; S^*(x, \bar{\lambda})\,r(x, \lambda), \tag{2.43}$$

where

$$r(x, \lambda) \;=\; \frac{\alpha(x) - \lambda}{\alpha(x) - \bar{\lambda}}.$$

But then, by virtue of relations (2.41), (2.43), (2.22), and (2.23), we have

$$h(x)\,JS^*(x, \lambda) \;=\; r(x, \lambda)(Kf)(x).$$

Since

$$r(x, \lambda) \;=\; 1 + \frac{\bar{\lambda} - \lambda}{\alpha(x) - \bar{\lambda}},$$

we get

$$h(x)JS^*(x, \lambda) = (Kf)(x) + (\bar{\lambda} - \lambda)(L^*Kf)(x). \tag{2.44}$$

By virtue of (2.13), we have $W(0, \bar{\lambda}) = E$. Therefore,

$$h(0)JS^*(x, \lambda) = \int_0^l f(t)S(t, \lambda)JS^*(x, \lambda)\,dt. \tag{2.45}$$

It follows from (2.43) that

$$S(t, \lambda)JW^*(t, \bar{\lambda})JW(t, \bar{\lambda}) = r(t, \bar{\lambda})S(t, \bar{\lambda}).$$

Consequently, the equality

$$r(t, \bar{\lambda}) = 1 + \frac{\lambda - \bar{\lambda}}{\alpha(t) - \lambda}$$

yields

$$g(x)JS^*(x, \lambda) = -(K^*f)(x) - (\lambda - \bar{\lambda})(K^*Lf)(x). \tag{2.46}$$

Thus, by setting $\lambda = -i$ in relations (2.40), (2.44), (2.45), and (2.46), we get

$$2K^*Kf = Kf + 2iL^*Kf + K^*f - 2iK^*Lf - \int_0^l f(t)S(t, -i)JS^*(x, -i)\,dt. \tag{2.47}$$

In view of (2.37), this equality implies that

$$\bar{B}f = \int_0^l f(t)S(t, -i)JS^*(x, -i)\,dt. \tag{2.48}$$

Note that, by virtue of (2.2), (2.12), (2.13), and (2.18), we have

$$S(x, -i) = P(x)JW(x, -i), \tag{2.49}$$

where

$$W(x, -i) = \int_0^{\hat{x}} \exp[-P^2(t)J\,dt]. \tag{2.50}$$

2.4. Δ-Basis of the Operator $\bar{A}$

Consider the set of vectors

$$h_k = (\delta_{k1}, \delta_{k2}, \dots, \delta_{kn}) \quad (k = \overline{1, n}), \tag{2.51}$$

where δ_{ki} is the Kronecker symbol. Since

$$\sum_{k=1}^{n} h_k^* h_k = E,$$

we have

$$S(t, -i)JS^*(x, -i) = S(t, -i) \sum_{k=1}^{n} h_k^* h_k J \sum_{s=1}^{n} h_s^* h_s S^*(x, -i)$$

and, consequently,

$$S(t, -i)JS^*(x, -i) = \sum_{k, \, s=1}^{n} g_k^* J_{ks} g_s, \tag{2.52}$$

where

$$g_k(x) = h_k S^*(x, -i), \quad J_{ks} = h_k J h_s^*. \tag{2.53}$$

Note that, in view of equality (2.19), we have

$$(g_k, g_m) = h_k \int_0^l S^*(x, -i)S(x, -i)dx h_m^* = -\frac{1}{2} h_k W^*(x, -i)JW(x, -i)\Big|_0^l h_m^*.$$

Moreover, since $W(l, -i) = \chi_A(-i)$ and $W(0, -i) = E$, we get

$$(g_k, g_m) = \frac{1}{2} h_k (J - \chi_A^*(-i)J\chi_A(-i)) h_m^*. \tag{2.54}$$

In particular, it follows from (2.54) that $g_k \in \vec{\mathcal{H}}$ $(k = \overline{1, n})$. Furthermore, in view of (2.48) and (2.52), we have

$$\bar{B}f = \sum_{k,\,s=1}^{n} (f, g_k) J_{ks} g_s. \tag{2.55}$$

Thus, the system of vectors $\{g_k\}_{k=1}^n$ forms the Δ-basis of the operator $\bar{A}$, and the corresponding coefficient matrix coincides with the matrix J.

2.5. Characteristic Matrix Function of the Operator $\bar{A}$

Let us calculate the characteristic matrix function $\chi_{\bar{A}}(\lambda)$ of the operator $\bar{A}$.

Recall that, for $\alpha = -i$ and $J^{-1} = J$, the characteristic matrix function $\chi_{\bar{A}}(\lambda)$ of the operator $\bar{A}$ is defined by the equality (Chapter 2, Subsection 1.2)

$$\chi_{\bar{A}}^*(-i)J\chi_{\bar{A}}(\lambda) = J + i(i-\lambda)\|((\bar{A}+iI)\bar{R}_\lambda g_k, g_m)\|,$$

where $\{g_k\}_{k=1}^n$ is a Δ-basis of the operator $\bar{A}$ and $\lambda \in \rho(\bar{A})$. Since $(\bar{A}+iI)\bar{R}_\lambda = I + (\lambda+i)\bar{R}_\lambda$, we have

$$\chi_{\bar{A}}^*(-i)J\chi_{\bar{A}}(\lambda) = J + i(i-\lambda)\|(g_k, g_m)\| + (\lambda+i)i(i-\lambda)\|(\bar{R}_\lambda g_k, g_m)\|. \tag{2.56}$$

It follows from (2.19) and (2.53) that

$$i(i-\lambda)g_k(t)S(t, \lambda) = h_k \frac{d}{d(t)}[W^*(t, -i)JW(t, \lambda)]. \tag{2.57}$$

Hence, in view of (2.17) and (2.57), we obtain

$$i(i-\lambda)(\bar{R}_\lambda g_k)(x) = i(i-\lambda)\frac{g_k(x)}{\alpha(x)-\lambda} - ih_k^*[W^*(t, -i)JW(t, \lambda)]\Big|_x^l \; JS^*(x, \bar{\lambda})$$

$$= i(i-\lambda)\frac{g_k(x)}{\alpha(x)-\lambda} - ih_k[\chi_{\bar{A}}^*(-i)J\chi_A(\lambda)$$

$$- W^*(x, -i)JW(x, \lambda)]JS^*(x, \bar{\lambda}). \tag{2.58}$$

Note that, in view of equalities (2.49) and (2.53), one can show, by analogy with the proof of equality (2.43), that

$$h_k W^*(x, -i) J W(x, \lambda) J S^*(x, \bar{\lambda}) = g_k(x) \frac{\alpha(x) - i}{\alpha(x) - \lambda}. \qquad (2.59)$$

But then, according to (2.58) and (2.59),

$$i(i - \lambda)(\bar{R}_\lambda g_k)(x) = i g_k(x) - i h_k \chi_A^*(-i) J \chi_A(\lambda) J S^*(x, \bar{\lambda}) \qquad (2.60)$$

and, hence,

$$(i - \lambda)(\bar{R}_\lambda g_k, g_m) = (g_k, g_m) - h_k \chi_A^*(-i) J \chi_A(\lambda) J \int_0^l S^*(x, \bar{\lambda}) S(x, -i) dx \, h_m^*.$$

$$(2.61)$$

By inserting (2.61) in (2.56), we obtain

$$\chi_A^*(-i) J \chi_{\bar{A}}(\lambda) = J - 2 \|(g_k, g_m)\|$$

$$- i(\lambda + i) \chi_A^*(-i) J \chi_A(\lambda) J \int_0^l S^*(x, \bar{\lambda}) S(x, -i) dx. \qquad (2.62)$$

In view of (2.54), we have

$$J - 2 \|(g_k, g_m)\| = \chi_A^*(-i) J \chi_A(-i);$$

therefore, relation (2.62) can be rewritten in the form

$$\chi_A^*(-i) J \chi_{\bar{A}}(\lambda) = \chi_A^*(-i) J \left[\chi_A(-i) - i(\lambda + i) \chi_A(\lambda) J \int_0^l S^*(x, \bar{\lambda}) S(x, -i) dx \right].$$

$$(2.63)$$

By using equality (2.19), we obtain

$$i(\lambda + i) \int_0^l S^*(x, \bar{\lambda}) S(x, -i) dx = W^*(x, \bar{\lambda}) J W(x, -i) \Big|_0^l = \chi_A^*(\bar{\lambda}) J \chi_A(-i) - J.$$

$$(2.64)$$

But then, according to (2.63), (2.64), and the equality $\chi_A(\lambda) J \chi_A^*(\bar{\lambda}) = J$, we get

$$\chi_{\bar{A}}^*(-i)J\chi_{\bar{A}}(\lambda) = \chi_A^*(-i)J\chi_A(\lambda) \tag{2.65}$$

and, hence, $\chi_{\bar{A}}(\lambda) = u\chi_A(\lambda)$, where the matrix $u = J\chi_{\bar{A}}^{-*}(-i)\chi_A^*(-i)J$ is J-unitary (which follows from (2.65) with $\lambda = -i$). Since characteristic matrix functions are determined up to the left J-unitary factor, we may suppose, without loss of generality, that $\chi_{\bar{A}}(\lambda) = \chi_A(\lambda)$. By virtue of Theorem 2.1 (Chapter 2), we conclude that L-simple parts of the operators A and $\bar{A}$ are unitary equivalent. In particular, if the operators A and $\bar{A}$ are L-simple (or simple), then these operators are unitary equivalent.

2.6. The Condition under Which $\mathfrak{D}_{A^*} = \mathfrak{D}_A$

Assume that A is a simple K'-operator with real spectrum. Then $L_A = \mathfrak{N}_{-i}$ (see Chapter 2, Subsection 2.2), i.e., a Δ-basis of the operator A is a basis of the defect space $\mathfrak{N}_{-i}$. Also assume that $\chi_A(\lambda)$ is the $r \times r$ (i.e., minimal) characteristic matrix function associated with the operator A, $\bar{A}$ is its triangular model defined by equality (2.6) and $\bar{A}_s$ is the simple part of the operator $\bar{A}$. Recall that the operators A and $\bar{A}_s$ are unitary equivalent. Therefore, the operator $\bar{A}_s$ is also a K'-operator; moreover, $\dim \bar{\mathfrak{N}}_{-i} = r$, where $\bar{\mathfrak{N}}_{-i}$ is the defect space of the Hermitian part $\bar{A}_0$ of the operator $\bar{A}$. Under these conditions, the equality $\mathfrak{D}_{A^*} = \mathfrak{D}_A$ holds if and only if $\mathfrak{D}_{\bar{A}_s^*} = \mathfrak{D}_{\bar{A}_s}$.

Assume that the last equality holds. Then, the corresponding defect space $\bar{\mathfrak{N}}_{-i}$ is a subset of $\mathfrak{D}_{\bar{A}_s}$ (Chapter 1, Theorem 2.7). In other words, the vectors g_k defined by relations (2.53) belong to the lineal $\mathfrak{D}_{\bar{A}_s}$ (and, hence, to the lineal $\mathfrak{D}_{\bar{A}}$).

Let us find the vector $\bar{A}g_k$. Since $\sigma(x) = \beta(x)\omega(x)$, in view of relations (2.53) and (2.49), we obtain

$$g_k(t)\sigma(t) = h_k W^*(t,-i)JP^2(t)(\alpha(t)+i)\omega(t)$$

$$= -ih_k W^*(t,-i)H(t)\omega(t) + ih_k W^*(t,-i)JP^2(t)\omega(t), \tag{2.66}$$

where $H(t) = i\alpha(t)JP^2(t)$. By virtue of (2.5), this equality yields

$$i\int_x^l g_k(t)\sigma(t)dt = h_k\int_x^l W^*(t,-i)d\omega(t) + h_k\int_x^l W^*(t,-i)JP^2(t)\omega(t)dt$$

$$= h_k[W^*(t,-i)\omega(t)]\Big|_x^l - h_k\int_x^l \frac{d}{dt}W^*(t,-i)\omega(t)dt$$

$$- h_k \int_x^l W^*(t, -i) JP^2(t) \omega(t) dt, \tag{2.67}$$

where, in view of (2.12) and (2.14),

$$\frac{d}{dx} W^*(x, -i) = -W^*(x, -i) JP^2(t). \tag{2.68}$$

Inserting (2.68) in (2.67) and taking the relation $W(l, -i) = \chi_A(-i)$ into account, we obtain

$$i \int_x^l g_k(t) \sigma(t) dt = h_k [\chi_A^*(-i) \omega(l) - W^*(x, -i) \omega(x)]. \tag{2.69}$$

Since

$$h_k W^*(x, -i) \omega(x) J \sigma^*(x) = h_k W^*(x, -i) \omega(x) J \omega^*(x) \beta^*(x)$$

$$= h_k W^*(x, -i) JP(x)(\alpha(x) - i) = g_k(x)(\alpha(x) - i),$$

we have

$$i \int_x^l g_k(t) \sigma(t) J \sigma^*(x) dt = h_k \chi_A^*(-i) \omega(l) J \sigma^*(x) - g_k(x)(\alpha(x) - i) \tag{2.70}$$

and, hence, by virtue of (2.6) and (2.70),

$$(\bar{A} g_k)(x) = i g_k(x) + \varphi_k(x),$$

where $\varphi_k(x) = h_k \chi_A^*(-i) \omega(l) J \sigma^*(x)$. Since $g_k \in \bar{\mathcal{H}}$ and $\bar{A} g_k \in \bar{\mathcal{H}}$, we conclude that $\varphi_k \in \bar{\mathcal{H}}$; thus

$$(\varphi_k, \varphi_k) = h_k \chi_A^*(-i) \omega(l) J \int_0^l \sigma^*(x) \sigma(x) J dx \, \omega^*(l) \chi_A(-i) h_k^* < \infty$$

for $k = 1, 2, \ldots, r$. This inequality and the fact the matrices $\chi_A^*(-i) \omega(l)$ and J are invertible imply that the matrix

$$\sigma = \int_0^l \sigma^*(x)\sigma(x)J\,dx$$

is finite (the elements of the matrix σ are finite numbers). Then the trace of this matrix is finite. Further, since

$$\sigma^*(x)\sigma(x)J = (\alpha^2(x)+1)\omega^*(x)P^2(x)\omega(x)J$$

and

$$\mathrm{tr}\,[\omega^*(x)P^2(x)\omega(x)J] = \mathrm{tr}\,[P^2(x)\omega(x)J\omega^*(x)] = \mathrm{tr}\,[P^2(x)J]$$

we obtain

$$|\mathrm{tr}\,\sigma| = \left| \int_0^l (\alpha^2(x)+1)\,\mathrm{tr}\,[P^2(x)J]\,dx \right| < \infty. \tag{2.71}$$

Moreover, since $\det e^A = e^{\mathrm{tr}\,A}$, relation (2.50) implies that

$$\det \chi_A(-i) = \det W(l,-i) = \exp\left\{ -\int_0^l \mathrm{tr}\,[P^2(t)J]\,dt \right\}$$

and, therefore,

$$\int_0^l \mathrm{tr}\,[P^2(x)J]\,dx$$

is finite. But then inequality (2.71) can be rewritten in the form

$$\left| \int_0^l \alpha^2(x)\,\mathrm{tr}\,[P^2(x)J]\,dx \right| < \infty \tag{2.72}$$

Thus, inequality (2.71) is true if $\mathcal{D}_{\tilde{A}_s^*} = \mathcal{D}_{\tilde{A}_s}$. The converse statement is also true, i.e., if inequality (2.71) is satisfied, then $g_k \in \mathcal{D}_{\tilde{A}_s}$, and hence, by Theorem 2.7 (Chapter 1), $\mathcal{D}_{\tilde{A}_s^*} = \mathcal{D}_{\tilde{A}_s}$.

Thus, we have proved the following assertion:

Theorem 2.1. *If A is a simple K'-operator with characteristic function of the form (4.1), then, for the validity of the equality $\mathcal{D}_{A^*} = \mathcal{D}_A$, it is necessary and* *sufficient that inequality (2.71) be satisfied.*

We note that if A is a dissipative operator $(J = E)$ satisfying the conditions of this theorem, then $\operatorname{tr} [P^2(x)J] = \operatorname{tr} P^2(x) = 1$. Hence, the following assertion is true:

Corollary 2.2. *If A is a simple dissipative operator with characteristic function of the form (4.1), then the equality $\mathfrak{D}_A = \mathfrak{D}_{A^*}$ holds if and only if $\alpha(\cdot) \in L_2(0, l)$.*

2.7. Scalar Case

In the scalar case $(n = 1)$,

$$J = \pm 1, \quad P(x) = 1, \quad \beta(x) = \alpha(x) + i, \quad \chi_A(\lambda) = \exp \int_0^l i\, \frac{1 + \lambda\alpha(t)}{\lambda - \alpha(t)} J\, dt,$$

$$\omega(x) = \exp \int_{x_0}^x i\alpha(t)J\, dt, \quad \sigma(x) = (\alpha(x) + i)\omega(x).$$

In this case, the triangular model has the form (2.6), where $\sigma(x)$ is defined by the last equality. The condition under which $\mathfrak{D}_{A^*} = \mathfrak{D}_A$ is formulated as Corollary 2.2.

Remark. Main results of this section were obtained by Kuzhel in [1–3].

3. K^r-Operators with Spectrum Concentrated at Infinity

3.1. Model

Assume that the characteristic matrix function $\chi_A(\lambda)$ of a K^r-operator A can be represented in the form (1.3). By using the same reasoning as in Subsection 2.1, we can rewrite $\chi_A(\lambda)$ as follows:

$$\chi_A(\lambda) = \int_0^{\overset{\frown}{l}} \exp[-i\lambda P^2(t)J\, dt]. \tag{3.1}$$

As in Subsection 2.1, we define the Hilbert space $\tilde{\mathcal{H}}$ and consider in it the operator P defined by the equality

$$(Pf)(x) = f(x)P(x). \tag{3.2}$$

Denote the space $\tilde{\mathcal{H}} \ominus \operatorname{Ker} P$ and the operator $P|_{\tilde{\mathcal{H}} \ominus \operatorname{Ker} P}$ by $\bar{\mathcal{H}}$ and $\bar{P}$, respectively. The operator P is thus supposed to be invertible.

Consider operators Q and J defined by the equalities

$$(Qf)(x) = \frac{1}{i}\frac{d}{dx}f(x), \quad (Jf)(x) = f(x)J, \tag{3.3}$$

where $f \in \mathfrak{D}_Q$ if the vector function f is absolutely continuous on $[0, l]$; $\{f, f'\} \subset \bar{\mathcal{H}}$, and $f(l) = 0$. The operator $\bar{A}$ defined in $\bar{\mathcal{H}}$ by the equality

$$\bar{A}f = \bar{P}^{-1}QJ\bar{P}^{-1}f \tag{3.4}$$

is called a model of a given K'-operator A.

Consider a matrix function

$$W(x, \lambda) = \int_0^{\overset{x}{\frown}} \exp\left[-i\lambda \bar{P}^2(t)J\,dt\right] \tag{3.5}$$

satisfying the differential equation

$$\frac{d}{dx}W(x, \lambda) = -i\lambda \bar{P}^2(x)JW(x, \lambda). \tag{3.6}$$

As in the previous section, we can show that

$$W^*(x, \lambda)JW(x, \bar{\lambda}) = J. \tag{3.7}$$

3.2. Resolvent and the Characteristic Matrix Function of a Model

Let $f \in \Delta_{\bar{A}-\lambda I}$. Then

$$f = (\bar{A} - \lambda I)g_\lambda = \bar{P}^{-1}QJ\bar{P}^{-1}g_\lambda - \lambda g_\lambda, \tag{3.8}$$

where g_λ is a vector from $\mathfrak{D}_{\bar{A}}$. Denote $\varphi = J\bar{P}^{-1}g_\lambda$; in this case, $g_\lambda = \bar{P}J\varphi$ and $f =$

$P^{-1}Q\varphi - \lambda PJ\varphi$, whence

$$Pf = Q\varphi - \lambda P^2 J\varphi. \tag{3.9}$$

Let $\psi(x) = \varphi(x)JW(x, \lambda)J$. Then $\varphi(x) = \psi(x)W^*(x, \bar{\lambda})$ by virtue of (3.7). Hence,

$$(Q\varphi)(x) = \frac{1}{i}\frac{d\psi(x)}{dx}\psi(x)W^*(x, \bar{\lambda}) + \lambda\psi(x)W^*(x, \bar{\lambda})JP^2(x). \tag{3.10}$$

Since $(P^2 J\varphi)(x) = \psi(x)W^*(x, \bar{\lambda})JP^2(x)$, it follows from (3.9) and (3.10) that

$$(Pf)(x) = \frac{1}{i}\frac{d\psi(x)}{dx}\psi(x)W^*(x, \bar{\lambda}) \tag{3.11}$$

and, in view of the equalities (3.7) and $(Pf)(x) = f(x)P(x)$,

$$\frac{d}{dx}\psi(x) = if(x)P(x)JW(x, \lambda)J.$$

Since $\psi(l) = 0$, we obtain

$$\psi(x) = -i\int_x^l f(t)S(t, \lambda)\,dt,$$

where $S(x, \lambda) = P(x)JW(x, \lambda)J$. We can now write the following expression for $\varphi(x)$:

$$\varphi(x) = -i\int_x^l f(t)S(t, \lambda)\,dt\,W^*(x, \bar{\lambda}).$$

At the same time, $(\bar{R}_\lambda f)(x) = g_\lambda(x) = (PJ\varphi)(x) = \varphi(x)JP(x)$. Therefore,

$$(\bar{R}_\lambda f)(x) = -i\int_x^l f(t)S(t, \lambda)JS^*(x, \bar{\lambda})\,dt. \tag{3.12}$$

We note that resolvent (3.12) can be formally obtained from resolvent (2.17) if we set $\alpha(x) \equiv 0$ in relation (2.17).

The boundedness of the resolvent $\bar{R}_\lambda$ can be established in the same way as the boundedness of the resolvent defined by relation (2.17). The auxiliary operator $\bar{B}$ can be found as in Subsection 2.3 and has the form

$$\bar{B} = \sum_{k,\,i=1}^{n} (\cdot, g_k) J_{ki} g_i,$$

where $g_k(x) = h_k S^*(x, -i)$. By using the same reasoning as in Subsection 2.5, we can show that $\chi_{\bar{A}}(\lambda) = u\chi_A(\lambda)$ and, hence, the L-simple part of the K'-operator A with spectrum concentrated at infinity is unitary equivalent to the L-simple part of the operator (model) $\bar{A}$ defined by (3.4).

Since $g_k(l) \neq 0$, we have $g_k \in \mathfrak{D}_{\bar{A}}$ and, in this case, $\mathfrak{D}_{A^*} \neq \mathfrak{D}_A$.

Remark. The results presented in this section were obtained for the first time by Kuzhel in [1–3].

4. *K'*-Operators with Discrete Spectrum

4.1. Model

Let A be a K'-operator with discrete spectrum. Then (see Section 1) the characteristic matrix function of the operator A can be represented in the form

$$\chi_A(\lambda) = \overset{N}{\underset{k=1}{\widehat{\prod}}} u_k [L(k) - \lambda E][L^*(k) - \lambda E]^{-1} \gamma(k) u_k^{-1}, \qquad (4.1)$$

where $N \leq \infty$ and u_k are J-unitary matrices,

$$L(k) = \begin{pmatrix} 0 & 0 \\ 0 & \lambda_k E_{q_k} \end{pmatrix}, \qquad \gamma(k) = \begin{pmatrix} E_{p_k} & 0 \\ 0 & \bar{\gamma}_k E_{q_k} \end{pmatrix} \qquad (4.2)$$

with $\mathrm{Im}\,\lambda_k > 0$ and $|\gamma_k| = 1$ (see Section 7, Chapter 2). We set

$$\pi(k) = \overset{k}{\underset{j=1}{\widehat{\prod}}} u_j \gamma^*(j) u_j^{-1} \qquad (k \leq N), \qquad (4.3)$$

$$\sigma(k) = \beta(k) u_k^* \pi^*(k-1), \qquad (4.4)$$

where $\beta(k)$ are nonnegative matrices defined by the condition $L(k) - L^*(k) =$

$-i\beta^2(k)J$. It is clear that

$$\pi(k)J\pi^*(k) = J, \qquad \sigma(k)J\sigma^*(k) = -2\,\text{Im}\,L(k). \tag{4.5}$$

Consider the Hilbert space $\vec{\mathcal{H}}$ of vector functions f of the form $f = (f(1), f(2), \ldots$ $f(k), \ldots)$ $(k \le N)$, where $f(k) \in \mathcal{E}_n$ ($\mathcal{E}_n$ is a Euclidean n-dimensional space). The scalar product in $\vec{\mathcal{H}}$ is defined by the equality

$$(f, g) = \sum_{k=1}^{N} f(k)g^*(k).$$

The operator $\bar{A}$ defined in $\vec{\mathcal{H}}$ by the equality

$$\bar{A}f = (\varphi(1), \varphi(2), \ldots, \varphi(k), \ldots), \tag{4.6}$$

where

$$\varphi(k) = f(k)L^*(k) + i \sum_{j=k+1}^{N} f(j)\sigma(j)J\sigma^*(k) \tag{4.7}$$

is called a (triangular) model of the operator A.

4.2. Resolvent of the Model

Assume that the characteristic matrix function $\chi_A(\lambda)$ is holomorphic and $\det \chi_A(\lambda) \ne 0$ at the point $\lambda \ne 0$. Let us show that $\lambda \in \rho(\bar{A})$. Indeed, assume that $f \in \Delta_{\bar{A}-\lambda I}$, i.e., $f = (\bar{A} - \lambda E)g_\lambda$, where g_λ is a vector from $\mathfrak{D}_{\bar{A}}$. Then

$$f(k) = g_\lambda(k)L^*(k) + i \sum_{j=k+1}^{N} g_\lambda(j)\sigma(j)J\sigma^*(k) - \lambda g_\lambda(k). \tag{4.8}$$

We set $y(N) = 0$ and

$$y(k) = i \sum_{j=k+1}^{N} g_\lambda(j)\sigma(j)J \quad (k < N). \tag{4.9}$$

Equalities (4.8) and (4.9) imply that

$$f(k) = g_\lambda(k)[L^*(k) - \lambda E] + y(k)\sigma^*(k)$$

and, in view of the invertibility of the matrix $L^*(k) - \lambda E$,

$$g_\lambda(k) = [f(k) - y(k)\sigma^*(k)][L^*(k) - \lambda E]^{-1}. \tag{4.10}$$

By using (4.9) and (4.10), we get

$$y(k-1) - y(k) = i[f(k) - y(k)\sigma^*(k)][L^*(k) - \lambda E]^{-1}\sigma(k)J$$

and, hence,

$$y(k-1) = y(k)v(k) + A(k), \tag{4.11}$$

where

$$v(k) = E - i\sigma^*(k)[L^*(k) - \lambda E]^{-1}\sigma(k)J, \tag{4.12}$$

$$A(k) = if(k)[L^*(k) - \lambda E]^{-1}\sigma(k)J. \tag{4.13}$$

In this case, since $\pi(k-1)u_k J u_k^* \pi^*(k-1) = J$, relations (4.4) and (4.12) yield

$$v(k)J = \pi(k-1)u_k(E - i\beta^2(k)J[L^*(k) - \lambda E]^{-1})Ju_k^*\pi^*(k-1).$$

But then, taking into account the equalities

$$-i\beta^2(k)J = L(k) - L^*(k), \quad Ju_k^*\pi^*(k-1)J = u_k^{-1}\pi^{-1}(k-1),$$

we get

$$v(k) = \pi(k-1)u_k[L(k) - \lambda E][L^*(k) - \lambda E]^{-1}u_k^{-1}\pi^{-1}(k-1). \tag{4.14}$$

Consider the matrices

$$W(0, \lambda) = E$$

and

$$W(m, \lambda) = \prod_{k=1}^{m} u_k[L(k) - \lambda E][L^*(k) - \lambda E]^{-1}\gamma(k)u_k^{-1}, \tag{4.15}$$

where $m \leq N$. It is easy to see that

$$W(m, \lambda) J W^*(m, \bar{\lambda}) = J \quad (m \leq N) \tag{4.16}$$

and

$$W(m, \lambda) = b(m, \lambda) W(m-1, \lambda), \tag{4.17}$$

where

$$b(m, \lambda) = u_m [L(m) - \lambda E][L^*(m) - \lambda E]^{-1} \gamma(m) u_m^{-1} \tag{4.18}$$

and $b(m, \lambda) J b^*(m, \bar{\lambda}) = J$. By comparing (4.18) with (4.14), we obtain

$$b(k, \lambda) = \pi^{-1}(k) v(k) \pi(k-1). \tag{4.19}$$

Let us show that the matrix

$$y(k) = \sum_{m=k+1}^{N} A(m) \pi(m-1) W(m-1, \lambda) J W^*(k, \bar{\lambda}) \pi^*(k) J \tag{4.20}$$

is a solution of Eq. (4.11). Indeed, by virtue of (4.5) and (4.16), we have

$$y(k) \pi(k) W(k, \lambda) = \sum_{m=k+1}^{N} A(m) \pi(m-1) W(m-1, \lambda).$$

But then

$$y(k-1) \pi(k-1) W(k-1, \lambda)$$

$$= A(k) \pi(k-1) W(k-1, \lambda) + \sum_{m=k+1}^{N} A(m) \pi(m-1) W(m-1, \lambda). \tag{4.21}$$

Multiplying (4.21) by the matrix $J W^*(k, \bar{\lambda}) \pi^*(k) J$ from the right, we get

$$y(k-1) S(k, \lambda) = A(k) S(k, \lambda) + y(k), \tag{4.22}$$

where $S(k, \lambda) = \pi(k-1) W(k-1, \lambda) J W^*(k, \bar{\lambda}) \pi^*(k) J$. It follows from equality (4.19) and the relation

$$JW^*(k, \bar{\lambda})\pi^*(k)J = W^{-1}(k, \lambda)\pi^{-1}(k) = W^{-1}(k-1)b^{-1}(k, \lambda)\pi^{-1}(k)$$

that

$$S(k, \lambda) = \pi(k-1)b^{-1}(k, \lambda)\pi^{-1}(k) = v^{-1}(k).$$

Multiplying now equality (4.22) by $v(k)$ from the right, we obtain (4.11).

Thus, the components $g_\lambda(k)$ of the vector g_λ are determined (for any f from $\bar{\mathcal{H}}$) by equalities (4.10) and (4.20). If, in addition, $f = 0$, then, by virtue of (4.13), we have $g_\lambda = 0$. Consequently, the operator $\bar{A} - \lambda I$ is invertible and $\bar{R}_\lambda f = g_\lambda$. Moreover, it follows from (4.10), (4.13), and (4.20) that

$$(\bar{R}_\lambda f)(k) = f(k)[L^*(k) - \lambda E]^{-1} - i \sum_{m=k+1}^{N} f(m)S(m, \lambda)J\hat{S}(k, \lambda),$$

where

$$S(m, \lambda) = [L^*(m) - \lambda E]^{-1}\sigma(m)J\pi(m-1)W(m-1, \lambda)$$

and

$$\hat{S}(k, \lambda) = W^*(k, \bar{\lambda})\pi^*(k)J\sigma^*(k)[L^*(k) - \lambda E]^{-1}.$$

By using equalities (4.3), (4.4), and (4.15), one can easily verify that

$$S(m, \lambda) = [L^*(m) - \lambda E]^{-1}\beta(m)u_m^* JW(m-1, \lambda) \qquad (4.23)$$

and $\hat{S}(k, \lambda) = S^*(k, \bar{\lambda})$. Finally, we get

$$(\bar{R}_\lambda f)(k) = f(k)[L^*(k) - \lambda E]^{-1} - i \sum_{m=k+1}^{N} f(m)S(m, \lambda)JS^*(k, \bar{\lambda}), \qquad (4.24)$$

where the matrix $S(m, \lambda)$ is defined by equality (4.23).

In what follows, we shall use the equality

$$i(\mu - \bar{\lambda})S^*(m, \lambda)S(m, \mu) = W^*(m-1, \lambda)JW(m-1, \mu) - W^*(m, \lambda)JW(m, \mu). \qquad (4.25)$$

To prove (4.25), we note that the equality $L(m) - L^*(m) = -i\beta^2(m)J$ and (4.18) imply that

$$J - b^*(m, \lambda) \, J b \, (m, \mu)$$

$$= i \, (\mu - \bar{\lambda}) \, J u_m \, \beta(m) \, [L \, (m) - \bar{\lambda}E]^{-1} \, [L^*(m) - \mu E]^{-1} \, \beta \, (m) \, u_m^* \, J. \qquad (4.26)$$

Multiplying the last equality by $W^*(m - 1, \lambda)$ from the left and by $W \, (m - 1, \mu)$ from the right and taking equalities (4.17) and (4.23) into account, we arrive at (4.25).

Let us prove now the boundedness of the operator $\bar{R}_\lambda$. Consider operators L and S defined by the equalities

$$(Lf)(k) = f(k) \, [L^*(k) - \lambda E]^{-1}, \qquad (Sf)(k) = \sum_{m = k+1}^{N} f(m) \, \Gamma \, (m, k; \lambda), \qquad (4.27)$$

where

$$\Gamma \, (m, k; \lambda) = S \, (m, \lambda) \, J S^*(k, \bar{\lambda}). \qquad (4.28)$$

Since $\lambda \neq 0$ and the determinant of the characteristic matrix function $\chi_A (\lambda)$ has no zeros and poles in a sufficiently small vicinity of the point λ, the boundedness of the operator L is evident.

Let us prove the boundedness of operator S. We set

$$\varphi \, (k) = \sum_{m = k+1}^{N} f(m) \, S \, (m, \lambda) \, J, \qquad A \, (k, \lambda) = S^*(k, \lambda) \, S \, (k, \lambda). \qquad (4.29)$$

Then $(Sf)(k) = \varphi \, (k) \, S^*(k, \bar{\lambda})$ and

$$\| Sf \|^2 = \sum_{k=1}^{N} \varphi \, (k) \, A \, (k, \bar{\lambda}) \, \varphi^*(k). \qquad (4.30)$$

For fixed k, we consider the operator S_λ defined in the Euclidean space $\mathcal{E}_n$ by the equality $S_\lambda \varphi \, (k) = \varphi \, (k) \, S^*(k, \bar{\lambda})$. Then

$$\varphi \, (k) \, A \, (k, \bar{\lambda}) \, \varphi^*(k) = \| S_\lambda \, \varphi(k) \|_{\mathcal{E}}^2 \le \text{tr} \, [A \, (k, \bar{\lambda})] \, \| \varphi(k) \|_{\mathcal{E}}^2$$

and, hence,

$$\| Sf \|^2 = \sum_{k=1}^{N} \text{tr} \, [A \, (k, \bar{\lambda})] \, \| \varphi(k) \|_{\mathcal{E}}^2. \qquad (4.31)$$

where, by virtue of the first equality in (4.29),

$$\| \varphi(k) \|_{\mathcal{E}} \le \sum_{m=1}^{N} \| f(m) S(m, \lambda) J \|_{\mathcal{E}} \tag{4.32}$$

By setting $\hat{A}(m, \lambda) = S(m, \lambda) S^*(m, \lambda)$, we get

$$\| f(m) S(m, \lambda) J \|_{\mathcal{E}}^2 = f(m) \hat{A}(m, \lambda) f^*(m) \le \mathrm{tr}\,[\hat{A}(m, \lambda)] \| f(m) \|_{\mathcal{E}}^2.$$

In other words, since $\mathrm{tr}\,[\hat{A}(m, \lambda)] = \mathrm{tr}\,[A(m, \lambda)]$, we have

$$\| f(m) S(m, \lambda) J \|_{\mathcal{E}} \le \mathrm{tr}^{1/2}[A(m, \lambda)] \| f(m) \|_{\mathcal{E}} \tag{4.33}$$

By virtue of (4.32) and (4.33), we have

$$\| \varphi(k) \|_{\mathcal{E}}^2 \le \left(\sum_{m=1}^{N} \mathrm{tr}^{1/2}[A(m, \lambda)] \| f(m) \|_{\mathcal{E}} \right)^2$$

$$\le \sum_{m=1}^{N} \mathrm{tr}\,[A(m, \lambda)] \sum_{m=1}^{N} \| f(m) \|_{\mathcal{E}}^2,$$

i.e.,

$$\| \varphi(k) \|_{\mathcal{E}}^2 \le \sum_{m=1}^{N} \mathrm{tr}\,[A(m, \lambda)] \| f \|^2.$$

Taking this inequality into account, we can rewrite (4.31) as

$$\| Sf \|^2 \le M_\lambda M_{\bar{\lambda}} \| f \|^2, \quad M_\lambda = \sum_{k=1}^{N} \mathrm{tr}\,[A(k, \lambda)]. \tag{4.34}$$

It follows from (4.29), (4.25), and the equalities $W(0, \lambda) = E$ and $W(N, \lambda) = \chi_A(\lambda)$ that

$$M_\lambda = \mathrm{tr}\, \sum_{k=1}^{N} S^*(k, \lambda) S(k, \lambda) = \mathrm{tr}\, \frac{\chi_A^*(\lambda) J \chi_A(\lambda) - J}{2 \, \mathrm{Im}\, \lambda} < \infty,$$

which proves the boundedness of the operators S and $\bar{R}_\lambda = L - iS$ at the point λ.

4.3. Operator $\bar{B}_\alpha$

Let α be a fixed nonreal point at which the characteristic matrix function $\chi_A(\alpha)$ is holomorphic and $\det \chi_A(\lambda) \neq 0$. Since $\bar{R}_\lambda = L - iS$, the operator $\bar{B}_\alpha = i\bar{R}_\alpha - i\bar{R}_\alpha^* + 2\operatorname{Im}\alpha \bar{R}_\alpha^* \bar{R}_\alpha$ can be represented in the form

$$\bar{B}_\alpha = iL - iL^* + 2\operatorname{Im}\alpha L^* L + S + S^* + 2\operatorname{Im}\alpha S^* S + 2i\operatorname{Im}\alpha(S^* L - L^* S).$$

$$(4.35)$$

For $\lambda = \alpha$, the operators L and L^* are determined by the following equalities:

$$(Lf)(k) = f(k)[L^*(k) - \alpha E]^{-1}, \qquad (L^* f)(k) = f(k)[L(k) - \bar{\alpha}E]^{-1}.$$

Hence,

$$iLf - iL^* f + 2\operatorname{Im}\alpha L^* Lf = (h(1), h(2), \ldots h(k), \ldots),$$

where

$$h(k) = if(k)[L^*(k) - \alpha E]^{-1}(L(k) - L^*(k))[L(k) - \bar{\alpha}E]^{-1}.$$

In view of the equality $L(k) - L^*(k) = -i\beta^2(k)J$, we obtain

$$h(k) = f(k)[L^*(k) - \alpha E]^{-1}\beta^2(k)J[L(k) - \bar{\alpha}E]^{-1}. \qquad (4.36)$$

Let us determine the vector $S^* Sf$. Taking expression (4.27) into account, we get

$$(S^* \varphi)(k) = \sum_{m=1}^{k-1} \varphi(m)\Gamma(m, k; \bar{\alpha}), \qquad (4.37)$$

where the matrix $\Gamma(m, k; \bar{\alpha})$ is defined by equality (4.28). Let $\varphi(m) = (Sf)(m)$, i.e.,

$$\varphi(m) = \sum_{j=m+1}^{N} f(j)\Gamma(j, m; \alpha).$$

Then

$$(S^* Sf)(k) = (S^* \varphi)(k) = \sum_{m=1}^{k-1} \sum_{j=m+1}^{N} f(j)\Gamma(j, m; \alpha)\Gamma(m, k; \bar{\alpha}). \qquad (4.38)$$

By using the obvious equality

$$\sum_{m=1}^{k-1} \sum_{j=m+1}^{N} a_{mj} = \sum_{j=1}^{N} \sum_{m=1}^{k-1} a_{mj} - \sum_{j=1}^{k-1} \sum_{m=j}^{k-1} a_{mj}$$

we rewrite (4.38) in the form

$$(S^* S f)(k) = \sum_{j=1}^{N} f(j) \sum_{m=1}^{k-1} \Gamma(j, m; \alpha) \Gamma(m, k; \bar{\alpha})$$

$$- \sum_{j=1}^{k-1} f(j) \sum_{m=j}^{k-1} \Gamma(j, m; \alpha) \Gamma(m, k; \bar{\alpha}). \qquad (4.39)$$

Furthermore, by virtue of (4.28) and (4.25), we have

$$2 \operatorname{Im} \alpha \sum_{m=j}^{k-1} \Gamma(j, m; \alpha) \Gamma(m, k; \bar{\alpha}) = S(j, \alpha) J [W^*(j-1, \bar{\alpha}) J W(j-1, \bar{\alpha})$$

$$- W^*(k-1, \bar{\alpha}) J W(k-1, \bar{\alpha})] J S^*(k, \alpha). \qquad (4.40)$$

By using equality (4.23), we obtain

$$S(j, \alpha) J [W^*(j-1, \bar{\alpha}) J W(j-1, \bar{\alpha}) = r(j, \alpha) S(j, \bar{\alpha}), \qquad (4.41)$$

where $r(j, \alpha) = [L^*(j) - \alpha E]^{-1} [L^*(j) - \bar{\alpha} E]$. Similarly, by virtue of (4.41),

$$W^*(k-1, \bar{\alpha}) J W(k-1, \bar{\alpha}) J S^*(k, \alpha) = S^*(k, \bar{\alpha}) r^*(k, \alpha). \qquad (4.42)$$

By inserting (4.41) and (4.42) in (4.40) and using equality (4.28), we get

$$2 \operatorname{Im} \alpha \sum_{m=j}^{k-1} \Gamma(j, m; \alpha) \Gamma(m, k; \bar{\alpha}) = r(j, \alpha) \Gamma(j, k; \bar{\alpha}) - \Gamma(j, k; \alpha) r^*(k, \alpha).$$

$$(4.43)$$

By analogy, one can establish that

$$2 \operatorname{Im} \alpha \sum_{m=1}^{k-1} \Gamma(j, m; \alpha) \Gamma(m, k; \bar{\alpha}) = S(j, \alpha) S^*(k, \alpha) - \Gamma(j, k; \alpha) r^*(k, \alpha). \qquad (4.44)$$

Therefore, by virtue of (4.39), (4.43), and (4.44), we have

$$2\operatorname{Im}\alpha(S^*Sf)(k) = \sum_{j=1}^{N} f(j)S(j,\alpha)JS^*(k,\alpha) - \sum_{j=1}^{N} f(j)\Gamma(j,k;\alpha)r^*(k,\alpha)$$

$$- \sum_{j=1}^{k-1} f(j)r(j,\alpha)\Gamma(j,k;\bar{\alpha}) + \sum_{j=1}^{k-1} f(j)\Gamma(j,k;\alpha)r^*(k,\alpha),$$

i.e.,

$$2\operatorname{Im}\alpha(S^*Sf)(k) = \sum_{j=1}^{N} f(j)S(j,\alpha)JS^*(k,\alpha) - f(k)\Gamma(k,k;\alpha)r^*(k,\alpha)$$

$$- \sum_{j=k+1}^{N} f(j)\Gamma(j,k;\alpha)r^*(k,\alpha) - \sum_{j=1}^{k-1} f(j)r(j,\alpha)\Gamma(j,k;\bar{\alpha}). \quad (4.45)$$

Since $r(j,\alpha) = E + (\alpha - \bar{\alpha})[L^*(j) - \alpha E]^{-1}$, we have

$$\sum_{j=1}^{k-1} f(j)r(j,\alpha)\Gamma(j,k;\bar{\alpha}) = (S^*f)(k) + (\alpha - \bar{\alpha})(S^*Lf)(k). \quad (4.46)$$

Similarly,

$$\sum_{j=k+1}^{N} f(j)\Gamma(j,k;\alpha)r^*(k,\alpha) = (Sf)(k) + (\bar{\alpha} - \alpha)(L^*Sf)(k). \quad (4.47)$$

Furthermore, it follows from (4.23) and (4.28) that

$$\Gamma(k,k;\alpha) = [L^*(k) - \alpha E]^{-1}\beta^2(k)J[L(k) - \alpha E]^{-1}$$

and, therefore, in view of (4.36), we get

$$f(k)\Gamma(k,k;\alpha)r^*(k,\alpha) = i(Lf)(k) - i(L^*f)(k) + 2\operatorname{Im}\alpha(L^*Lf)(k). \quad (4.48)$$

Finally, equalities (4.45), (4.46), (4.47), and (4.48) yield

$$2\operatorname{Im}\alpha(S^*Sf)(k) = \sum_{j=1}^{N} f(j)S(j,\alpha)JS^*(k,\alpha) - i(Lf)(k) + i(L^*f)(k)$$

$$- 2 \operatorname{Im} \alpha (L^* L f)(k) - (S f)(k) - (\bar{\alpha} - \alpha)(L^* S f)(k)$$

$$- (S^* f)(k) - (\alpha - \bar{\alpha})(S^* L f)(k). \tag{4.49}$$

By using equalities (4.35) and (4.49), we obtain

$$(\bar{B}_\alpha f)(k) = \sum_{j=1}^{N} f(j) S(j, \alpha) J S^*(k, \alpha). \tag{4.50}$$

By analogy with Subsection 2.4, we can establish that

$$\bar{B}_\alpha f = \sum_{m,\, s=1}^{n} (f, g_m) J_{ms} g_s,$$

where

$$g_m(k) = h_m S^*(k, \alpha) \quad (m = \overline{1, n}) \tag{4.51}$$

and the vectors h_m are defined by equalities (2.51).

Thus, the system of vectors $\{g_m\}_{m=1}^{n}$ forms a Δ-basis of the operator $\bar{A}$ defined by equalities (4.6) and (4.7), and the corresponding coefficient matrix coincides with J.

4.4. Characteristic Matrix Function of the Operator $\bar{A}$

Recall that, for $J^{-1} = J$, the characteristic matrix function $\chi_{\bar{A}}(\lambda)$ of the operator $\bar{A}$ is defined by the equality (Chapter 2, Subsection 1.2)

$$\chi_{\bar{A}}^*(\alpha) J \chi_{\bar{A}}(\lambda) = J + i(\bar{\alpha} - \lambda) \| ((\bar{A} - \alpha I) \bar{R}_\lambda g_j, g_m) \|,$$

where $\{g_k\}_{k=1}^{n}$ is a Δ-basis of the operator $\bar{A}$ and $\lambda \in \rho(\bar{A})$. Since $(\bar{A} - \alpha I) \bar{R}_\lambda = I + (\lambda - \alpha) \bar{R}_\lambda$, we have

$$\chi_{\bar{A}}^*(\alpha) J \chi_{\bar{A}}(\lambda) = J + i(\bar{\alpha} - \lambda) \| (g_j, g_m) \| + (\lambda - \alpha) i (\bar{\alpha} - \lambda) \| (\bar{R}_\lambda g_j, g_m) \|. \tag{4.52}$$

By virtue of (4.24),

$$(\bar{R}_\lambda g_j)(k) = g_j(k) [L^*(k) - \lambda E]^{-1} - i \sum_{m=k+1}^{N} g_j(m) S(m, \lambda) J S^*(k, \bar{\lambda}).$$

Note that, in view of (4.25), (4.51), and the equality $W(N, \lambda) = \chi_A(\lambda)$, we have

$$\sum_{m=k+1}^{N} g_j(m) S(m, \lambda) = \frac{1}{i(\lambda - \overline{\alpha})} h_j[W^*(k, \alpha) J W(k, \lambda) - \chi_A^*(\alpha) J \chi_A(\lambda)]. \quad (4.53)$$

Thus, by virtue of the last three equalities, we get

$$(\overline{\alpha} - \lambda)(\overline{R}_\lambda g_j)(k) = (\overline{\alpha} - \lambda) g_j(k)[L^*(k) - \lambda E]^{-1}$$

$$+ h_j W^*(k, \alpha) J W(k, \lambda) J S^*(k, \overline{\lambda}) - h_j \chi_A^*(\alpha) J \chi_A(\lambda) J S^*(k, \overline{\lambda}). \quad (4.54)$$

Note that, in view of equalities (4.17) and (4.23), one can show, by analogy with the proof of equality (4.41), that

$$W^*(k, \alpha) J W(k, \lambda) J S^*(k, \overline{\lambda})$$

$$= W^*(k-1, \alpha) b^*(k, \alpha) J b(k, \lambda) u_k \beta(k)[L(k) - \lambda E]^{-1}. \quad (4.55)$$

Note that, by virtue of (4.18), we have

$$b^*(k, \alpha) J b(k, \lambda) u_k$$

$$= u_k^{-*}[L(k) - \overline{\alpha} E]^{-1}[L^*(k) - \overline{\alpha} E] J [L(k) - \lambda E][L^*(k) - \lambda E]^{-1}. \quad (4.56)$$

It follows from equalities (4.23), (4.51), (4.55), and (4.56) that

$$h_j W^*(k, \alpha) J W(k, \lambda) J S^*(k, \overline{\lambda}) = g_j(k)[L^*(k) - \overline{\alpha} E][L^*(k) - \lambda E]^{-1}. \quad (4.57)$$

By inserting (4.57) in (4.54), we obtain

$$(\overline{\alpha} - \lambda)(\overline{R}_\lambda g_j)(k) = g_j(k) - \varphi_j(k), \quad (4.58)$$

where

$$\varphi_j(k) = h_j \chi_A^*(\alpha) J \chi_A(\lambda) J S^*(k, \overline{\lambda}). \quad (4.59)$$

Therefore, by virtue of (4.52) and (4.58), we get

$$\chi_A^*(\alpha) J \chi_{\overline{A}}(\lambda) = J + 2 \operatorname{Im} \alpha \| (g_j, g_m) \| - i(\lambda - \alpha) \| (\varphi_j, g_m) \|. \quad (4.60)$$

According to (4.25), (4.51), and the equalities $W(0, \alpha) = E$ and $W(N, \alpha) = \chi_A(\alpha)$, we have

$$(g_j, g_m) = -\frac{1}{2 \operatorname{Im} \alpha} h_j [J - \chi_A^*(\alpha) J \chi_A(\alpha)] h_m^*,$$

i.e.,

$$2 \operatorname{Im} \alpha \| (g_j, g_m) \| = \chi_A^*(\alpha) J \chi_A(\alpha) - J. \qquad (4.61)$$

Therefore, by virtue of (4.60) and (4.61),

$$\chi_{\bar{A}}^*(\alpha) J \chi_{\bar{A}}(\lambda) = \chi_A^*(\alpha) J \chi_A(\alpha) - i(\lambda - \alpha) \| (\varphi_j, g_m) \|. \qquad (4.62)$$

By using equalities (4.25), (4.51), and (4.59), we get

$$i(\lambda - \alpha) \| (\varphi_j, g_m) \| = \chi_A^*(\alpha) J \chi_A(\lambda) J [\chi_A^*(\bar{\lambda}) J \chi_A(\alpha) - J]. \qquad (4.63)$$

Finally, by inserting (4.63) in (4.62) and taking the equality $\chi_A(\lambda) J \chi_A^*(\bar{\lambda}) = J$ into account, we obtain

$$\chi_{\bar{A}}^*(\alpha) J \chi_{\bar{A}}(\lambda) = \chi_A^*(\alpha) J \chi_A(\lambda).$$

By analogy with Subsection 2.5, one can now easily verify that L-simple parts of the operators A and $\bar{A}$ are unitary equivalent.

4.5. Spectrum of the Operator A

It is easy to see that the numbers $\bar{\lambda}_k$ in (4.2) (i.e., the nonreal poles of the characteristic matrix function $\chi_A(\lambda)$) are eigenvalues of the operator $\bar{A}$ defined by equalities (4.6) and (4.7); consequently, these poles are eigenvalues of the operator A. If the point λ_0 $(\operatorname{Im} \lambda_0 \neq 0)$ is not a pole of the characteristic matrix function $\chi_A(\lambda)$, then the matrix function $\chi_A(\lambda)$ is holomorphic at λ_0. Consequently, λ_0 cannot belong to the spectrum of the operator A.

Thus, the nonreal spectrum of the operator A coincides with the set of poles of the characteristic matrix function $\chi_A(\lambda)$.

4.6. The Condition under Which $\mathfrak{D}_{A^*} = \mathfrak{D}_A$

Let A be a simple K^r-operator with discrete spectrum (see Subsection 4.1). As in Subsection 2.6, one can establish that the equality $\mathfrak{D}_{A^*} = \mathfrak{D}_A$ holds if and only if $\mathfrak{D}_{\bar{A}_s^*} = \mathfrak{D}_{\bar{A}_s}$, where $\bar{A}_s$ is the simple part of the operator $\bar{A}$. Moreover, if the latter equality holds, then the vectors g_m defined by (4.51) belong to the lineal $\mathfrak{D}_{\bar{A}}$.

Let us find the vectors $\varphi_m = \bar{A} g_m$. By virtue of (4.7) and (4.51), we have

$$\varphi_m(k) = g_m(k)L^*(k) + ih_m \sum_{j=k+1}^{N} S^*(j, \alpha)\sigma(j)J\sigma^*(k), \qquad (4.64)$$

where, in view of (4.4) and (4.23),

$$S^*(j, \alpha)\sigma(j) = W^*(j-1, \alpha)u_j^{-*}\beta^2(j)J[L(j)-\bar{\alpha}E]^{-1}u_j^*\pi^*(j-1).$$

By using (4.3), (4.15), and the equality

$$i\beta^2(j)J[L(j)-\bar{\alpha}E]^{-1} = [L^*(j)-L(j)][L(j)-\bar{\alpha}E]^{-1}$$

$$= [L^*(j)-\bar{\alpha}E][L(j)-\bar{\alpha}E]^{-1} - E,$$

we get

$$iS^*(j, \alpha)\sigma(j) = W^*(j, \alpha)\pi^*(j) - W^*(j-1, \alpha)\pi^*(j-1).$$

But then

$$i\sum_{j=k+1}^{N} S^*(j, \alpha)\sigma(j) = \chi_A^*(\alpha)\pi^*(N) - W^*(k, \alpha)\pi^*(k).$$

It follows from relations (4.3), (4.4), (4.15), and (4.23) that

$$W^*(k, \alpha)\pi^*(k)J\sigma^*(k) = S^*(k, \alpha)[L^*(k)-\bar{\alpha}E]$$

and, hence,

$$ih_m \sum_{j=k+1}^{N} S^*(j, \alpha)\sigma(j)J\sigma^*(k)$$

$$= h_m \chi_A^*(\alpha)\pi^*(N)J\sigma^*(k) - g_m(k)[L^*(k)-\bar{\alpha}E].$$

By using the last equality, we can rewrite (4.64) as follows:

$$\varphi_m(k) = \bar{\alpha}\, g_m(k) + ih_m \chi_A^*(\alpha)\, \pi^*(N)\, J\sigma^*(k).$$

Thus, $\varphi_m = \bar{\alpha}\, g_m + \psi_m$, where ψ_m is a vector with the components $\psi_m(k) = ih_m \chi_A^*(\alpha)\, \pi^*(N)\, J\sigma^*(k)$. Furthermore, since $g_m \in \vec{\mathcal{H}}$ and $\varphi_m = \vec{A} g_m \in \vec{\mathcal{H}}$, this vector ψ_m belongs to $\vec{\mathcal{H}}$. Hence,

$$(\psi_m, \psi_m) = h_m \chi_A^*(\alpha)\, \pi^*(N)\, J\sigma\pi(N)\chi_A(\alpha)\, h_m^* < \infty, \quad m = 1, 2, \ldots, r, \qquad (4.65)$$

where

$$\sigma := \sum_{k=1}^{N} \sigma^*(k)\,\sigma(k)\, J.$$

It follows from inequality (4.65) and the invertibility of the matrices $\chi_A^*(\alpha)\, \pi^*(N)$ and J that the matrix σ is finite. But then the trace of this matrix is also finite; moreover,

$$\mathrm{tr}\,\sigma = \sum_{k=1}^{N} \mathrm{tr}\,[\sigma^*(k)\,\sigma(k)\, J].$$

Note that, by virtue of equality (4.5), we get

$$\mathrm{tr}\,[\sigma^*(k)\,\sigma(k)\, J] = \mathrm{tr}\,[\sigma(k)\, J\sigma^*(k)] = -2\,\mathrm{tr}\,[\mathrm{Im}\, L(k)].$$

By virtue of (4.2), we have $\mathrm{tr}\,[\mathrm{Im}\, L(k)] = q_k\, \mathrm{Im}\,\lambda_k$ and, therefore,

$$\mathrm{tr}\,\sigma = -2\sum_{k=1}^{N} q_k\, \mathrm{Im}\,\lambda_k. \qquad (4.66)$$

We have already mentioned that the trace of the matrix σ is finite; consequently, the right-hand side of (4.66) is convergent (for $N = \infty$). Since $\mathrm{Im}\,\lambda_k > 0$, we have

$$\sum_{k=1}^{N} |\mathrm{Im}\,\bar{\lambda}_k| = \sum_{k=1}^{N} \mathrm{Im}\,\lambda_k \leq \sum_{k=1}^{N} q_k\, \mathrm{Im}\,\lambda_k < \infty.$$

Conversely, if $\sum_{k=1}^{N} |\mathrm{Im}\,\bar{\lambda}_k| < \infty$, then

$$\sum_{k=1}^{N} q_k \operatorname{Im} \lambda_k \le r \sum_{k=1}^{N} |\operatorname{Im} \bar{\lambda}_k| < \infty$$

and $\operatorname{tr} \sigma$ is finite. But then $\psi_m \in \vec{\mathcal{H}}$ and $g_m \in \mathfrak{D}_{\bar{A}}$.

Hence, we have proved the following theorem:

Theorem 4.1. *If A is a simple K^r-operator whose characteristic matrix function can be represented in the form (4.1), then the equality $\mathfrak{D}_{A^\bullet} = \mathfrak{D}_A$ holds if and only if $\sum_{k=1}^{N} |\operatorname{Im} \bar{\lambda}_k| < \infty$, where $\{\bar{\lambda}_k\}$ is the set of nonreal eigenvalues of the operator A.*

Corollary 4.2. *If a simple K^r-operator of the indicated type has finitely many nonreal eignevalues, then $\mathfrak{D}_{A^\bullet} = \mathfrak{D}_A$.*

4.7. Remark

In the previous subsections, we have studied K^r-operators A whose characteristic matrix functions $\chi_A(\lambda)$ can be represented in the form (4.1). Similar results can also be established in the case where the characteristic matrix function $\chi_A(\lambda)$ of the operator A admits the representation

$$\chi_A(\lambda) = \prod_{k=1}^{N} V_k [M^*(k) - \lambda E][M(k) - \lambda E]^{-1} \gamma(k) V_k^{-1}, \qquad (4.67)$$

where

$$M(k) = \begin{pmatrix} \lambda_k E_{p_k'} & 0 \\ 0 & 0 \end{pmatrix}, \quad \gamma(k) = \begin{pmatrix} \gamma_k E_{p_k'} & 0 \\ 0 & E_{q_k'} \end{pmatrix},$$

$\operatorname{Im} \lambda_k > 0$, and $|\gamma_k| = 1$. By using the same reasoning as in Subsections 4.1–4.6, we can construct the model operator $\vec{A}$ for this case. This operator is defined in $\vec{\mathcal{H}}$ by the equalities

$$(\vec{A} f)(k) = f(k) M(k) + i \sum_{j=k+1}^{N} f(j) \sigma(j) J \sigma^*(k), \qquad (4.68)$$

where

$$\sigma(k) = \beta(k)\, V_k^* \pi^*(k-1), \qquad \pi(k) = \prod_{j=1}^{\hat{k}} V_j \gamma^*(j) V_j^{-1},$$

and $\beta(k)$ are nonnegative matrices defined by the relations $M(k) - M^*(k) = 2i\beta^2(k)J$. We can repeat the procedure used in the previous subsections and show that the simple part of the operator A is unitary equivalent to the simple part of the operator $\bar{A}$. In this case, the numbers λ_k (the poles of the characteristic matrix function $\chi_A(\lambda)$) are eigenvalues of the operator $\bar{A}$ and, hence, of A.

By analogy with Subsection 4.6, we can also prove that if A is a simple K^r-operator of the indicated type, then the equality $\mathcal{D}_{A^*} = \mathcal{D}_A$ holds if and only if

$$\sum_{k=1}^{N} \operatorname{Im}\lambda_k < \infty.$$

Remark. The principal results of this section were obtained by Kuzhel in [1–3].

5. General Case

5.1. The Model of an Arbitrary K^r-Operator

Recall that, in the general case, the characteristic matrix function $\chi_A(\lambda)$ of an arbitrary K^r-operator A can be represented as (see Chapter 2, Subsection 7.6)

$$\chi_A(\lambda) = W_5(\lambda)\, W_4(\lambda)\, W_3(\lambda)\, W_2(\lambda)\, W_1(\lambda), \tag{5.1}$$

where

$$W_k(\lambda) = \int_0^{\hat{l}_k} \exp\left[-i\lambda P_k^2(t) J\, dt\right] \quad (k = 3; 5),$$

$$W_4(\lambda) = \int_0^{\hat{l}_4} \exp\left[i\,\frac{1+\lambda\alpha(t)}{\lambda-\alpha(t)}\, P_4^2(t)\, J\, dt\right]$$

and the matrices $W_1(\lambda)$ and $W_2(\lambda)$ are defined by equalities (7.27) and (7.28).

By using the representations of these matrix functions we can construct the model operators $\bar{A}_k$ in the corresponding spaces $\bar{\mathcal{H}}_k$ $(k = \overline{1,\,5})$; thus, in particular,

the operators $\bar{A}_5$ and $\bar{A}_3$ can be constructed in the same way as the model operator $\bar{A}$ in Section 3 (relation (3.4));

the operator $\bar{A}_4$ can be constructed in the same way as the operator $\bar{A}$ in Section 2 (relation (2.6));

the operator $\bar{A}_2$ can be constructed by analogy with $\bar{A}$ in Section 4 (relations (4.6) and (4.7));

the operator $\bar{A}_1$ can be constructed by analogy with the operator $\bar{A}$ in Section 4 (relation (4.68)).

Then we construct the coupling of these operators in the space $\bar{\mathcal{H}} = \bar{\mathcal{H}}_1 \oplus \bar{\mathcal{H}}_2 \oplus \bar{\mathcal{H}}_3 \oplus \bar{\mathcal{H}}_4 \oplus \bar{\mathcal{H}}_5$, i.e.,

$$\bar{A} = (((\bar{A}_1 \gamma \bar{A}_2) \gamma \bar{A}_3) \gamma \bar{A}_4) \gamma \bar{A}_5.$$

By the multiplication theorem (Theorem 6.4 in Chapter 2), we can show that $\chi_{\bar{A}}(\lambda) = \chi_A(\lambda)$ and, thus, the simple parts of the operators A and $\bar{A}$ are unitary equivalent.

According to the results of Subsection 6.1 (Chapter 2), the model of the operator A can be written in the form

$$\bar{A}f = S_0 S^{-1} f \quad \left(f \in \mathfrak{D}_{\bar{A}}, \ \mathfrak{D}_{\bar{A}} = S[\mathfrak{D}_{\bar{A}_1} \oplus \mathfrak{D}_{\bar{A}_2} \oplus \mathfrak{D}_{\bar{A}_3} \oplus \mathfrak{D}_{\bar{A}_4} \oplus \mathfrak{D}_{\bar{A}_5}]\right)$$

where

$$S = \begin{pmatrix} I & K_{12} & K_{13} & K_{14} & K_{15} \\ 0 & I & K_{23} & K_{24} & K_{25} \\ 0 & 0 & I & K_{34} & K_{35} \\ 0 & 0 & 0 & I & K_{45} \\ 0 & 0 & 0 & 0 & I \end{pmatrix}, \quad S_0 = \begin{pmatrix} \bar{A}_1 & A_{12} & A_{13} & A_{14} & A_{15} \\ 0 & \bar{A}_2 & A_{23} & A_{24} & A_{25} \\ 0 & 0 & \bar{A}_3 & A_{34} & A_{35} \\ 0 & 0 & 0 & \bar{A}_4 & A_{45} \\ 0 & 0 & 0 & 0 & \bar{A}_5 \end{pmatrix}.$$

The operators K_{mi} and A_{mi} in the expressions for the matrices S and S_0 can be found explicitly (by using the corresponding relations in Section 6 of Chapter 2). However, it is not necessary to do this because the spectrum of the operator $\bar{A}$ is determined by the diagonal elements of the matrix S_0.

In many special cases (dissipative operators, K_I^r-operators, K^I-operators, etc.), this model can be substantially simplified.

Sometimes, it is convenient to represent the model $\bar{A}$ of the operator A as $\bar{A} = \bar{D} \gamma \bar{C}$, where $\bar{D} = \bar{A}_1 \gamma \bar{A}_2$ and $\bar{C} = (\bar{A}_3 \gamma \bar{A}_4) \gamma \bar{A}_5$. In this notation, the operator $\bar{D}$ describes the nonreal point spectrum of the original K^r-operator A, while the operator $\bar{C}$ characterizes the continuous spectrum of the operator A. In this connection, it is naturally to call the operators $\bar{D}$ and $\bar{C}$ the *discrete* and *continuous components of the triangular model* $\bar{A}$, respectively.

5.2. Completeness Criterion

The vector $\varphi \neq 0$ is called the *root vector* of the operator A corresponding to the number λ_0 if there exists natural n such that

$$(A - \lambda_0 I)^n \varphi = 0. \tag{5.2}$$

If n is the least natural number such that equality (5.2) is satisfied, then it is clear that the vector $f = (A - \lambda_0 I)^{n-1} \varphi$ is the eigenvector of the operator A corresponding to the eigenvalue λ_0.

Assume that $\bar{A}$ is a model of the simple K^r-operator A. Then $A = u^{-1} \bar{A}_s u$, where $\bar{A}_s$ is the simple part of the operator $\bar{A}$, and u is a unitary operator. If in addition φ is the root vector of the operator A corresponding to the nonreal eigenvalue λ_0, then the vector $u\varphi$ is the root vector of the operator $\bar{A}_s$ (and hence, of $\bar{A}$) corresponding to the same eigenvalue λ_0. The converse statement is also true: If $u\varphi$ is the root vector of the operator $\bar{A}$ and $\text{Im}\,\lambda_0 \neq 0$, then φ is the root vector of the operator A.

The collection of root vectors of the operator A acting in $\mathcal{H}$ is called complete if the closure of the linear span of these vectors coincides with $\mathcal{H}$.

It is clear that if A is a simple K^r-operator and $\bar{A}$ is its model, then the collection of root vectors of the operator A is complete if and only if the system of root vectors of the simple part $\bar{A}_s$ of the operator $\bar{A}$ is complete.

Theorem 5.1 *(completeness criterion). If A is a simple dissipative K^r-operator, then*

$$|\det \chi_A (-i)| \leq \prod_{k=1}^{N} \left| \frac{\bar{\lambda}_k + i}{\lambda_k + i} \right|^{p_k}, \tag{5.3}$$

where λ_k $(\text{Im}\,\lambda_k > 0)$ are poles of the characteristic matrix function[7] $\chi_A (\lambda)$ and

---

[7] Or nonreal eigenvalues of the operator A (this is the same).

p_k is the multiplicity of the pole λ_k. The collection of root vectors of the operator A is complete if and only if (5.3) turns into the equality.

Proof. By the condition of the theorem, A is a simple dissipative operator and, thus, $J = E$. Under this condition, the factor $W_1(\lambda)$ in (5.1) is absent. But then

$$\det \chi_A(-i) = \det W_2(-i)\det W_3(-i)\det W_4(-i)\det W_5(-i), \qquad (5.4)$$

where

$$\det W_2(-i) = \prod_{k=1}^{N}\left(\frac{\bar\lambda_k + i}{\lambda_k + i}\right)^{p_k} \qquad (N \leq \infty).$$

To find $\det W_3(-i)$, we use the equalities

$$W_3(-i) = \int_0^{l_3} \exp[-P_3^2(t)dt] \qquad \text{and} \qquad \det e^M = e^{\mathrm{tr}M},$$

where M is an arbitrary square matrix. As a result, by using the equality $\mathrm{tr}\,P_3^2(t) = 1$, we obtain

$$\det W_3(-i) = \exp\left\{\int_0^{l_3} -\mathrm{tr}\,P_3^2(t)dt\right\} = e^{-l_3}.$$

Similarly, $\det W_4(-i) = e^{-l_3}$ and $\det W_5(-i) = e^{-l_5}$ and, hence,

$$|\det \chi_A(-i)| \leq \prod_{k=1}^{N}\left|\frac{\bar\lambda_k + i}{\lambda_k + i}\right|^{p_k} e^{-(l_3 + l_4 + l_5)} \qquad (5.5)$$

and we arrive at inequality (5.3).

In the case under consideration, the model $\bar A$ of the operator A is defined as the coupling of the operators $\bar A_2$ and $\bar C = (\bar A_3 \gamma \bar A_4)\gamma \bar A_5$ in the space $\vec{\mathcal{H}} = \vec{\mathcal{H}}_2 \oplus \vec{\mathcal{H}}_c$, where $\vec{\mathcal{H}}_c = \vec{\mathcal{H}}_3 \oplus \vec{\mathcal{H}}_4 \oplus \vec{\mathcal{H}}_5$. Hence,

$$\bar A = \bar A_2 P_2 S^{-1} + \bar C P_c + \Gamma P_c,$$

where P_2 and P_c are the orthoprojectors onto the subspaces $\vec{\mathcal{H}}_2$ and $\vec{\mathcal{H}}_c$, respectively, in $\vec{\mathcal{H}}$ and Γ is an operator acting from $\vec{\mathcal{H}}_c$ into $\vec{\mathcal{H}}_2$.

Assume that φ is the root vector of the operator $\vec{A}$ corresponding to the number λ_0 and n is the least natural number for which $(\vec{A} - \lambda_0 I)^n \varphi = 0$. We set $\varphi_k = (\vec{A} - \lambda_0 I)^k \varphi$. Since φ_{n-1} is the eigenvector of the operator $\vec{A}$ corresponding to the eigenvalue λ_0, we have

$$0 = (\vec{A} - \lambda_0 I) \varphi_{n-1}$$

$$= \vec{A}_2 P_2 S^{-1} \varphi_{n-1} + \Gamma P_c \varphi_{n-1} - \lambda_0 P_2 \varphi_{n-1} + (\vec{C} - \lambda_0 I) P_c \varphi_{n-1}.$$

All vectors on the right-hand side of this equality except $(\vec{C} - \lambda_0 I) P_c \varphi_{n-1}$ belong to the space $\vec{\mathcal{H}}_2$. Consequently, $(\vec{C} - \lambda_0 I) P_c \varphi_{n-1} = 0$. In view of the fact that the operator $\vec{C}$ does not have nonreal eigenvalues, we have $P_c \varphi_{n-1} = 0$, i.e., $\varphi_{n-1} \in \vec{\mathcal{H}}_2$.

By applying the same procedure to the vector $\varphi_{n-1} = (\vec{A} - \lambda_0 I) \varphi_{n-2}$, we find that $\varphi_{n-2} \in \vec{\mathcal{H}}_2$. Similarly, we can prove that the vectors $\varphi_{n-3}, \ldots, \varphi_2$, and φ_1 lie in $\vec{\mathcal{H}}_2$.

Thus, all root vectors corresponding to nonreal points of the spectrum of the operator $\vec{A}$ lie in the subspace $\vec{\mathcal{H}}_2$. But then the closure $\vec{\mathcal{H}}_0$ of the linear span of the indicated root vectors is a subspace of the space $\vec{\mathcal{H}}_2$.

Let the system of root vectors of the operator A be complete in $\mathcal{H}$. Then the simple part of the model $\vec{A}$ coincides with the simple part of the model $\vec{A}_2$ and, hence, $\chi_A(\lambda) = \chi_{\vec{A}_2}(\lambda) \ (= W_2(\lambda))$. But then $\det \chi_A(-i) = \det W_2(-i)$. Thus, all the other factors in (5.4) are equal to one, i.e., $l_3 = l_4 = l_5 = 0$. As a result, equality (5.5) takes the form of (5.3).

Conversely, if (5.3) turns into the equality, then we have $l_3 = l_4 = l_5 = 0$ in (5.5) and, hence, $\chi_A(\lambda) = W_2(\lambda)$. But then the operator $\vec{A}_2$ is the model of the operator A. If the closure $\mathcal{H}_0$ of the system of root vectors of the operator A did not coincide with $\mathcal{H}$, then this operator would be representable as a coupling of the operator $A_0 = A|_{\mathcal{H}_0 \cap \mathcal{D}_A}$ and an operator A_1 that acts in the subspace $\vec{\mathcal{H}}_1 = \mathcal{H} \ominus \mathcal{H}_0$. In this case, A_1 would have no nonreal eigenvectors (this can be easily verified by analogy with the proof of the inclusion $\vec{\mathcal{H}}_0 \subset \vec{\mathcal{H}}_2$). Therefore, the operator A_1 should be asociated with the continuous component of the model, which is impossible because $l_3 = l_4 = l_5 = 0$. Therefore, if (5.3) turns into the equality, then $\mathcal{H}_0 = \mathcal{H}$, i.e., the system of root vectors of the operator A is complete. $\blacksquare$

Theorem 5.2. *If a dissipative K^r-operator A has no continuous spectrum and $\mathfrak{D}_{A^*} = \mathfrak{D}_A$, then the system of root vectors of the operator A is complete.*

Proof. Since $\mathfrak{D}_{A^*} = \mathfrak{D}_A$, the operator A can be represented as $A = H + iK$, where H is self-adjoint and K is a finite-dimensional nonnegative operator. Assume that λ_0 is a real eigenvalue of the operator A and x is the corresponding eigenvector. Then $(Hx, x) + i(Kx, x) = \lambda_0(x, x)$ and, hence, $(Kx, x) = 0$. Since $K \geq 0$, we have $Kx = 0$. As a result, we obtain $Hx = \lambda_0 x$.

Let $\mathcal{L}$ be the linear span of all eigenvectors corresponding to real eigenvalues of the operator A. Since the operator K vanishes on the lineal $\mathcal{L}$, it also vanishes on the subspace $\mathcal{H}_1 = \overline{\mathcal{L}}$. Consequently, the subspace $\mathcal{H}_1$ reduces the operator K.

Let us show that the subspace $\mathcal{H}_1$ reduces the operator H, as well. For this purpose, we consider the symmetric operator $A_1' = A|_{\mathcal{L}}$ and assume that A_1 is its closure. Since A is closed and $\mathcal{H}_1 = \overline{\mathcal{L}}$, then $\mathfrak{D}_{A_1} = \mathcal{H}_1 \cap \mathfrak{D}_A$.

Let us show that A_1 is a self-adjoint operator in $\mathcal{H}_1$. For this purpose, it suffices to show that its defect numbers are zero. We assume that $\mathfrak{M}_\lambda = (A_1 - \lambda I)\mathfrak{D}_{A_1}$ and φ is an eigenvector of the operator A corresponding to the eigenvalue λ_0. Then $(A_1 - \lambda I)\varphi = (\lambda_0 - \lambda)\varphi$ and, hence, $\mathcal{L} \subset \mathfrak{M}_\lambda$. Since $\overline{\mathcal{L}} = \mathcal{H}_1$ and $\mathfrak{M}$ is a subspace of the space $\mathcal{H}$, we have $\mathfrak{M}_\lambda = \mathcal{H}_1$. Therefore, the defect numbers of the operator A_1 are zero and, consequently, A_1 is self-adjoint (in the space $\mathcal{H}_1$). Furthermore, $K|_{\mathcal{H}_1} = 0$ and, hence, $A_1 = H|_{\mathfrak{D}_A \cap \mathcal{H}_1}$.

Thus, the self-adjoint operator H induces the self-adjoint operator A_1 in the subspace $\mathcal{H}_1$. Thus, $\mathcal{H}_1$ reduces the operator H and, hence, the operator A. But then $\mathfrak{D}_A = \mathfrak{D}_{A_1} \oplus \mathfrak{D}_{A_2}$, where $A_2 = A|_{\mathfrak{D}_A \cap \mathcal{H}_2}$ and $\mathcal{H}_2 = \mathcal{H} \ominus \mathcal{H}_1$. In this case, $\mathfrak{D}_{A_2^*} = \mathfrak{D}_{A_2}$, i.e., A_2 belongs to the class of K_1^r-operators and, by construction, has no real spectrum. Therefore, the model of the operator A_2 cannot contain the operators $\bar{A}_3$, $\bar{A}_4$, and $\bar{A}_5$, i.e., $l_3 = l_4 = l_5 = 0$. Moreover, A_2 is a simple operator (because it cannot induce a self-adjoint operator in any invariant subspace). By using Theorem 5.1, we complete the proof.

∎

5.3. Condition under Which $\mathfrak{D}_{A^*} = \mathfrak{D}_A$

Assume that A is a simple K^r-operator and $\mathfrak{D}_{A^*} = \mathfrak{D}_A$. By analogy with Subsection 2.6, we can show that the defect space $\bar{\mathfrak{N}}_{-i}$ belongs to $\mathfrak{D}_{\bar{A}}$, where $\bar{A}$ is the model of the operator A. In view of the results in Subsection 5.1, we conclude that defect spaces of the model operators $\bar{A}_k$ $(k = \overline{1, 5})$ are also contained in the corresponding lineals $\mathfrak{D}_{\bar{A}_k}$. However, this is impossible for the operators $\bar{A}_3$ and $\bar{A}_5$ (see Section 3). Therefore, if $\mathfrak{D}_{A^*} = \mathfrak{D}_A$, these operators are absent in the general model of the operator A.

Thus, if $\mathfrak{D}_{A^*} = \mathfrak{D}_A$, then the model $\bar{A}$ of the operator A is a coupling of the model operators $\bar{A}_1$, $\bar{A}_2$, and $\bar{A}_4$. By combining the results of Subsections 2.6, 4.6, and 4.7, we conclude that

$$\sum_k |\operatorname{Im} \bar{\mu}_k| + \sum_k \operatorname{Im} \lambda_k + \left| \int_0^l \alpha^2(x) \operatorname{tr} [P^2(x) J] dx \right| < \infty, \qquad (5.6)$$

where $\{\bar{\mu}_k\}$ $(\operatorname{Im} \mu_k > 0)$ and $\{\lambda_k\}$ $(\operatorname{Im} \lambda_k > 0)$ form the set of nonreal eigenvalues of the operator A.

In the case of dissipative operators, inequality (5.6) can be rewritten as follows:

$$\sum_k \operatorname{Im} \lambda_k + \int_0^l \alpha^2(x) dx < \infty.$$

Remark. The basic results of this section were obtained by the author in [1 – 3]. In the case of bounded nonself-adjoint operators, a criterion of completeness was obtained by Livsic [3] in the following form:

$$\sum_k \operatorname{Im} \lambda_k = \operatorname{tr} \frac{A - A^*}{2i}.$$

Polyatskii [1] suggested a criterion of completeness of a system of eigenvectors and adjoint vectors of a quasiunitary contraction T of rank r in the following form:

$$|\det \tau| = \prod_k |\lambda_k|;$$

here, λ_k are eigenvalues of the operator T, $\tau = \| (Te_k, e_j') \|_{k,\,j=1}^r$, $\{e_k\}_1^r$ and $\{e_k'\}_1^r$ are orthonormal bases in the subspaces $\mathfrak{D}_T = (I - T^*T) \mathcal{H}$ and $\mathfrak{D}_T' = (I - TT^*) \mathcal{H}$, respectively.

6. Models of Linear Operators (A Brief Survey)

Operators from a class $\mathcal{M}$ are called *model operators* for operators from a given class $\mathcal{K}$ if each operator A from $\mathcal{K}$ is unitary equivalent (or similar) to some operator S from $\mathcal{M}$. It is natural to focus our attention mainly on the situations where the operators from $\mathcal{M}$ are simpler than the corresponding operators from K.

In the case of normal (e.g., self-adjoint or unitary) operators, their spectral decompositions can be regarded as model operators.

For linear operators acting in finite-dimensional spaces, the matrices of model operators are always triangular or have normal Jordan form.

Note that models of various classes of nonunitary or nonself-adjoint linear operators were constructed by different methods in last decades. Sometimes, it is possible to construct different models of the same class of operators adapted to the solution of various particular problems.

The fact that a simple operator T is unitary equivalent to the simple part of its triangular model $\vec{T}$ (or to the operator $\vec{T}$ if the latter is simple) is an important property of model operators. Note that a nonunitary (nonself-adjoint) operator T is called simple if it does not induce a unitary (self-adjoint) operator in any nontrivial invariant subspace.

The investigation of the characteristics of the original operator T can be significantly simplified if one knows the corresponding characteristics of the model operator $\vec{T}$ (its spectrum, invariant subspaces, etc.).

6.1. Friedrichs Model

One of the first models of linear operators was studied by Friedrichs [1]. In this work, Friedrichs suggested to consider an operator

$$H_\varepsilon = H_0 + \varepsilon V$$

as a model in the theory of perturbations of continuous spectrum. Here, H_0 and V are defined in the space $L_2(-1, 1)$ by the equalities

$$(H_0 f)(x) = xf(x), \qquad (Vf)(x) = \int_{-1}^{1} v(x, y) f(y) dy.$$

The self-adjoint operator H_0 has absolutely continuous spectrum located on the interval $[-1, 1]$. Friedrichs showed that the operators H_0 and H_ε are unitary equivalent ($H_\varepsilon = UH_0 U^*$) for sufficiently small ε under certain restrictions imposed on the kernel $v(x, y)$. The unitary operator U is defined explicitly in terms of the solution of the integral equation of the second kind.

Later, Friedrichs [2] proposed a more general model. Unlike the first model where the investigation was carried out for the interval $[-1, 1]$, in this work, the corresponding interval l on the real axis is not necessarily finite, f are vector functions with values in a Hilbert space $\mathcal{H}$, and the kernel $v(x, y)$ is a bounded operator in $\mathcal{H}$.

In this paper, Friedrichs established the existence of strong limits (of wave operators)

$$U^{(\pm)} = \lim_{t \to \pm\infty} [\exp{(iH_e t)}\exp{(-iH_0 t)}]$$

and proved the fact that the scattering operator (the S-matrix) $S = U^{(+)*} U^{(-)}$ is unitary and commutes with H_0. In more details, the Friedrichs model and its further generalization are considered, e.g., in Faddeev [1].

6.2. Livsic Method

6.2.1. *Bounded Nonself-Adjoint Operators.* The first general model for a broad class of bounded nonself-adjoint operators was constructed by Livsic [3] (see also Brodskii and Livsic [1]).

Below, we present main ideas and results of this work. We say that a bounded operator A acting in $\mathcal{H}$ belongs to a class (Ω) if $A - A^*$ is a finite-dimensional operator. In this case, the operator $(A - A^*) i^{-1}$ can be represented in the form

$$\frac{A - A^*}{i} f = \sum_{\alpha, \beta=1}^{r} (f, g_\alpha) J_{\alpha\beta} g_\beta \quad (r < \infty),$$

where $\{g_k\}_1^r$ is a set of vectors, which are called channels, and $J = \| J_{ki} \|$ is a unitary Hermitian matrix. The matrix function of a complex variable λ

$$W_A(\lambda) = I - i\|((A - \lambda I)^{-1} g_\alpha, g_\beta)\| J \quad (\lambda \in \rho(A)) \tag{6.1}$$

is called the characteristic matrix of the operator A. The characteristic matrix function thus defined is J-nonexpanding in the lower plane, J-noncontracting in the upper plane, and J-unitary on the real axis. This matrix admits a multiplicative representation

$$W_A(\lambda) = \int_0^l \exp\left[\frac{i dE(t) J}{\lambda - \alpha(t)}\right] \prod_{k=1}^{N} \left(E + \frac{i}{\lambda - \mu_k} \Pi_k^* \Pi_k J\right), \tag{6.2}$$

where $N \le \infty$, $l < \infty$, $\alpha(t)$, and $E(t)$ are nondecreasing functions bounded on $[0, l]$, $\{\mu_k\}_1^N$ is the set of all nonreal points of the spectrum of the operator A, and Π_k are one-row matrices satisfying the condition $\Pi_k J \Pi_k^* = 2 \operatorname{Im} \mu_k$. Moreover, the number of copies of each eigenvalue μ_k in (6.2) is equal to its multiplicity. In addition,

$$E(t) = \int_0^t \Pi^*(x)\Pi(x)dt, \quad \operatorname{tr} \Pi^*(x)\Pi(x) \equiv 1$$

where $\Pi(x)$ is a rectangular or square matrix whose rows are linearly independent at every point of a certain set of positive measure.

The model $\bar{A}$ of the operator A is defined in the space $\bar{\mathcal{H}} = \bar{\mathcal{H}}_1 \oplus \bar{\mathcal{H}}_2$ by the equality

$$\bar{A} = \begin{pmatrix} \bar{A}_1 & \Gamma \\ 0 & \bar{A}_2 \end{pmatrix}, \tag{6.3}$$

where

$$\bar{\mathcal{H}}_1 = \begin{cases} \mathbb{C}_2^N & (N < \infty), \\ l_2 & (N = \infty) \end{cases}$$

and $\bar{\mathcal{H}}_2 = L_2(0, l; \mathbb{C}^r)$. The operators $\bar{A}_1$, $\bar{A}_2$, and Γ are defined by the following equalities:

$$(\bar{A}_1 f)_k = f_k \mu_k + i \sum_{s=k+1}^{N} f_s \Pi_s J \Pi_k^*,$$

$$(\bar{A}_2 f)(x) = f(x)\alpha(x) + i \int_x^l f(t)\Pi(t)J\Pi^*(x)dt,$$

$$(\Gamma f)_k = i \int_0^l f(t)\Pi(t)J \Pi_k^* dt.$$

The operator $\bar{A}$ defined by equality (6.3) is called the triangular model of the operator A and the operators $\bar{A}_1$ and $\bar{A}_2$ are called discrete and continuous part of the triangular model, respectively.

It was showed that the characteristic matrix function of operator A coincides with the characteristic matrix function of its model $\bar{A}$, i.e., $W_{\bar{A}}(\lambda) = W_A(\lambda)$. This means that the E-simple parts of the operators $\bar{A}$ and A are unitary equivalent. The E-simple part A_E (it is also called the kernel of the operator A) is defined by the equality $A_E = A|_{\mathcal{H}_E}$, where $E = \langle g_1, g_2, \dots, g_r \rangle$ and $\mathcal{H}_E$ is the closure of the linear span of all vectors of the form $A^n g_k$ $(k = \overline{1, r};\ n \in \mathbb{N} \cup \{0\})$.

In particular, if A is a simple dissipative operator from the class (Ω), then

$$\sum_{k=1}^{N} \text{Im}\, \mu_k \le \text{tr}\, \frac{A - A^*}{2i}. \tag{6.4}$$

The set $\{\mathcal{H}_{\mu_k}\}$ of all proper subspaces of the operator A is full in the space $\mathcal{H}$ if and only if this relation turns into the equality.

Note that Mukminov [1] obtained this result without using the method of characteristic functions.

Later, Livsic applied the results obtained to the investigation of some physical problems (the theory of scattering matrix, intermediate systems and their decay, the scattering of photons by free electrons, etc.; for more details, see Livsic [4]).

6.2.2. *Reduction of Nonself-Adjoint Operators to the Diagonal Form.* The results presented in the previous subsection can be regarded as an infinite-dimensional analogue of the Schur–Toeplitz theorem on the unitary equivalence of a (finite) square matrix and a triangular matrix. On the other hand, an arbitrary square matrix is similar (not necessarily unitarily similar) to a matrix in the Jordan normal form, which becomes diagonal under certain conditions. In this connection, it is interesting to find the simplest form of a linear operator A acting in an infinite-dimensional space. Spectral theory solves this problem in the case of self-adjoint operators. If A is a nonself-adjoint operator without continuous spectrum, then the problem of finding its simplest form is quite similar to the corresponding problem for finite-dimensional matrices (however, in the infinite-dimensional case, one meets certain specific difficulties connected with the fact that the reducing operator and its inverse may be unbounded).

In this connection, it is quite interesting to study the problem of reduction of nonself-adjoint operators with continuous spectra to the simplest form. For a broad class of operators, this problem was solved by Sahnovich [1, 3].

Here, we sketch the results of his second work.

Consider bounded dissipative operators whose imaginary components are nuclear. The class of such operators is denoted by (Ω^+). The characteristic matrix function $W_A(\lambda)$ of an operator A from (Ω^+) has the form

$$W_A(\lambda) = I - i \| ((A - \lambda I)^{-1} g_\alpha, g_\beta) \|,$$

where $\{g_\alpha\}$ is an orthogonal system of eigenvectors of the operator $(A - A^*) i^{-1}$ and

$$\frac{A - A^*}{i} g_\alpha = (g_\alpha, g_\alpha) g_\alpha \qquad (1 \leq \alpha \leq m, \quad m \leq \infty).$$

We say that the operator A from (Ω^+) has absolutely continuous spectrum if $\det W_A(\lambda)$ is the outer function, i.e., the following representation is true:

$$\det W_A(\lambda) = \exp\left[-i \int_a^b \frac{P^2(s)}{s - \lambda} ds \right] \qquad (\lambda \in [a, b])$$

The corresponding operator class is denoted by (Ω_1). As follows from the results of Livsic, the simple part of an operator A from (Ω_1) is unitary equivalent to the triangular model

$$(\bar{A}f)(x) = xf(x) + i \int_a^x f(t)\beta(t)dt\,\beta(x) \qquad (f \in L_2(a, b, \mathbb{C}_2^m)), \qquad (6.5)$$

where $\beta(x) \geq 0$ and

$$\int_a^b \beta^2(x)dx < \infty.$$

In this case,

$$W_{\bar{A}}(\lambda) = \int_a^b \exp\left[-i\,\frac{dE(t)}{t-\lambda}\right], \qquad E(t) = \int_a^t \beta^2(s)\,ds.$$

Thus, in order to reduce a simple operator A from the class (Ω_1) to the simplest form, it suffices to solve this problem for its model $\bar{A}$. Sahnovich showed that if the matrix $\beta^2(x)$ is measurable and

$$\text{Vrai max} \, \|\beta^2(x)\| = M < \infty \qquad (a \leq x \leq b),$$

then the simple part $\bar{A}_s$ of the operator $\bar{A}$ defined by equality (6.5) is linearly equivalent to the operator Q (i.e., $\bar{A}_s = BQB^{-1}$), where Q is the operator of multiplication by an independent variable (which, clearly, can be regarded as a continuous analogue of a diagonal matrix):

$$(Qf)(x) = xf(x) \qquad (f \in L_2(a, b, \mathbb{C}_2^m)).$$

In this case, it is also interesting to determine the explicit form of the transformation B. We have

$$(B\varphi)(x) = \frac{1}{\sqrt{2\pi}}\frac{d}{dx}\int_a^x \varphi(\sigma)\sqrt{G(\sigma)}\,U(x, \sigma)d\sigma\,\beta^{-1}(x),$$

where $U(x, \sigma)$ and $G(\sigma)$ are unitary and nonnegative matrices, respectively, defined by the following equalities:

$$U(x, \sigma) = s - \lim_{\varepsilon \to +0} W(\sigma - \varepsilon, \sigma) W(x, \sigma) W^{-1}(\sigma + \varepsilon, \sigma),$$

$$G(\sigma) = s - \lim_{\varepsilon \to +0} W(\sigma - \varepsilon, \sigma) [e^{\pi \beta^2(\sigma)} - e^{-\pi \beta^2(\sigma)}] W^{-1}(\sigma + \varepsilon, \sigma),$$

where, in turn,

$$W(x, \lambda) = \int_a^{\overset{x}{\frown}} \exp\left[-i \frac{dE(t)}{t - \lambda}\right].$$

Thus,

$$W(b, \lambda) = \chi_{\bar{A}}(\lambda) \ (= \chi_A(\lambda)).$$

By using these results, Sahnovich proved the existence of the limits

$$W_{\pm}(A^*, A) = s - \lim_{t \to \pm\infty} e^{iA^* t} e^{-iAt}.$$

In particular, these limits exist if $\beta^2(t)$ satisfies the Hölder condition

$$\|\beta^2(t_2) - \beta^2(t_1)\| \leq L |t_2 - t_1|^{\alpha} \qquad (0 < \alpha \leq 1)$$

In this case, the operators B and B^{-1} are bounded.

By analogy with the self-adjoint case, the operators $W_{\pm}(A^*, A)$ are called the wave operators and

$$S = W_-(A^*, A) W_+(A^*, A)$$

is called the scattering operator.

The similarity problem was studied for various classes of operators by many authors and by different methods. In particular, this problem was investigated by Sz.-Nagy and Foias [4], Gokhberg and Krein [2], Sahnovich [2], Naboko [5], Malamud [1], and others. As a rule, these authors use only the corresponding characteristic function or the resolvent. The transformation B is not determined but its existence is established.

6.2.3. *One-Cellular Operators.* An operator A is called *one-cellular* if, in any pair of its invariant subspaces, one is a subspace of the other. In the case of finite-dimensional subspaces, this is equivalent to the existence of one Jordan cell.

The concept of one-cellular operators was introduced by Brodskii. Operators of this type were studied in detail by Brodskii and Kiselevskii (e.g., see the monograph of Brodskii [1] and the papers of Kiselevskii [1, 2]).

The principal results obtained in this field can be formulated as follows: Let (Ω_0^+) be the class of simple dissipative operators A such that $\sigma(A) = \{0\}$. Then an operator $A \in (\Omega_0^+)$ is one-cellular if

$$\lim_{z \to 0} \sup \{|z| \ln \|(A + iI)^{-1}\|\} = \operatorname{tr} \frac{A - A^*}{i}. \tag{6.6}$$

Moreover, it was shown by Brodskii and Kiselevskii [1] that equality (6.6) is also a necessary condition for the operator A to be one-cellular.

For example, the integral operator

$$(Af)(x) = \int_0^x f(t)\,dt$$

is one-cellular.

For more detailed information about one-cellular operators, see the monographs of Gohberg and Krein [1] and Nikolsky [1].

6.2.4. *Bounded Nonunitary Operators.* Polyatskii [1] constructed a model for invertible operators T such that the operator $I - T^* T$ is finite-dimensional. The char- acteristic function of the operator T is given by equality (9.2) (Chapter 2) and, according to the Potapov theorem, can be represented in the form (7.4) (Chapter 2). On the basis of this representation, the model operator $\vec{T}$ is constructed in the space $\mathcal{H} = \mathcal{H}_1 \oplus \mathcal{H}_2$, where $\mathcal{H}_1$ and $\mathcal{H}_2$ are spaces of vector functions of discrete and continuous arguments, respectively, with values in $\mathbb{C}_2^r$. This operator has the form

$$\vec{T} = \begin{pmatrix} \vec{T}_1 & \Gamma \\ 0 & \vec{T}_2 \end{pmatrix},$$

where

$$(\vec{T}_1 f)(k) = f(k)\, t(k)\, e^{i\varphi(k)} - \sum_{j=k+1}^N f(j) P(j) U^*(j) \pi^*(j-1) \pi^{-*}(k) U^{-*}(k) JP(k)\, e^{i\varphi(k)},$$

$N \le \infty$, $\{t(k)\}$ is a sequence of real diagonal $r \times r$-matrices that have a single eigen- value not equal to one, $P^2(k) = |E - t^2(k)|$, $\{\varphi(k)\}$ is a scalar non-decreasing sequence (function),

$$\pi(m) \;=\; \prod_{j=1}^{m} U(j)t(j)U^{-1}(j),$$

$$(\bar{T}_2 f)(x) = f(x)e^{i\theta(x)} - 2\int_x^l f(t)P(t)\pi^*(t)\pi^{-*}(x)JP(x)e^{i\theta(x)}\,dt, \quad l < \infty,$$

$$P^2(x) = E'(x), \quad \pi(x) = \pi(N)\int_0^{\hat{x}} e^{-P^2(t)J}\,dt.$$

The operator $\Gamma: \vec{\mathcal{H}}_2 \to \vec{\mathcal{H}}_1$ is defined as follows:

$$(\Gamma f)(k) = -\sqrt{2}\int_0^l f(t)P(t)\pi^*(t)\pi^{-*}(k)U^{-*}(k)JP(k)\widetilde{}e^{i\phi(k)}\,dt.$$

It was established that the discrete spectrum of the operator T consists of the eigenvalues of the matrices $t(k)$ that are not equal to one, and the continuous spectrum coincides with the range of the function $e^{i\theta(x)}$ $(0 \le x \le e)$.

In the case of simple contractions, the criterion of completeness of a system of finite-dimensional invariant subspaces is also formulated.

6.3. Sz.-Nagy–Foias Method

6.3.1. *Functional Model for Contractions.* Assume that T is a contraction in the Hilbert space $\mathcal{H}$, and $\mathcal{H}_0$ is the maximal invariant subspace in which the operator T induces a unitary operator. By virtue of the theorem on canonical representation (Appendix 1), the subspace $\mathcal{H}_0$ reduces the operator T. In the case where $\mathcal{H}_0 = \{0\}$, the contraction T is called completely nonunitary.

If T is a contraction in $\mathcal{H}$, then the operators

$$D_T = (I - T^*T)^{1/2}, \quad D_{T^*} = (I - TT^*)^{1/2}$$

and the subspaces $\mathcal{D}_T = \overline{D_T \mathcal{H}}$, $\mathcal{D}_{T^*} = \overline{D_{T^*} \mathcal{H}}$ are called *defect operators* and *defect spaces*, respectively, for the operator T.

Recall that the characteristic function $\theta_T(\lambda)$ of the contraction T is defined by the following equality (see Chapter 2, Subsection 9.1):

$$\theta_T(\lambda) = [-T + \lambda D_{T^*}(I - \lambda T^*)^{-1}D_T]\Big|_{D_T}.$$

In the case of completely nonunitary operators, this characteristic function is pure in the sense that

$$\| \theta_T(0)f \| < \|f\| \quad (\forall f \in D_T, \ f \neq 0).$$

Further, let E be a separable Hilbert space and let $L_2(E)$ be the space of measurable square-summable vector functions defined on the unit circle and taking values from E. We also denote the Hardy subspace of the space $L_2(E)$ by $H^2(E)$ and set

$$\Delta_T(t) = [I - \theta_T^*(e^{it})\theta_T(e^{it})]^{1/2}.$$

Below, we present the Sz.-Nagy–Foias basic theorem about a model operator.

Theorem. *An arbitrary completely nonunitary contraction T acting in the separable Hilbert space $\mathcal{H}$ is unitary equivalent to the operator $\bar{T}$ acting in the functional space*

$$\vec{\mathcal{H}} = \left[H^2(\mathcal{D}_{T^*}) \oplus \overline{\Delta_T L_2(\mathcal{D}_T)} \right] \ominus \{\theta_T U \oplus \Delta_T U \,|\, U \in H^2(\mathcal{D}_T)\}$$

and is defined on the vectors $u \oplus v \in \vec{\mathcal{H}}$ by the following equality:

$$\vec{T}^*(u \oplus v) = e^{-it}[u(e^{it}) - u(0)] \oplus e^{-it}v(t).$$

In the case where $T \in C_{\cdot 0}$ (and only in this case), the model of the operator T can be simplified and rewritten as follows [8]:

$$\vec{\mathcal{H}} = H^2(\mathcal{D}_{T^*}) \ominus \theta_T H^2(\mathcal{D}_T), \quad (\vec{T}^* u)(\lambda) = \lambda^{-1}[u(\lambda) - u(0)].$$

The model $\bar{T}$ of the operator T thus constructed is a completely nonunitary contraction and its characteristic function $\theta_{\bar{T}}(\lambda)$ coincides with the characteristic function $\theta_T(\lambda)$. This implies the unitary equivalence of the operators T and $\bar{T}$.

In the case where $T \in C_{\cdot 0}$, a model for the operator T was obtained and used in Rota [1] and Helson [1].

For more detailed information about this model, see the monograph of Sz.-Nagy and Foias [4] and the paper of Sz.-Nagy [1]. We only note that, in the latter work, the model operator is constructed under the assumption that $\theta(\lambda)$ is a purely contracting analytic function, which, in particular, may coincide with the characteristic function $\theta_T(\lambda)$.

---

[8] We note that the class $C_{\cdot 0}$ consists of contractions T for which $T^{*n}h \to 0$ ($\forall h \in \mathcal{H}$).

6.3.2. *Operators with Bounded Characteristic Function.* McEnnis [3] generalized some results of Sz.-Nagy and Foias (presented in the previous subsection) for the case of bounded linear operators. According to McEnnis [3], the characteristic function of the operator $T \in \mathcal{B}[\mathcal{H}]$ is defined by the equality

$$\theta_T(\lambda) = [-TJ_T + \lambda J_{T^*} Q_{T^*} (I - \lambda T^*)^{-1} J_T Q_T] \Big|_{\mathcal{D}_T}, \qquad (6.7)$$

where

$$Q_T = |I - T^* T|^{1/2}, \quad J_T = \text{sign}(I - T^* T),$$

$$Q_{T^*} = |I - TT^*|^{1/2}, \quad J_{T^*} = \text{sign}(I - TT^*)$$

and $\mathcal{D}_T = J_T \mathcal{H}$. In the case where

$$\sup_{\lambda \in D} \| \theta_T(\lambda) \| < \infty$$

(D is an open unit circle), McEnnis constructed a functional model, which has the same form as the Sz.-Nagy–Foias model. The only difference is that, in this case,

$$\Delta_T(t) = [J_T(I - \theta_T^*(e^{it})\theta_T(e^{it}))]^{1/2}.$$

In conclusion, we note that, by virtue of (6.7) and the equalities

$$J_T Q_T = Q_T J_T, \quad TJ_T|_{\mathcal{D}_T} = J_{T^*} T|_{\mathcal{D}_T}$$

(for the proof of the latter, see, e.g., Kuzhel [20]), we have

$$J_{T^*} \theta_T(\lambda) J_T = [-TJ_T + \lambda Q_{T^*} (I - \lambda T^*)^{-1} Q_T] \Big|_{\mathcal{D}_T}.$$

Thus, the operator function $J_{T^*} \theta_T(\lambda) J_T$ differs from the characteristic function $\theta_T(\lambda)$ defined by equality (9.5) (Chapter 2) only by sign.

6.3.3. *Dissipative Operators.* Shvartsman [1] used the ideas and methods of Sz.-Nagy and Foias to construct and investigate the model of an arbitrary bounded dissipative operator. After this, Karpenko and Tikhonov [1] generalized these results to the case of unbounded dissipative operators with nonempty set of regular points.

Below, we outline the principal methods and results of these works.

Assume that A is a dissipative operator acting in the Hilbert space $\mathcal{H}$ and $\rho(A) \neq \varnothing$.

Also let $\alpha \in \rho(A)$ be a fixed point in the lower half plane. Then in equality (8.2) (Chapter 2), one can set $J = I$ and write the characteristic function $\theta_A(\lambda)$ of the operator A as

$$\theta_A(\lambda)\,\theta_A^*(\alpha) = I + i(\bar\alpha - \lambda)\,Q^* T_{\alpha\lambda}\,Q \qquad (\lambda \in \rho(A)),$$

where $T_{\alpha\lambda} = (A - \alpha I)(A - \lambda I)^{-1}$, the operator Q is determined from the condition $B_\alpha = QQ^*$, and $B_\alpha = iR_\alpha - iR_\alpha^* + 2\,\mathrm{Im}\,\alpha\,R_\alpha^*\,R_\alpha$. Note that the operator Q is defined not uniquely and acts from a Hilbert space $\mathfrak{G}$ into $\mathcal{H}$. In particular, we can take $Q = \sqrt{B_\alpha}$. Then $\mathfrak{G}$ is a subspace of the space $\mathcal{H}$ and contains the defect space $\mathfrak{N}_\alpha$ of the operator A (or coincides with $\mathfrak{N}_\alpha$).

In the case under consideration, the first equality in (8.14) (Chapter 2) can be rewritten in the form

$$\theta_A^*(\lambda)\theta_A(\lambda) \le I \qquad (\mathrm{Im}\,\lambda \le 0,\ \lambda \in \rho(A)).$$

Denote by $\Phi_{\mathfrak{G}}^+$ the class of operator functions $\chi(\cdot)$ holomorphic in the lower half plane taking values in $\mathfrak{G}$ and satisfying the following condition:

$$\chi^*(\lambda)\chi(\lambda) \le I \qquad (\mathrm{Im}\,\lambda \le 0).$$

Therefore, $\theta_A(\cdot) \in \Phi_{\mathfrak{G}}^+$.

Further, we assume that the Hilbert space $\mathfrak{G}$ is separable and consider the Hilbert space $L_2(\mathfrak{G})$ of vector functions φ defined and measurable on $\mathbb{R}$ with values in $\mathfrak{G}$ which satisfy the following condition:

$$\|\varphi\|^2 = \int_{-\infty}^{\infty} \|\varphi(t)\|_{\mathfrak{G}}^2\, dt < \infty. \tag{6.8}$$

In this case, the scalar product in $L_2(\mathfrak{G})$ is defined as

$$(f, \varphi) = \int_{-\infty}^{\infty} (f(t), \varphi(t))_{\mathfrak{G}}\, dt. \tag{6.9}$$

(In (6.8) and (6.9), $\|\cdot\|_{\mathfrak{G}}$ and $(\cdot,\cdot)_{\mathfrak{G}}$ denote, respectively, the norm and scalar product in $\mathfrak{G}$.)

Denote by $H^2(\mathfrak{G})$ the class of functions φ from $L_2(\mathfrak{G})$ analytically extendable to the lower half plane and satisfying the following condition:

$$\sup_{\tau<0} \int_{-\infty}^{\infty} \| \varphi(t+i\tau) \|_{\mathfrak{G}}^2 \, dt < \infty.$$

Further, let $\Delta(t) = (I - \theta_A^*(t)\theta_A(t))^{1/2}$, where $\theta_A(t)$ is a nontangent limit of the characteristic function $\theta_A(\lambda)$ (in the sense of strong convergence).

We set

$$\vec{\mathcal{H}} = \left[\mathcal{H}^2(\mathfrak{G}) \oplus \overline{\Delta_A L_2(\mathfrak{G})} \right] \ominus \{ \theta_A u \oplus \Delta_A u \, | \, u \in \mathcal{H}^2(\mathfrak{G}) \}$$

and define in $\vec{\mathcal{H}}$ an operator $\vec{A}$ as follows:

$$\mathfrak{D}_{\vec{A}} = \left\{ \frac{u(t)-u(\alpha)}{t-\alpha} \oplus \frac{v(t)}{t-\alpha} \, \middle| \, u(t) \oplus v(t) \in \vec{\mathcal{H}} \right\},$$

$$\vec{A}[\varphi(t) \oplus \psi(t)] = \tilde{\varphi}(t) \oplus \tilde{\psi}(t),$$

where

$$\varphi(t) = \frac{u(t)-u(\alpha)}{t-\alpha}, \quad \tilde{\varphi}(t) = \frac{tu(t)-\alpha u(\alpha)}{t-\alpha}, \quad \psi(t) = \frac{v(t)}{t-\alpha}, \quad \tilde{\psi}(t) = t\psi(t).$$

It is easy to see that the operator $\vec{A}$ thus defined is the maximal simple dissipative operator and its characteristic function $\theta_{\vec{A}}(\lambda)$ is related to the characteristic function $\theta_A(\lambda)$ of the operator A by equality $\theta_{\vec{A}}(\lambda) = \theta_A(\lambda) U$, where U is a nonunitary operator. Consequently, the simple part of the operator A is unitary equivalent to the operator $\vec{A}$.

6.4. Unified Model

In this section, we present basic ideas of the unified (coordinateless) method for the construction of a functional model of contractions. This method was anticipated by Sz.-Nagy and Foias [4], Douglas [1], Pavlov [1], suggested by Vasyunin [1] and Makarov and Vasyunin [1], and developed by Nikolsky and Vasyunin [1]. A detailed exposition of this approach to the construction of the functional model can be found in the survey by Nikolsky and Khrushchev [1].

Assume that T is a contraction in the Hilbert space $\mathcal{H}_0$ and U is its minimal unitary dilation. According to the Sarason theorem (see Section 5 in Chapter 4), the space $\mathcal{H}$, in which the operator U acts, can be represented in the form

$$\mathcal{H} = \mathcal{G}_* \oplus \mathcal{H}_0 \oplus \mathcal{G}, \tag{6.10}$$

where $U\mathcal{G} \subset \mathcal{G}$, $U^{*}\mathcal{G}_{*} \subset \mathcal{G}_{*}$, and the spaces

$$L = \mathcal{G} \ominus U\mathcal{G}, \quad L_{*} = \mathcal{G}_{*} \ominus U^{*}\mathcal{G}_{*}$$

are wandering subspaces of the operator U in the sense that

$$U^{n} L \perp U^{k} L, \quad U^{n} L_{*} \perp U^{k} L_{*} \quad (\forall \{n, k\} \in \mathbb{Z}, \quad n \neq k).$$

Moreover, the following equalities hold:

$$\mathcal{G} = \sum_{n \geq 0} \oplus U^{n} L, \quad \mathcal{G}_{*} = \sum_{n \geq 0} \oplus U^{*n} L_{*}.$$

Let E and E_{*} be fixed spaces of the same dimensionality as the spaces L and L_{*}, respectively. Consider unitary mappings

$$V: E \to L, \quad V_{*}: E_{*} \to L_{*}$$

and a mapping

$$\Pi = (\pi, \pi_{*}): L_{2}(E) \oplus L_{2}(E_{*}) \to \mathcal{H},$$

where π and π_{*} are defined by the following equalities:

$$\pi\left(\sum_{k \in \mathbb{Z}} z^{k} e_{k}\right) = \sum_{k \in \mathbb{Z}} U^{k} V e_{k} \quad (e_{k} \in E),$$

$$\pi_{*}\left(\sum_{k \in \mathbb{Z}} z^{k} e_{k}^{*}\right) = \sum_{k \in \mathbb{Z}} U^{k+1} V_{*} e_{k}^{*} \quad (e_{k}^{*} \in E_{*}).$$

Taking the properties of the spaces L and L_{*} into account, one can show that the mappings π and π_{*} are isometric:

$$\pi^{*}\pi = I_{L_{2}(E)}, \quad \pi_{*}^{*}\pi_{*} = I_{L_{2}(E_{*})}.$$

Furthermore, $\pi Z = U\pi$, $\pi_{*} Z = U\pi_{*}$, and, hence,

$$\Pi Z = U\Pi \qquad (6.11)$$

(Z denotes both the identical mapping of a part of the plane into itself $(Z(f) = \xi)$ and the multiplication operator $f \to Zf$).

One can easily check that

$$\pi \, \mathcal{H}^2 (E) \perp \pi_* \, \mathcal{H}^2_- (E_*), \tag{6.12}$$

where

$$\mathcal{H}^2 (E) = \left\{ \sum_{n \geq 0} Z^n e_n \Big| e_n \in E, \quad \sum_{n \geq 0} \| e_n \|^2 < \infty \right\}$$

$$= \{ f \in L_2(E) | \hat{f}(k) = 0, \quad k < 0 \}$$

($\hat{f}(k)$ is the kth Fourier coefficient of the function f) and

$$\mathcal{H}^2_- (E) := L_2(E) \ominus \mathcal{H}^2 (E).$$

In addition, we have

$$\pi_*^* \pi : \; L_2(E) \to L_2(E_*), \qquad \pi_*^* \pi \, \mathcal{H}^2 (E) \subset \mathcal{H}^2 (E_*).$$

This allows us to conclude that there is a function $\theta \in \mathcal{H}^\infty (E \to E_*)$ such that

$$\pi_*^* \pi = \theta \quad (\| \theta \|_\infty = \| \pi_*^* \pi \| \leq 1).$$

By using this procedure, every nonunitary contraction T can be associated with a contracting operator function $\theta = \theta_T$. In this case, the correspondence $T \to \theta_T$ is an isomorphism to within a unitary equivalence. The operator function θ_T defined in this way is called the characteristic function of the contraction T and, under a proper choice of the spaces $E (= \mathcal{D}_T)$ and $E_* (= \mathcal{D}_{T^*})$ and mappings V and V_*, can be represented in the standard form

$$\theta_T (\lambda) = [-T + \lambda D_{T^*} (I - \lambda T^*)^{-1} D_T] \big|_{\mathcal{D}_T}.$$

The model M_θ of the operator T is defined by the equality

$$M_\theta = P_\theta U \big|_{\mathcal{K}_T},$$

where

$$\mathcal{K}_\theta = \mathcal{H} \ominus [\pi_* \mathcal{H}^2_- (E_*) \oplus \pi \mathcal{H}^2 (E)],$$

the projector $P_\theta = P_{\mathcal{K}_\theta}$ has the form

$$P_\theta = I - \pi P_+ \pi^* - \pi_* P_- \pi_*^*$$

and $P_\pm$ are the orthoprojectors in L^2 to $\mathcal{H}^2$ and $\mathcal{H}_-^2$, respectively.

In conclusion, we note that different realizations (transcriptions) of the universal model M_θ depending on the choice of the corresponding "parameters" (i.e., spaces E and E_*, mappings V and V_*, and a spectral representation of the unitary dilation U) were studied by Nikolsky and Vasyunin [1] and Nikolsky and Khrushchev [1]. As a result, we arrive at the models of Sz.-Nagy–Foias (Subsection 6.3.1), de Branges–Rovnyak [2], and Pavlov [1]. In these works, one can also find the characteristic of the spectrum of the operator M_θ and the description of its invariant subspaces.

An important role in various applications is played by the "commutant-lifting theorem". It states that each operator X in the commutant

$$\{M_\theta\}' := \{X \mid XM_\theta = M_\theta X\}$$

can be represented in the form

$$X = P_\theta Y \big|_{\mathcal{K}_\theta},$$

where Y is an operator in the space of isometric dilation $\mathcal{H}^2(E_*) \oplus \Delta L_2(E)$, which commutes with the shift Z and leaves the subspace

$$\{\theta f \oplus \Delta f \mid f \in \mathcal{H}^2(E)\}$$

invariant.

6.5. Clark Method

6.5.1. *Bounded Nonunitary Operators.* An original method for constructing model operators was proposed by Clark [2]. It can be described as follows: Assume that $T \in \mathcal{B}[\mathcal{H}]$, the operators J_T, J_{T^*} and Q_T, Q_{T^*} are defined as in Subsection 6.3.2, and the characteristic function θ_T of an operator T is defined by equality (9.5) (Chapter 2)

$$\theta_T(z) = TJ_T - zQ_{T^*}(I - zT^*)^{-1}Q_T.$$

One can easily prove that (see Kuzhel [17])

$$J_{T^*} - \theta_T(z) J_T \theta_T^*(\lambda) = (1 - z\bar{\lambda}) T_z T_\lambda^*, \tag{6.13}$$

where $T_z = Q_{T^*}(I - zT^*)^{-1}$. Relation (6.13) implies that the operator function

$$b(\lambda, z) = (1 - z\bar{\lambda})^{-1}[J_{T^*} - \theta_T(z) J_T \theta_T^*(\lambda)]$$

is positive definite in a certain vicinity D of the origin $(0 \in D \subset \{z \mid |z| < 1\})$, in the sense that, for any points $z_1, z_2, \dots, z_n$ from the set D and any nonzero vectors from $\mathcal{H}_2 = \mathcal{D}_{T^*} (= D_{T^*}\mathcal{H})$, we have

$$\sum_{i,\, j=1}^{n} (b(z_i, z_j)x_i, x_j) > 0.$$

Further, let $\tilde{D} = \{\bar{z}^{-1} \mid z \in D\}$ and let $\mathcal{H}_0$ be a set of elements of the form $K_z f$, where $z \in D \cup \tilde{D}$ and $f \in \mathcal{H}_2$. Denote by $\mathcal{H}_1$ the set of all finite linear combinations of elements from $\mathcal{H}_0$ and define a scalar product in $\mathcal{H}_1$ by the equality

$$(K_z f, K_\lambda \varphi) = (b(z, \lambda)f, \varphi). \tag{6.14}$$

By completing $\mathcal{H}_1$ with respect to the corresponding norm, we obtain the Hilbert space $\bar{\mathcal{H}}$.

On the lineal $\mathcal{H}_1$, we define an operator S by the equality

$$SK_z f = \bar{z} K_z f - h_0 J_T \theta_T^*(z) f, \tag{6.15}$$

where

$$h_0 \varphi = \lim_{R_0 \leq R \to \infty} RK_R J_{T^*} \theta_T(0) f,$$

where, in turn, R_0 is a sufficiently large number such that $[R_0, \infty) \subset \tilde{D}$.

Then we prove that the operator S is bounded and show that the characteristic function θ_{S^*} of the operator S^* is equal to the characteristic function θ_T. Moreover, the operator S^* is defined by the equality

$$S^* h_0 f = -K_0 J_* \theta_T(0) f.$$

Hence, it can be regarded as a model of the given operator T.

It is worth noting that Clark in [2] used not the characteristic function $\theta_T(z)$ but an analytic operator function $B(z)$ satisfying certain conditions. This approach enabled him to solve the inverse problem for the operator class under consideration, namely, the problem of reconstruction of a bounded nonunitary operator for a given operator function.

6.5.2. *Unbounded Operators.* Karpenko [2] applied the Clark method to the construction of the model of an arbitrary linear operator A with nonempty set of regular points. She used the following procedure:

Assume that A is a closed linear operator acting in the Hilbert space $\mathcal{H}$ and $\rho(A) \neq \varnothing$. The operator A is not supposed to be bounded or densely defined. Let α (Im $\alpha \neq 0$) be a fixed point from $\rho(A)$. As stated above, the operator

$$B_\alpha = iR_\alpha - iR_\alpha^* + 2\operatorname{Im}\alpha\, R_\alpha^* R_\alpha \quad (R_\alpha = (A - \alpha I)^{-1})$$

can be represented in the form $B_\alpha = QJQ^*$, where Q is a linear operator acting from the auxiliary space $\mathfrak{S}$ into $\mathcal{H}$ and J is a unitary self-adjoint operator acting in $\mathfrak{S}$. Without loss of generality, we can assume that $\mathfrak{S} = \mathfrak{N}_\alpha$, where $\mathfrak{N}_\alpha$ is the defect space of the Hermitian part H_0 of the operator A and the operator Q is self-adjoint in $\mathfrak{S}$. In this case, by virtue of equality (8.2) (Chapter 2), the characteristic function $\theta_A(\lambda)$ of the operator A is defined by the functional equation

$$\theta_A(\lambda)J\theta_A^*(\alpha) = J + i(\bar{\alpha} - \lambda)QT_{\alpha\lambda}Q \quad (\lambda \in \rho(A)),$$

where $T_{\alpha\lambda} = (A - \alpha I)(A - \lambda I)^{-1}$. Furthermore, as shown in Subsection 8.3 (Chapter 2), we have

$$Q[\theta_A^*(\mu)J\theta_A(\lambda) - J]Q = i(\bar{\mu} - \lambda)Q\,T_{\bar{\alpha}\mu}\,T_{\bar{\alpha}\lambda}\,Q. \tag{6.16}$$

Since $\mathfrak{S} = \mathfrak{N}_\alpha$, we conclude that $\operatorname{Ker} Q = \{0\}$. Therefore, taking equality (6.16) into account, we can show that the operator function

$$b(\lambda, \mu) = \frac{i}{\bar{\mu} - \lambda}[J - \theta_A^*(\mu)J\theta_A(\lambda)] \tag{6.17}$$

is positive definite in a sufficiently small neighbourhood D_α of the point α in the subspace $\mathfrak{S}$.

As in the previous subsection, we consider the set $\mathcal{H}_0$ of elements of the form $K_\lambda f$, each of which is defined by the pair $\langle \lambda, f \rangle$, where $\lambda \in D = D_\alpha \cup D_{\bar{\alpha}}$, $f \in \mathfrak{S}$, and $D_{\bar{\alpha}}$ is a sufficiently small neighbourhood of the point $\bar{\alpha}$. The lineal $\mathcal{H}_1$ and the Hilbert space $\vec{\mathcal{H}}$ are defined just as in the previous subsection (taking expression (6.17) for $b(\lambda, \mu)$ into account). The operator $\bar{A}$ is defined in the space $\vec{\mathcal{H}}$ (on the elements of $\mathcal{H}_1$) by the equality

$$\bar{A} m_\lambda f = \lambda K_\lambda f - \alpha K_\alpha J \theta_A^*(\bar{\alpha}) J \theta_A(\lambda) f, \tag{6.18}$$

where

$$m_\lambda f = K_\lambda f - K_\alpha J \theta_A^*(\bar{\alpha}) J \theta_A(\lambda) f \quad (\lambda \in D, \ f \in \mathfrak{G}). \tag{6.19}$$

The definition of the operator $\bar{A}$ is correct. This operator is simple and its resolvent (at the point $\bar{\alpha}$) can be represented in the form

$$\bar{R}_\alpha K_\lambda f = \frac{1}{\lambda - \alpha} [K_\lambda f - K_\alpha J \theta_A^*(\bar{\alpha}) J \theta_A(\lambda)].$$

One can also show that

$$\bar{B}_\alpha K_\lambda f = \frac{1}{\lambda - \alpha} K_{\bar{\alpha}} J [J - \theta_A^*(\bar{\alpha}) J \theta_A(\lambda)] f$$

and the operators $\bar{Q}$ and $\bar{J}$ in the representation $\bar{B}_\alpha = \bar{Q} J \bar{Q}$ have the form (for $\mathrm{Im}\, \alpha < 0$)

$$\bar{Q} K_\lambda f = \frac{i \sqrt{-2\,\mathrm{Im}\,\alpha}}{\lambda - \alpha} K_{\bar{\alpha}} J \Phi^{-1} [J - \theta_A^*(\bar{\alpha}) J \theta_A(\lambda)] f,$$

$$\bar{J} K_{\bar{\alpha}} f = K_{\bar{\alpha}} \Phi^{-1} J \Phi f \quad (\bar{J} : K_{\bar{\alpha}} \mathfrak{G} \to K_{\bar{\alpha}} \mathfrak{G}),$$

where $\Phi = [J - \theta_A^*(\bar{\alpha}) J \theta_A(\bar{\alpha})]^{1/2}$ and the operator $\bar{J}$ is unitary and self-adjoint.

These relations enable us to find the characteristic function $\theta_{\bar{A}}$ of the operator $\bar{A}$, namely,

$$\theta_{\bar{A}}(\lambda) = L^* \theta_A(\lambda) L,$$

where the unitary operator L acting from $K_{\bar{\alpha}} \mathfrak{G}$ into $\mathfrak{G}$ is given by the equality

$$L K_{\bar{\alpha}} f = \frac{1}{\sqrt{-2\,\mathrm{Im}\,\alpha}} J \Phi f.$$

Thus, the simple part of the operator A is unitary equivalent to the model operator

$\bar{A}$ defined by equalities (6.18) and (6.19).

Remarks. Here, we do not consider the models of de Branges–Rovnyak [2] and Pavlov [1] It has already been noted above that these models can be obtained by proper realizations of the unified model.

Ball [2] used the canonical model of de Branges–Rovnyak to study the unitary parts of contractions and perturbations of unitary operators.

Kriete [1] established similar results for the self-adjoint parts of dissipative operators.

Tsekanovskii [1] constructed a model, which coincides in form with the continuous component of the Livsic model (Subsection 6.2.1) for a class (rather narrow) of unbounded accretive operators in the space $L_2((0, \infty); E)$.

Solomyak [1] constructed a universal model of dissipative operators. This model was constructed by using the Cayley transformation of the universal model of contractions.

4. Dilations of Linear Operators

1. A Brief Survey

The study of nonunitary and nonself-adjoint operators is largely based on the methods of characteristic functions and the method of dilations (i.e., the method for finding an "extension" of a given operator to a unitary or self-adjoint operator). The theory of unitary dilations of contractions was thoroughly developed in a great number of works written by different authors (for detailed information, see, e.g., the monographs by Sz.-Nagy and Foias [1] and Nikolsky [1] and the survey by Nikolsky and Khrushchev [1]).

Later Davis [1] used the method of Sz.-Nagy–Foias to construct and investigate a J-unitary dilation of an arbitrary closed and densely defined operator T. In the case of bounded operators, the J-unitary dilation was constructed by different methods by Davis and Foias [1], Sahnovich [4], McEnnis [2], and Kuzhel [26].

For dissipative operators, the results outlined above lead to J-unitary dilations, which is inconvenient. At the same time, by very simple argument, one can show that self-adjoint dilations of dissipative operators exist. To show this, it suffices to apply the Cayley transformation.

Thus, in the case of dissipative operators, the problem is reduced to the construction of a self-adjoint dilation in the explicit form. First, this problem was solved by Pavlov [2] for the Schrödinger operator $L = -\Delta + q + i\rho$, where q and ρ are real functions continuous in $\mathbb{R}^3$, $0 \leq \rho \leq \text{const} < \infty$, and the operator $A = -\Delta + q$ is supposed to be self-adjoint.

Kuzhel and Kudryashov (see Kuzhel and Kudryashov [1], Kuzhel [25], and Kudryashov [1]) constructed symmetric and self-adjoint dilations of an arbitrary dissipative operator with nonempty set of regular points and a J-selfadjoint dilation of an arbitrary linear operator A with $\rho(A) \neq 0$ (see Sections 2–4). They suggested two methods for the construction of a self-adjoint (J-selfadjoint) dilation one of which is similar to the Sz.-Nagy–Foias method, while the second one is based on the method of Pavlov [2].

Solomyak [1] constructed a self-adjoint dilation of a dissipative operator as the Cayley transformation of a standard unitary dilation of a contraction.

173

2. Self-Adjoint Dilation of a Dissipative Operator
(First Method)

2.1. The Notion of Dilation in the Case of Unbounded Operators

If operators are bounded, then an operator B acting in the Hilbert space $\mathcal{H}$ is called a dilation of the operator $A \in \mathcal{B}[\mathcal{H}_0]$ if $\mathcal{H}_0$ is a subspace of the space $\mathcal{H}$ and

$$A^n = PB^n|_{\mathcal{H}_0} \quad (\forall\, n \in \mathbf{N}), \tag{2.1}$$

where P is an orthoprojector from $\mathcal{H}$ onto $\mathcal{H}_0$. Condition (2.1) is equivalent to each of the following conditions:

(i) $(A^n x, y) = (B^n x, y) \quad (\forall\, \{x, y\} \subset \mathcal{H}_0, \ \forall\, n \in \mathbf{N})$. $\tag{2.2}$;

(ii) the equality

$$(A - \lambda I)^{-1} = P(B - \lambda I)^{-1}|_{\mathcal{H}_0} \tag{2.3}$$

holds in a neughbourhood of a fixed point λ_0 from $\rho(A) \cap \rho(B)$;

(iii) for a fixed point α from $\rho(A) \cap \rho(B)$ and any $n \in \mathbf{N}$,

$$R^n(A, \alpha) = PR^n(B, \alpha) \quad (R(T, \lambda) = (T - \lambda I)^{-1}). \tag{2.4}$$

Indeed, the fact that conditions (2.1) and (2.2) are equivalent is obvious. In order to establish the equivalence of conditions (2.1) and (2.3) or (2.3) and (2.4), it suffices to use the expansion

$$R(T, \lambda) = - \sum_{k=0}^{\infty} \frac{1}{\lambda^{k+1}} T^k \quad (|\lambda| > \|T\|)$$

or

$$R(T, \lambda) = \sum_{k=0}^{\infty} (\lambda - \alpha)^k R^{k+1}(T, \alpha) \quad (|\lambda - \alpha| < \|R(T, \alpha\|^{-1}),$$

respectively.

Conditions (2.3) and (2.4) have are also meaningful for the case of unbounded operators. Thus, each of them can be used to define a dilation of an arbitrary linear operator with nonempty resolvent set.

2.2. Auxiliary Constructions

Assume that A is a dissipative operator acting in the Hilbert space $\mathcal{H}_0$ and $-i \in \rho(A)$. Consider the operators

$$B_{-i} = iR_{-i} - iR_{-i}^* - 2R_{-i}^* R_{-i} \quad \text{and} \quad \tilde{B}_{-i} = iR_{-i} - iR_{-i}^* - 2R_{-i}R_{-i}^*.$$

As shown in Section 2 (Chapter 1), these operators map the space $\mathcal{H}_0$ into the defect spaces $\mathfrak{N}_{-i}$ and $\mathfrak{N}_i$ of the Hermitian part A_0 of the operator A, respectively. Furthermore, equalities (2.14) and (2.15) in Chapter 1 hold. For $\alpha = -i$ and $y = x$, they take the form

$$(B_{-i}(A + iI)x, \ (A + iI)x) = 2 \operatorname{Im}(Ax, x), \tag{2.5}$$

$$(\tilde{B}_{-i}(A^* - iI)x, \ (A^* - iI)x) = 2 \operatorname{Im}(-A^*x, x). \tag{2.6}$$

These equalities imply that the operators B_{-i} and $\tilde{B}_{-i}$ are nonnegative (for $\tilde{B}_{-i}$, this follows from the fact that the operator $-A^*$ is dissipative by virtue of Theorem 3.4 in Chapter 1).

Consider the operators $Q = \sqrt{B_{-i}}$ and $D = Q(A + iI)$. Then, taking equality (2.14) in Chapter 1 into account, we obtain

$$(Dx, Dy) = i(x, Ay) - i(Ax, y) \quad (\{x, y\} \subset \mathcal{D}_A) \tag{2.7}$$

and, in particular, $\| Dx \|^2 = 2 \operatorname{Im}(Ax, x)$.

Now consider the operators $\tilde{Q} = \sqrt{\tilde{B}_{-i}}$ and $T = I - 2iR_{-i}$. Then $TB_{-i} = \tilde{B}_{-i}T$ and, hence,

$$TQ = \tilde{Q}T. \tag{2.8}$$

2.3. Self-Adjoint Dilation

As before, let A be a dissipative operator acting in the Hilbert space $\mathcal{H}_0$ and let $\mathfrak{N}_\lambda$ be the defect space of the Hermitian part A_0 of the operator A. We construct a Hilbert space $\mathcal{H}$ of vectors of the form

$$f = (\ldots, f_{-2}, f_{-1}, \boxed{f_0}, f_1, f_2, \ldots), \qquad \sum_{k=-\infty}^{\infty} \|f_k\|^2 < \infty,$$

where $f_0 \in \mathcal{H}_0$, $f_n \in \mathfrak{N}_{-i}$, $f_{-i} \in \mathfrak{N}_i$ $(n \geq 1)$; the frame denotes the zero position in the corresponding sequence.

We define a scalar product in $\mathcal{H}$ by the equality

$$(f, g) = \sum_{k=-\infty}^{\infty} (f_k, g_k). \qquad (2.9)$$

By identifying the vector $\varphi = (\ldots, 0, 0, \boxed{f_0}, 0, 0, \ldots)$ with f_0, we obtain an inclusion of $\mathcal{H}_0$ in $\mathcal{H}$. In this case, the operator P is defined by the equality

$$Pf = f_0 \qquad (f = (\ldots, f_{-2}, f_{-1}, \boxed{f_0}, f_1, f_2, \ldots))$$

and is the orthoprojector onto $\mathcal{H}_0$ in $\mathcal{H}$.

In the space $\mathcal{H}$, we consider the operators

$$S_n^+ f = \sum_{k=n}^{\infty} f_k - \frac{1}{2} f_n, \qquad S^+ f = \sum_{k=1}^{\infty} f_k, \qquad (2.10)$$

$$S_n^- f = \sum_{k=n}^{\infty} f_{-k} - \frac{1}{2} f_{-n}, \qquad S^- f = \sum_{k=1}^{\infty} f_{-k}, \qquad (2.11)$$

where $n \geq 1$. By using these operators, we define the operator B in $\mathcal{H}$ as follows:

1. The vector $f = (\ldots, f_{-1}, \boxed{f_0}, f_1, \ldots) \in \mathcal{D}_B$ if and only if the following conditions hold:

   (a) $f \in \mathcal{D}_{S^+} \cap \mathcal{D}_{S^-}$;

   (b) $\varphi = \frac{1}{\sqrt{2}} f_0 + \tilde{Q} S^- f \in \mathcal{D}_A$;

   (c) $\sum_{k=-\infty}^{\infty} \|g_k\|^2 < \infty$,

where $g_0 = -if_0 + \sqrt{2}\,(A + iI)\varphi$, $g_{-n} = 2i\,S_n^- f$, and $g_n = -2i\,S_n^+ f$ $(n \geq 1)$;

(d) $S^+ f = T^* S^- f + iD\varphi$, where $T = I - 2iR_{-i}$, $D = Q(A + iI)$.

2. The operator B is defined by the equality

$$Bf = (\ldots,\ g_{-2},\ g_{-1},\ \boxed{g_0},\ g_1,\ g_2,\ldots)) \qquad (f \in \mathfrak{D}_B).$$

Theorem 2.1. *The operator B defined by Conditions 1 and 2 is a self-adjoint dilation of the operator A.*

Proof. We first prove the Hermitian property of the operator B. Let $\{f, h\} \subset \mathfrak{D}_B$. Then

$$(Bf, h) - (f, Bh) = (g_0, h_0) - (f_0, \tilde{g}_0) + a + b, \qquad (2.12)$$

where

$$a = \sum_{n=1}^{\infty} [(g_{-n}, h_{-n}) - (f_{-n}, \tilde{g}_{-n})], \qquad b = \sum_{n=1}^{\infty} [(g_n, h_n) - (f_n, \tilde{g}_n)].$$

The components g_k in this relation are defined by Conditions (c) and (d) in 1; the vectors $\tilde{g}_k$ are defined by similar relations

$$\tilde{g}_0 = -ih_0 + \sqrt{2}\,(A + iI)\,\tilde{\varphi}, \qquad \tilde{\varphi} = \frac{1}{\sqrt{2}}\,h_0 + \tilde{Q}S^- h \in \mathfrak{D}_A,$$

$$\tilde{g}_{-n} = 2i\,S_n^- h, \qquad \tilde{g}_n = -2i\,S_n^+ h \quad (n \geq 1).$$

But then

$$a = 2i\left[\sum_{n=1}^{\infty}\sum_{k=n}^{\infty} (f_{-k},\ h_{-n}) + \sum_{n=1}^{\infty}\sum_{k=n+1}^{\infty} (f_{-n},\ h_{-k}) \right].$$

The equality

$$\sum_{n=1}^{\infty}\sum_{k=n+1}^{\infty} a_{nk} = \sum_{k=1}^{\infty}\sum_{n=1}^{\infty} a_{nk} - \sum_{k=1}^{\infty}\sum_{n=k}^{\infty} a_{nk}$$

and the expression for the operator S^- yields

$$a = 2i \sum_{n,\,k=1}^{\infty} (f_{+k}, h_{-n}) = 2i\,(S^- f,\, S^- h).$$

By the same argument, we establish that $b = -2i\,(S^+ f,\, S^+ h)$. These two equalities and Condition (d) in 1 imply that

$$a + b = 2i\,[((I - TT^*)S^- f,\, S^- h) + i\,(S^- f,\, TD\tilde{\varphi}) - i\,(TD\varphi,\, S^- h) - (D\varphi,\, D\tilde{\varphi})],$$

$$(2.13)$$

where $I - TT^* = 2\tilde{Q}^2$. By virtue of equality (2.8) and the expressions for the operators T and D, we obtain $TD = \tilde{Q}(A - iI)$. We insert the expressions for the operators $I - TT^*$ and TD in (2.13). Then, in view of equality (2.7), we have

$$a + b = 2i\,[2(\tilde{Q}S^- f,\, \tilde{Q}S^- h) + i\,(\tilde{Q}S^- f,\, (A - iI)\tilde{\varphi})$$

$$- i\,((A - iI)\varphi,\, \tilde{Q}S^- h) - i\,(\varphi,\, A\tilde{\varphi}) + i\,(A\varphi,\, \tilde{\varphi})].$$

$$(2.14)$$

By virtue of Condition (b) in 1, we get

$$\tilde{Q}S^- f = \varphi - \frac{1}{\sqrt{2}} f_0, \qquad \tilde{Q}S^- h = \tilde{\varphi} - \frac{1}{\sqrt{2}} h_0$$

and

$$a + b = 2i\,(f_0, h_0) - i\sqrt{2}\,(\varphi, h_0) - i\sqrt{2}\,(f_0, \tilde{\varphi}) + \sqrt{2}\,(f_0, A\tilde{\varphi}) - \sqrt{2}\,(A\varphi, h_0).$$

$$(2.15)$$

On the other hand, by using the expressions for g_0 and $\tilde{g}_0$ (Condition (c) in 1), we obtain

$$(g_0, h_0) - (f_0, \tilde{g}_0) = -2i\,(f_0, h_0) + \sqrt{2}\,(A\varphi, h_0)$$

$$- \sqrt{2}\,(f_0, A\tilde{\varphi}) + i\sqrt{2}\,(\varphi, h_0) + i\sqrt{2}\,(f_0, \tilde{\varphi}) \qquad (2.16)$$

and, hence, by virtue of (2.12), (2.15), and (2.16),

$$(Bf, h) = (f, Bh) \qquad (\forall\, \{f, h\} \subset \mathfrak{D}_B).$$

The operator B is thus Hermitian.

To prove that the operator B is self-adjoint, it suffices to show that the equation

$$(B - \lambda I)f = h \quad (\lambda \in \{-i, i\}) \tag{2.17}$$

is solvable for any h from $\mathcal{H}$. This equation is equivalent to the equations

$$g_k - \lambda f_k = h_k \quad (k \in \mathbb{Z}), \tag{2.18}$$

where the vectors g_k are defined by Condition (c) in 1. Thus, we obtain the following relations for the components f_k of the vector f:

$$\sqrt{2}(A + iI)\varphi = (\lambda + i)f_0 + h_0, \tag{2.19}$$

$$2i \sum_{k=n}^{\infty} f_{-k} = (\lambda + i)f_{-n} + h_{-n} \quad (n \geq 1), \tag{2.20}$$

$$-2i \sum_{k=n}^{\infty} f_k = (\lambda - i)f_n + h_n \quad (n \geq 1). \tag{2.21}$$

Moreover, equations (2.20) and (2.21) are equivalent to

$$(\lambda + i)f_{-(n+1)} + (i - \lambda)f_{-n} = h_{-n} - h_{-(n+1)} \tag{2.22}$$

and

$$(i - \lambda)f_{n+1} + (\lambda + i)f_n = h_{n+1} - h_n, \tag{2.23}$$

respectively.

Let $\lambda = -i$. Then, by virtue of (2.22) and (2.23), we get

$$f_{-n} = \frac{i}{2}(h_{-(n+1)} - h_{-n}), \quad f_{n+1} = \frac{i}{2}(h_n - h_{n+1}) \quad (n \geq 1). \tag{2.24}$$

Thus, to find the vector f, it remains to determine the last two components, namely f_0 and f_1.

In (2.20), we now set $\lambda = -i$ and $n = 1$ and take into account the expressions for the operators S^+ and S^- (equalities (2.10) and (2.11)); this gives

$$S^-f = -\frac{i}{2}h_{-1}, \quad S^+f = f_1 + \frac{i}{2}h_1. \tag{2.25}$$

By applying the operator Q to both sides of equality (2.19), we obtain (for $\lambda = -i$)

$$D\varphi = \frac{1}{\sqrt{2}} Q h_0. \qquad (2.26)$$

In view equalities (2.25) and (2.26), we can rewrite Condition (d) in 1 in the form

$$f_1 + \frac{i}{2} h_1 = -\frac{i}{2} T^* h_{-1} + \frac{i}{\sqrt{2}} Q h_0,$$

whence

$$f_1 = -\frac{i}{2} T^* h_{-1} + \frac{i}{\sqrt{2}} Q h_0 - \frac{i}{2} h_1. \qquad (2.27)$$

By using equalities (2.19) and (2.25) and Condition (b) in 1, we can find f_0. Indeed,

$$f_0 = R_{-i} h_0 + \frac{i}{\sqrt{2}} \tilde{Q} h_{-1}. \qquad (2.28)$$

Hence, for any $h \in \mathcal{H}$ and $\lambda = -i$, equation (2.17) is uniquely solvable. Taking the expressions (2.24), (2.27), and (2.28) for the components of the vector f into account, we conclude that $f \in \mathcal{H}$.

Now consider the case where $\lambda = i$. Under this condition, (2.22) and (2.23) imply that

$$f_{-(n+1)} = \frac{i}{2} (h_{-(n+1)} - h_{-n}), \quad f_n = \frac{i}{2} (h_n - h_{n+1}) \quad (n \geq 1). \qquad (2.29)$$

It remains to find the components f_{-1} and f_0. As before, it follows from (2.20) and (2.21) with $n = 1$ and $\lambda = i$ that

$$S^- f = f_{-1} - \frac{i}{2} h_{-1}, \quad S^+ f = \frac{i}{2} h_1. \qquad (2.30)$$

Let $\psi = (A + iI)\varphi$. Then

$$T\psi = (A - iI)\varphi = (A + iI)\varphi - 2i\varphi.$$

By using equality (2.19) for $\lambda = i$ and the expression for the vector φ (Condition (b) in 1), we obtain

$$T\psi = \frac{1}{\sqrt{2}} h_0 - 2i \tilde{Q} S^- f. \qquad (2.31)$$

By virtue of (2.30), (2.31) and Condition (d) in 1, we arrive at the following system for finding the vectors ψ and $S^- f$:

$$\begin{cases} T^* S^- f + iQ\psi = \dfrac{i}{2} h_1, & (2.32) \\[2mm] 2\tilde{Q} S^- f - iT\psi = -\dfrac{i}{\sqrt{2}} h_0. & (2.33) \end{cases}$$

Let us apply the operators T and $\tilde{Q}$ to both sides of equalities (2.32) and (2.33), respectively. By virtue of the equality $2\tilde{Q}^2 = I - TT^*$, this enables us to rewrite this system in the form

$$\begin{cases} TT^* S^{-1} f + iTQ\psi = \dfrac{i}{2} Th_1, \\[2mm] (I - TT^*)S^- f - i\tilde{Q}T\psi = -\dfrac{i}{\sqrt{2}} \tilde{Q}h_0, \end{cases}$$

whence, in view of the equality $TQ = \tilde{Q}T$,

$$S^- f = \dfrac{i}{2} Th_1 - \dfrac{i}{\sqrt{2}} \tilde{Q}h_0. \qquad (2.34)$$

By inserting (2.30) in (2.34), we find f_{-1}:

$$f_{-1} = \dfrac{i}{2} h_{-1} - \dfrac{i}{\sqrt{2}} \tilde{Q}h_0 + \dfrac{i}{2} Th_1. \qquad (2.35)$$

Similarly, by applying the operators $2Q$ and T^*, respectively, to both sides of equalities (2.32) and (2.33) and taking into account that $2Q^2 = I - T^* T$ and $QT^* = T^* \tilde{Q}$, we find ψ, namely

$$\psi = Qh_1 + \dfrac{1}{\sqrt{2}} T^* h_0. \qquad (2.36)$$

Since $\psi = (A + iI)\varphi$ and $T^* = I + 2i R_{-i}^*$, by virtue of (2.19) and (2.36), we conclude that

$$f_0 = R_{-i}^* h_0 - \dfrac{i}{\sqrt{2}} Qh_1. \qquad (2.37)$$

Thus, in the case where $\lambda = i$, the vector f whose components are defined by relations (2.29), (2.35), and (2.37) belongs to the space $\mathcal{H}$ and is a solution of equation (2.17). This proves that the operator B is self-adjoint.

It remains to show that the operator B is a dilation of the operator A. Assume that $V(-i, \varepsilon)$ is an ε-neighbourhood of the point $-i$ such that

$$V(-i, \varepsilon) = \rho(A) \cap \rho(B); \quad h_0 = (\dots, 0, \boxed{h_0}, 0, \dots) \in \mathcal{H}$$

and $f = (B - \lambda I)^{-1} h_0$, where $\lambda \in V(-i, \varepsilon)$. Then relations (2.19), (2.22), and (2.23) hold. Moreover, the right-hand sides of (2.22) and (2.23) are equal to zero. Consequently,

$$f_{-(n+1)} = \left(\frac{\lambda - i}{\lambda + i}\right)^n f_{-1}; \quad f_{n+1} = \left(\frac{\lambda + i}{\lambda - i}\right)^n f_1. \tag{2.38}$$

Since $|(\lambda - i)(\lambda + i)^{-1}| > 1$ for $\mathrm{Im}\,\lambda < 0$, we have $f_{-n} = 0$ for any natural n. This means that $S^- f = 0$ and $\varphi = 2^{-1/2} f_0$. But then $f_0 = (A - \lambda I)^{-1} h_0$ according to (2.19) and, hence,

$$(B - \lambda I)^{-1} h_0 = \left(\dots, 0, \boxed{(A - \lambda I)^{-1} h_0}, f_1, \left(\frac{\lambda + i}{\lambda - i}\right) f_1, \dots \right).$$

This yields

$$P(B - \lambda I)^{-1} h_0 = (A - \lambda I)^{-1} h_0 \quad (\forall\, h_0 \in \mathcal{H}_0).$$

Thus, B is a self-adjoint dilation of the operator A. ∎

Theorem 2.2. *If B is a self-adjoint dilation of a dissipative operator A, then $\sigma_p(B) \subset \sigma_p(A)$.*

Proof. Indeed, let $Bf = \lambda f$ $(f \neq 0)$. Then, by virtue of Condition 2, we have

$$g_k = \lambda f_k \quad (k \in \mathbb{Z}). \tag{2.39}$$

Taking Condition (c) in 1 and the expressions (2.10) and (2.11) for the operators S_n^+ and S_n^- into account, we obtain

$$g_n - g_{n+1} = -i(f_n + f_{n+1}), \quad g_{-n} - g_{-(n+1)} = i(f_{-n} + f_{-(n+1)}). \tag{2.40}$$

By virtue of (2.39) and (2.40), we can write

$$f_{n+1} = \left(\frac{\lambda+i}{\lambda-i}\right)f_n, \quad f_{-(n+1)} = \frac{\lambda-i}{\lambda+i}f_{-n} \quad (n \ge 1). \tag{2.41}$$

Since $\bar{\lambda} = \lambda$, we have $|(\lambda+i)(\lambda-i)^{-1}| = 1$ and, hence, these equalities are possible only in the case where $f_n = f_{-n} = 0 \ (n \in \mathbb{N})$. But then $S^- f = 0$ and, therefore, (Condition (b) in 1), $\varphi = 2^{-1/2}f_0$. Substituting this value of φ in the expression for g_0 (Condition (c) in 1) and taking (2.39) into account, we find that $Af_0 = \lambda f_0$.

∎

In conclusion, we note that, in the general case, the dilation B is not an extension of the operator A but is an extension of its Hermitian part A_0. Moreover, $\mathcal{D}_A \cap \mathcal{D}_B = \mathcal{9}_A$ ($\mathcal{9}_A$ is the domain where the operator A is Hermitian). In particular, if the operator A is Hermitian, then B is a self-adjoint extension of the operator A (which may lead out of the space $\mathcal{H}_0$). This is, in fact, a new method for constructing self-adjoint extensions of Hermitian operators.

2.4. Symmetric Dilations of Dissipative Operators

In some cases, it is necessary to construct symmetric dilations of dissipative operators. This problem is much simpler than the previous one and can be solved similarly. Therefore, we only outline the main stages of the corresponding construction.

Assume that A is a dissipative operator acting in $\mathcal{H}_0$ and $-i = \rho(A)$. Consider the nonnegative operator $B_{-i} = iR_{-i} - iR_{-i}^* - 2R_{-i}^* R_{-i}$ and the operators $Q = \sqrt{B_{-i}}$ and $D = Q(A + iI)$. Let $\mathcal{H}$ be the Hilbert space composed of the vectors $f = (f_0, f_1, f_2, \ldots)$, where $f_0 \in \mathcal{H}_0$, $f_k \in \mathfrak{N}_{-i}$, and

$$\sum_{k=0}^{\infty} \|f_k\|^2 < \infty$$

($\mathfrak{N}_{-i}$ is the defect space of the Hermitian part A_0 of the operator A). We define a scalar product in $\mathcal{H}$ by the equality

$$(f, g) = \sum_{k=0}^{\infty} (f_k, g_k).$$

By identifying the vector $f = (f_0, 0, 0, \ldots)$ with the vector f_0, we obtain the isometric inclusion of $\mathcal{H}_0$ in $\mathcal{H}$.

In the space $\mathcal{H}$, we consider the operators S_n^+ and S_n^- defined by (2.10) and an operator B defined in $\mathcal{H}$ as follows:

1. $f = (f_0, f_1, f_2, \ldots) \in \mathfrak{D}_B$ if and only if $f_0 \in \mathfrak{D}_A$, $f \in \mathfrak{D}_{S^+}$, $\| S_n^+ f \| < \infty$, and
    $Sf = (i/\sqrt{2}) D f_0$;

2. if $f \in \mathfrak{D}_B$, then $Bf = (g_0, g_1, g_2, \ldots)$, where $g_0 = A f_0$, $g_n = -2i S_n^+ f$ $(n \geq 1)$.

The operator B thus defined is closed, densely defined, and symmetric; moreover, it is a dilation of the operator A.

Remarks. The results in this section belong to the author and were published in Kuzhel and Kudryashov [1] and Kuzhel [25].

3. Symmetric and Self-Adjoint Dilations of Dissipative Operators (Second Method)

3.1. Preliminary Remarks

Consider the spaces of vector functions $\mathcal{H}_+ = L_2(0, \infty; \mathcal{H})$ and $\mathcal{H}_- = L_2(-\infty, 0; \mathcal{H})$, where $\mathcal{H}$ is a Hilbert space. Scalar products in these spaces are defined, respectively, by the equalities

$$(f_+, h_+)_+ = \int_0^\infty (f_+(t), h_+(t))_{\mathcal{H}} dt \quad \text{and} \quad (f_-, h_-)_- = \int_{-\infty}^0 (f_-(t), h_-(t))_{\mathcal{H}} dt.$$

The derivative of the function $f(\cdot)$ from $\mathcal{H}_+$ (or $\mathcal{H}_-$) is defined as follows:

$$\frac{d}{dt} f(t) = s - \lim_{\Delta t \to 0} \frac{f(t + \Delta t) - f(t)}{\Delta t}.$$

The Sobolev class $W_2^1(0, \infty; \mathcal{H})$ consists of the functions $f \in \mathcal{H}_+$ for which $df(t)/dt$ exists and belongs to the space $\mathcal{H}_+$. The Sobolev class $W_2^1(-\infty, 0; \mathcal{H})$ is defined similarly. These classes are dense in $\mathcal{H}_+$ and $\mathcal{H}_-$, respectively.

The operator of differentiation possesses the following properties:

$$\left(\frac{d}{dt} f_\pm, f_\pm \right) + \left(f_\pm, \frac{d}{dt} f_\pm \right)_\pm = \mp \| f_\pm(0) \|_{\mathcal{H}}^2 \qquad (3.1)$$

$$\frac{d}{dt} Kf(t) = K\frac{d}{dt}f(t) \quad (K \in \mathcal{B}[\mathcal{H}]).$$ (3.2)

Relation (3.1) follows from the generalized formula of integration by parts and the condition

$$\lim_{t \to \pm\infty} \|f_{\pm}(t)\|_{\mathcal{H}} = 0;$$

relation (3.2) is a consequence of the definition of derivative and the fact that the operator K is continuous.

3.2. Symmetric Dilations

Assume that A is a closed dissipative operator acting in the Hilbert space $\mathcal{H}_0$ and $-i \in \rho(A)$. Consider the space of vector functions $\mathcal{H}_- = L_2(-\infty, 0, \mathcal{H}_1)$, where $\mathcal{H}_1 = \overline{Q\mathcal{H}_0}$, $Q = \sqrt{B_{-i}}$, and $B_{-i} = iR_{-i} - iR^*_{-i} - 2R^*_{-i}R_{-i}$. We construct the Hilbert space $\mathcal{H} = \mathcal{H}_- \oplus \mathcal{H}_0$ with the scalar product

$$(f, g) = (f_-, g_-)_- + (f_0, g_0)_0, \quad f = \begin{pmatrix} f_- \\ f_0 \end{pmatrix}, \quad g = \begin{pmatrix} g_- \\ g_0 \end{pmatrix},$$

and define an operator L in $\mathcal{H}$ as follows:

1. The vector

$$f = \begin{pmatrix} f_- \\ f_0 \end{pmatrix} \in \mathfrak{D}_L$$

if and only if

(a) $f_- \in W_2^1(-\infty, 0; \mathcal{H}_1)$;

(b) $f_0 \in \mathfrak{D}_A$;

(c) $f_-(0) = iDf_0$, where $D = Q(A + iI)$.

2. If $f \in \mathfrak{D}_L$, then

$$Lf = \begin{pmatrix} \Gamma_- f_- \\ A f_0 \end{pmatrix},$$

where

$$\Gamma_- f_- = -i \frac{d}{dt} f_-(t).$$

Theorem 3.1. *The operator* L *defined by Conditions 1 and 2 is a symmetric dilation of the operator* A.

Proof. Let $f \in \mathfrak{D}_L$. By using the equality

$$\| D f_0 \|_0^2 = 2 \operatorname{Im} (A f_0 \; f_0)_0 \qquad (3.3)$$

proved in Subsection 2.2, the conditions imposed on $\mathfrak{D}_L$, and relation (3.1), we obtain

$$(Lf, \; f) - (f, \; Lf) = 2i \operatorname{Im} (A f_0 \; f_0)_0 - i \| f_-(0) \|_0^2 = 0,$$

i.e., the operator L is Hermitian. Moreover, the operator L is closed and densely defined. This follows directly from the fact that the operator A is closed and densely defined.

Let us show that the range of the operator $L + iI$ is dense in $\mathcal{H}$. Indeed, assume that $g \perp \Delta_{L+iI}$, where

$$g = \begin{pmatrix} g_- \\ g_0 \end{pmatrix} \in \mathcal{H}.$$

Let $f_-(0) = f_0 = 0$. Consider the operator $\Gamma = \Gamma_- |_{\mathfrak{D}}$, where

$$\mathfrak{D} = \{ f_- \in \mathfrak{D}_{\Gamma_-} \,|\, f_-(0) = 0 \}.$$

Then, since $-i \in \rho(\Gamma)$, we conclude that $g = 0$. Furthermore, since $-i \in \rho(A)$, we have $g_0 = 0$ and, hence, $g = 0$.

By using the expression for the operator L, we find that

$$(L - \lambda I)^{-1} f = \begin{pmatrix} (\Gamma - \lambda I)^{-1} f_- + i e^{i\lambda t} D R_\lambda f_0 \\ (A - \lambda I)^{-1} f_0 \end{pmatrix}.$$

This yields $P(L - \lambda I)^{-1} f_0 = (A - \lambda I)^{-1} f_0$, where P is the orthoprojector onto $\mathcal{H}_0$ in $\mathcal{H}$ and the vector $\begin{pmatrix} 0 \\ f_0 \end{pmatrix}$ is identified with the vector f_0. Consequently, L is a symmetric dilation of the operator A.

∎

3.3. Self-Adjoint Dilations

Consider the space of vector functions $\mathcal{H}_+ = L_2(0, \infty; \mathcal{H}_2)$, where $\mathcal{H}_2 = \overline{\tilde{Q} \mathcal{H}_0}$, $\tilde{Q} = \sqrt{\tilde{B}_{-i}}$, and $\tilde{B}_{-i} = i R_{-i} - i R_{-i}^* - 2R_{-i} R_{-i}^*$. We construct the Hilbert space $\mathcal{H} = \mathcal{H}_- \oplus \mathcal{H}_0 \oplus \mathcal{H}_+$ (the space $\mathcal{H}_-$ is defined as in Subsection 3.2). A scalar product in $\mathcal{H}$ is defined by the equality

$$(f, g) = (f_- g_-)_- + (f_0 g_0)_0 + (f_+ g_+)_+, \quad f = \begin{pmatrix} f_- \\ f_0 \\ f_+ \end{pmatrix} \in \mathcal{H}, \quad g = \begin{pmatrix} g_- \\ g_0 \\ g_+ \end{pmatrix} \in \mathcal{H}.$$

In $\mathcal{H}$, we construct an operator S as follows:

1. The vector

$$f = \begin{pmatrix} f_- \\ f_0 \\ f_+ \end{pmatrix}$$

belongs to $\mathfrak{D}_S$ if and only if

(a) $f_- \in W_2^1(-\infty, 0; \mathcal{H}_1)$, $f_+ \in W_2^1(0, \infty; \mathcal{H}_2)$;

(b) $\varphi = f_0 + \tilde{Q} f_+(0) \in \mathfrak{D}_A$;

(c) $f_-(0) = T^* f_+(0) + i D \varphi$, where $T = I - 2i R_{-i}$, $D = Q(A + iI)$.

2. If

$$f = \begin{pmatrix} f_- \\ f_0 \\ f_+ \end{pmatrix} \in \mathfrak{D}_S,$$

then

$$Sf = \begin{pmatrix} \Gamma_- f_- \\ -i f_0 + (A + iI)\varphi \\ \Gamma_+ f_+ \end{pmatrix},$$

where

$$\Gamma_\pm f_\pm = -i \frac{d}{dt} f_\pm(t).$$

Theorem 3.2. *The operator S defined by Conditions 1 and 2 is a self-adjoint dilation of the operator A.*

Proof. Let us prove that the operator S is Hermitian. By using (3.1), Condition (c) in 1, (3.3), and relation $TD = \tilde{Q}(A - iI)$ (see Section 2), we get

$(Sf, f) - (f, Sf)$

$$= i \| f_+(0) \|_0^2 - i \| f_-(0) \|_0^2 + 2i \operatorname{Im} (A \varphi, f_0)_0 + 2i \operatorname{Im} i (\tilde{Q} f_+(0), f_0)_0$$

$$= -2i \operatorname{Im} [2 (f_+(0), R_{-i}^* f_+(0))_0 + (T^* f_+(0), D\varphi)_0 + 2i \| R_{-i}^* f_+(0) \|_0^2$$

$$- i (\tilde{Q} f_+(0), f_0)_0 + (A \varphi, \tilde{Q} f_+(0))_0]$$

$$= -2i \operatorname{Im} [2 (f_+(0), R_{-i}^* f_+(0))_0 + i (\tilde{B}_{-i} f_+(0), f_+(0))_0$$

$$+ 2i (R_{-i} R_{-i}^* f_+(0), f_+(0))_0] = 0,$$

where the last equality is obtained by using the expression for the operator $\tilde{B}_{-i}$.

Let us show that $\overline{\Delta}_{S+iI} = \mathcal{H}$. Indeed, assume that

$$g = \begin{pmatrix} g_- \\ g_0 \\ g_+ \end{pmatrix} \in \mathcal{H}$$

and $g \perp \Delta_{S+iI}$. We set $f_-(0) = f_0 = 0$ and $f_+ = 0$. Since $-i \in \rho(\Gamma)$, where the operator Γ is defined as in Subsection 3.2, we have $g_- = 0$.

We set $f_+(0) = 0$. Then, by virtue of Conditions (b) and (c) in 1, $f_-(0) = iQ(A + iI)f_0$, where f_0 is an arbitrary vector from $\mathfrak{D}_A$. Since $-i \in \rho(A)$, we have $g_0 = 0$.

Assume that $f_+(0)$ is an arbitrary vector from $\mathcal{H}_0$. Then, since $\overline{\Delta_{\Gamma_+ + iI}} = \mathcal{H}_+$, we conclude that $g_+ = 0$.

Consequently, $g = 0$.

We now show that $\overline{\Delta_{S-iI}} = \mathcal{H}$. As above, let $g \perp \Delta_{S-iI}$, where

$$ g = \begin{pmatrix} g_- \\ g_0 \\ g_+ \end{pmatrix} $$

and $g \in \mathcal{H}$. We set $f_0 = f_+(0) = 0$ and $f_- = 0$. Since $-i \in \rho(\Gamma_+|_{\mathfrak{D}'})$, where $\mathfrak{D}' = \{f_+ \in \mathfrak{D}_{\Gamma_+} | f_+(0) = 0\}$, we have $g_+ = 0$.

In view of Condition (c) in 1 and equality $T^* \tilde{Q} = QT^*$, we find that

$$ (A + iI)\varphi - (A^* - iI)\psi = 2if_0 \tag{3.4} $$

where $\psi = f_0 + Qf_-(0) \in \mathfrak{D}_{A^*}$. We set $f_- = 0$. Then, by virtue of (3.4),

$$ (A + iI)\varphi - 2if_0 = (A^* - iI)f_0. \tag{3.5} $$

Let us show that, for any vector f_0 from $\mathfrak{D}_{A^*}$, there are vectors $\varphi \in \mathfrak{D}_A$ and $f_+(0) \in \mathcal{H}_2$ such that $\varphi = f_0 + \tilde{Q}f_+(0)$. Indeed, let $f_1 = -i(A^* - iI)f_0$. We set $\varphi = (iR_{-i} + 2R_{-i}R_{-i}^*)f_1$. Then $\varphi \in \mathfrak{D}_A$ and $\varphi = f_0 + \tilde{Q}f_+(0)$, where $f_+(0) = \tilde{Q}f_1$. The vectors $f_+(0)$, φ, and f_0 satisfy equality (3.5). Therefore, $g_0 = 0$.

Let $f_-(0)$ be an arbitrary vector in $\mathcal{H}_1$. Since $\overline{\Delta_{\Gamma_- - iI}} = \mathcal{H}_-$, we have $g_- = 0$. Consequently, $g = 0$.

It is easy to see that the operator S is closed. Thus, S is a self-adjoint operator.

It remains to show that S is a dilation of the operator A. Indeed, by direct calculation, we can show that, for $\lambda \in \rho(S) \cap \rho(A)$,

$$ (S - \lambda I)^{-1}f = \begin{pmatrix} (\Gamma - \lambda I)^{-1} f_- + e^{i\lambda t} T^* \psi_+(0) + iDR_\lambda (f_0 + (\lambda + i)\tilde{Q}\psi_+(0)) \\ R_\lambda f_0 - (A + iI)R_\lambda \tilde{Q}\psi_+(0) \\ (\Gamma_+ - \lambda I)^{-1} f_+ \end{pmatrix}, $$

where $\psi_+(0) = [(\Gamma_+ - \lambda I)^{-1}f_+(t)]_{t=0}$ and, hence, $P(S - \lambda I)^{-1}f_0 = (A - \lambda I)^{-1}f_0$.

Remarks. The results presented in this section are due to Kudryashov (Kuzhel and Kudryashov [1] and Kudryashov [1]). As already mentioned, for a dissipative operator A with $\mathfrak{D}_{A^*} = \mathfrak{D}_A$, a similar self-adjoint dilation was constructed by Pavlov [2]. Petrov [1] used the results of Pavlov and Kudryashov to construct spectral projectors of arbitrary dissipative densely defined operators.

4. *J*-Unitary and *J*-Selfadjoint Dilations

4.1. *J*-Unitary Dilations

Assume that T is a linear bounded operator acting in the Hilbert space $\mathcal{H}_0$. Consider the operators

$$D_T = |I - T^*T|^{1/2}, \quad D_{T^*} = |I - TT^*|^{1/2} \tag{4.1}$$

and

$$V_T = \text{sign}\,(I - T^*T), \quad V_{T^*} = \text{sign}\,(I - TT^*) \tag{4.2}$$

and the subspaces $\mathfrak{D} = \overline{D_T \mathcal{H}_0}$ and $\mathfrak{D}_* = \overline{D_{T^*} \mathcal{H}_0}$. Let $\mathcal{H}$ be the Hilbert space composed of the vectors x of the form

$$x = (\dots,\, x_{-2},\, x_{-1},\, \boxed{x_0},\, x_1,\, x_2,\, \dots), \tag{4.3}$$

where $x_0 \in \mathcal{H}_0$, $x_n \in \mathfrak{D}$, $x_{-n} \in \mathfrak{D}_*$ $(n \in \mathbb{N})$, and

$$\sum_{k=-\infty}^{\infty} \| x_k \|^2 < \infty.$$

A scalar product in $\mathcal{H}$ is defined by equality

$$(x, y) = \sum_{k=-\infty}^{\infty} (x_k, y_k)_0, \tag{4.4}$$

where $(\cdot, \cdot)_0$ is a scalar product in $\mathcal{H}_0$. As in Section 2, we imbed the space $\mathcal{H}_0$ in $\mathcal{H}$.

In the space $\mathcal{H}$, we consider an operator J defined by the equality

$$Jx = (\ldots, V_{T^*}x_{-2}, V_{T^*}x_{-1}, \boxed{x_0}, V_T x_1, V_T x_2, \ldots)$$

so that $J^* = J$ and $J^2 = I$. In terms of the operator J, we define in $\mathcal{H}$ a new scalar product: $[x, y] = (Jx, y)$. The notions of J-metric, J-unitary property, and J-self-adjointness are defined as usual (see, e.g., Azizov and Iokhvidov [1]).

In $\mathcal{H}$, we introduce linear operators U and U'

$$Ux = (\ldots, x_{-3}, x_{-2}, \boxed{V_T^* D_{T^*} x_{-1} + Tx_0}, D_T x_0 - T^* x_{-1}, x_1, x_2, \ldots),$$

$$U'x = (\ldots, x_{-2}, x_{-1}, D_{T^*} x_0 - Tx_1, \boxed{T^* x_0 + VD_T x_1}, x_2, x_3, \ldots).$$

By using the relations $D_T V_T = V_T D_T$, $D_{T^*} V_{T^*} = V_{T^*} D_{T^*}$; $TD_T = D_{T^*} T$, and $TV_T = V_{T^*} T$, we establish that $UU' = U'U = I$ and, thus, $U' = U^{-1}$. In addition, $[Ux, Uy] = [x, y]$ for any x and y from $\mathcal{H}$. Hence, the operator U is J-unitary. In this case, $T^n = PU^n|_{\mathcal{H}_0}$ ($\forall n \in \mathbb{N}$), where P is the orthoprojector onto $\mathcal{H}_0$ in $\mathcal{H}$. Thus, U is a J-unitary dilation of the operator T. Similarly, we can show that the operator U' is a J-unitary dilation of the operator T^*.

In the case where T is a contraction, the operators V_T and V_{T^*} are equal to the identity operators in $\mathfrak{D}$ and $\mathfrak{D}_*$, respectively, and the dilation just constructed coincides with the unitary dilation of the contraction T. Furthermore, J is the identity operator in $\mathcal{H}$ and, hence, the J-metric introduced above coincides with the original one.

4.2. J-Selfadjoint Dilation (First Method)

Assume that A is a linear closed densely defined operator acting in $\mathcal{H}_0$ and $\rho(A) \neq \varnothing$. As before, we suppose that $-i \in \rho(A)$. Let the operators B_{-i} and $\tilde{B}_{-i}$ be defined as in Subsection 2.2. Consider the following operators:

$$Q = \sqrt{|B_{-i}|}, \quad \tilde{Q} = \sqrt{|\tilde{B}_{-i}|}, \quad V = \operatorname{sign} B_{-i}, \quad \tilde{V} = \operatorname{sign} \tilde{B}_{-i}, \quad D = VQ(A + iI).$$

$$(4.5).$$

The space $\mathcal{H}$ is defined as in Subsection 2.3. In $\mathcal{H}$, we define the operator B just as in Subsection 2.3, taking into account the fact that the operators Q, $\tilde{Q}$, and D are defined by (4.5). A *J*-metric in $\mathcal{H}$ is defined by the equality $[x, y] = (Jx, y)$, where

$$Jx = (\dots, \tilde{V} x_{-2}, \tilde{V} x_{-1}, \boxed{x_0}, Vx_1, Vx_2, \dots).$$

By the same reasoning as in Subsection 2.3, we establish that the operator B is *J*-self-adjoint. Moreover, for any λ from some neighbourhood of the point $-i$, we have

$$(A - \lambda I)^{-1} = P(B - \lambda I)^{-1}|_{\mathcal{H}_0}$$

and, therefore, the operator B is a *J*-self-adjoint dilation of the operator A.

In the case where A is a dissipative operator, V and $\tilde{V}$ are the identity operators in the respective subspaces. Hence, we arrive at the results presented in Subsection 2.3.

4.3. *J*-Selfadjoint Dilation (Second Method)

As in Subsection 4.2, we assume that A is a linear closed operator acting in the Hilbert space $\mathcal{H}_0$ and $-i \in \rho(A)$. The space $\mathcal{H}$ and the operator S in $\mathcal{H}$ are constructed as in Subsection 3.3. The operators Q, $\tilde{Q}$, and D are defined by (4.5). By the same argument as in Subsection 3.3, we show that the operator S is a *J*-selfadjoint dilation of the operator A.

Remarks. *J*-selfadjoint dilations of arbitrary linear operators were constructed by Kuzhel [24–26]. The second method for constructing *J*-selfadjoint dilations was suggested by Kudryashov [2].

5. Coordinate-Free Form of Dilations

As mentioned above, a bounded operator $B\colon \mathcal{H} \to \mathcal{H}$ is called a dilation of the operator $T \in \mathcal{B}[\mathcal{H}_0]$ if $\mathcal{H}_0$ is a subspace of the space $\mathcal{H}$ and

$$T^n = P_0 B^n|_{\mathcal{H}_0} \qquad (\forall\, n \in \mathbb{N}), \tag{5.1}$$

where P_0 is the orthoprojector onto $\mathcal{H}_0$ in $\mathcal{H}$.

Sarason [1] established the following criterion for the operator B to be a dilation of the operator $T \overset{\text{def}}{=} P_0 B|_{\mathcal{H}_0}$.

Lemma 5.1. *The operator* $B: \mathcal{H} \to \mathcal{H}$ *is a dilation of the operator* $T \overset{\text{def}}{=} P_0 B |_{\mathcal{H}_0}$ *if and only if the space* $\mathcal{H}$ *can be represented in the form*

$$\mathcal{H} = \mathcal{G}_* \oplus \mathcal{H}_0 \oplus \mathcal{G} \tag{5.2}$$

and

$$B\mathcal{G} \subset \mathcal{G}, \quad B^* \mathcal{G}_* \subset \mathcal{G}_*. \tag{5.3}$$

Proof. Assume that B is a dilation of the operator T and L is a linear span of the lineals $B^n \mathcal{H}_0$ $(n \geq 0)$. We set

$$\mathcal{G}' = \overline{L}, \quad \mathcal{G} = \mathcal{G}' \ominus \mathcal{H}_0, \quad \mathcal{G}_* = \mathcal{H} \ominus \mathcal{G}'. \tag{5.4}$$

Therefore, $\mathcal{H}$ is representable in the form (5.2).

It is clear that $B\mathcal{G}' \subset \mathcal{G}'$. Hence, $B^* \mathcal{G}_* \subset \mathcal{G}_*$. Let us show that the first inclusion in (5.3) is true. For this purpose, we first note that, in view of relation (5.1) and the definition of the lineal L, the equality

$$P_0 B g = T P_0 g \quad (g \in \mathcal{G}') \tag{5.5}$$

holds on this lineal and, consequently, in the entire space $\mathcal{G}'$. Further, since the space $\mathcal{H}$ is representable in the form (5.2), the inclusion $B\mathcal{G} \subset \mathcal{G}$ will be established if we show that $B^* (\mathcal{G}_* \oplus \mathcal{H}_0) \subset \mathcal{G}_* \oplus \mathcal{H}_0$. In view of the inclusion $B^* \mathcal{G}_* \subset \mathcal{G}_*$, for this purpose, it suffices to show that

$$B^* \mathcal{H}_0 \perp \mathcal{G}. \tag{5.6}$$

Let $x_0 \in \mathcal{H}_0$ and $g \in \mathcal{G}$. Then, by virtue of the equalities $\mathcal{G} = \mathcal{G}' \oplus \mathcal{H}_0$ and (5.5),

$$(B^* x_0, g) = (x_0, P_0 B g) = (T^* x_0, g) = 0.$$

This proves relation (5.6) and the inclusion

$$B\mathcal{G} \subset \mathcal{G}.$$

Conversely, assume that relations (5.2) and (5.3) are satisfied. Then it is easy to see that

$$P_0 B^n P = 0 \quad (n \geq 0), \quad P' B P_0 = B P_0, \tag{5.7}$$

where P and $P' = P + P_0$ are the orthoprojectors onto the spaces 9 and $9' = 9 \oplus \mathcal{H}_0$, respectively.

Let us show that equalities (5.1) take place. By the condition, we have $TP_0 = P_0 BP_0$. Thus, taking equalities (5.7) and $P_0 = P' - P$ into account and arguing by induction, we conclude that

$$T^{n+1}P_0 = T^n P_0 TP_0 = P_0 B^n (P' - P) BP_0 = P_0 B^n P'BP_0 = P_0 B^{n+1} P_0,$$

which proves (5.1).

One can easily check that the operator matrix of the dilation B has the following form with respect to expansion (5.2):

$$B = \begin{pmatrix} P_* B|_9_* & 0 & 0 \\ A & T & 0 \\ F & C & B|_9 \end{pmatrix},$$

where $A = P_0 B|_{9_*}$, $F = PB|_{9_*}$, $C = PB|_{\mathcal{H}_0}$, P_* is the orthoprojector onto 9_* in $\mathcal{H}$, and the operators P_0 and P have the same meaning as above.

In particular, if T is a contraction and $U = B$ is its minimal unitary dilation, then

$$U = \begin{pmatrix} P_* U|_9_* & 0 & 0 \\ D_T \cdot V_*^* & T & 0 \\ -VT^* V_*^* & VD_T & U|_9 \end{pmatrix}, \qquad (5.8)$$

where D_T and D_{T^*} are determined as in Subsection 6.3.1 (Chapter 3) and V and V_* are partial isometries from $\mathcal{H}_0$ to 9 and 9_*, respectively.

It can be also shown that, under certain additional conditions, the operator U defined in the space $\mathcal{H} = 9_* \oplus \mathcal{H}_0 \oplus 9$ by (5.8) is the minimal unitary dilation of the contraction $T \in \mathcal{B}[\mathcal{H}_0]$ (for more details, see Nikolsky and Khrushchev [1]).

Remarks. In Solomyak [1], the representation of a unitary dilation U of the contraction T in the form (5.8) is used to obtain (by using the Cayley transformation) a self-adjoint dilation S of the maximal dissipative operator L.

According to Solomyak, his dilation insignificantly differs from the dilation constructed by Kudryashov. He also indicated that "Kuzhel's and Kudryashov's dilations can be obtained from each other by a conformal change combined with the Fourier transformation."

Appendix 1. Contractions. Triangulation of Contractions

1. General Properties of Contractions

A linear operator $T: \mathcal{H} \to \mathcal{H}'$ is called a *contraction* (from $\mathcal{H}$ to $\mathcal{H}'$) if $\|T\| \leq 1$. If, in addition, $\mathcal{H}' = \mathcal{H}$, then we say that T is a contraction in $\mathcal{H}$.

Let T be a contraction from $\mathcal{H}$ to $\mathcal{H}'$. Then T^* is a contraction from $\mathcal{H}'$ to $\mathcal{H}$ (because $\|T^*\| = \|T\| \leq 1$).

Consider the operators

$$R = I - T^*T \quad \text{and} \quad R_* = I' - TT^*,$$

where T is a contraction from $\mathcal{H}$ to $\mathcal{H}'$ and I and I' are the identity operators in $\mathcal{H}$ and $\mathcal{H}'$, respectively. Since

$$(Rx, x) = \|x\|^2 - \|Tx\|_1^2 \geq 0 \quad (\forall x \in \mathcal{H}) \tag{1}$$

($\|\cdot\|_1$ denotes a norm in $\mathcal{H}'$), the operator R is nonnegative. By the same argument, we prove that $R_* \geq 0$. Hence, we can consider the self-adjoint nonnegative operators

$$\mathcal{D} = (I - T^*T)^{1/2}, \quad \mathcal{D}_* = (I' - TT^*)^{1/2}, \tag{2}$$

the subspaces

$$\mathfrak{D} = \overline{D\mathcal{H}}, \quad \mathfrak{D}_* = \overline{D_*\mathcal{H}'}, \tag{3}$$

and the numbers $m = \dim \mathfrak{D}$, $m_* = \dim \mathfrak{D}_*$, which are called the *defect operators*, *defect spaces*, and *deficiency indices* of the operator T, respectively.

Defect operators play an important role in the investigation of contractions. In particular, if $\mathcal{D} = 0$ (or, equivalently, the deficiency index $m = 0$), then the operator T is isometric. If, in addition, $\mathcal{D}_* = 0$ (i.e., $m_* = 0$), then the operator T is unitary.

Further, since $TR = R_*T$ and $RT^* = T^*R_*$, we have

$$T\mathcal{D} = \mathcal{D}_*T, \quad \mathcal{D}T^* = T^*\mathcal{D}_*. \tag{4}$$

By using these equalities, we can show that

195

$$T\mathfrak{D} \subset \mathfrak{D}_*, \quad T^*\mathfrak{D}_* \subset \mathfrak{D}.$$

In what follows, we need the following assertion:

Proposition 1. *If T is a contraction, then the equality* $Tx = y$, *where* $\|y\| = \|x\|$, *holds if and only if* $T^*y = x$.

Proof. If $Tx = y$, then $(T^*y, x) = (x, T^*y) = \|y\|^2$ and, hence (in view of the equality $\|x\| = \|y\|$),

$$\|T^*y - x\|^2 = \|T^*y\|^2 - \|y\|^2 \leq 0,$$

which is possible only if $T^*y = x$. Similarly, if $T^*y = x$, then $Tx = y$.

■

2. Canonical Representation. Completely Nonunitary Contractions

Lemma 2. *If T is a contraction in* $\mathcal{H}$, *then the sets*

$$\mathcal{H}_+ = \{x \in \mathcal{H} \mid \|T^n x\| = \|x\| \quad (\forall n \in \mathbf{N})\}, \tag{5}$$

$$\mathcal{H}_- = \{x \in \mathcal{H} \mid \|T^{*n} x\| = \|x\| \quad (\forall n \in \mathbf{N})\} \tag{6}$$

are subspaces of the space $\mathcal{H}$.

Proof. Assume that $\{x, y\} \subset \mathcal{H}_+$. Let us show that

$$(Tx, Ty) = (x, y). \tag{7}$$

Indeed, if $h = Tx$, then $\|h\| = \|x\|$ and, hence, due to Proposition 1, $T^*h = x$. But then

$$(Tx, Ty) = (T^*h, y) = (x, y)$$

proving relation (7). By using this equality, we obtain

$$\|T(x + y)\|^2 = \|Tx\|^2 + (Tx, Ty) + (Ty, Tx) + \|Ty\|^2$$

$$= \|x\|^2 + (x, y) + (y, x) + \|y\|^2 = \|x+y\|^2,$$

i.e., $\|T(x+y)\| = \|x+y\|$.

Denote $x' = Tx$, $y' = Ty$. Then $\{x', y'\} \subset \mathcal{H}_+$ and, taking the already established equality into account, we get

$$\|T^2(x+y)\| = \|T(x'+y')\| = \|x'+y'\| = \|T(x+y)\| = \|x+y\|.$$

Similarly, by induction, we prove that

$$\|T^n(x+y)\| = \|x+y\| \quad (\forall n \in \mathbf{N})$$

and, hence, $x+y \in \mathcal{H}_+$, i.e., $\mathcal{H}_+$ is a lineal. Assume that n is a fixed natural number and let $\{x_k\}$ be a sequence of elements in $\mathcal{H}_+$ strongly convergent to an element x. Passing in the equality $\|T^n x_k\| = \|x_k\|$ to the limit as $k \to \infty$, we obtain $\|T^n x\| = \|x\|$ ($\forall n \in \mathbf{N}$). Thus, the lineal $\mathcal{H}_+$ is closed. Similarly, we establish that the set $\mathcal{H}_-$ is also linear and closed.

◼

Proposition 3. *The operator T induces a unitary operator in the space*

$$\mathcal{H}_0 = \mathcal{H}_+ \cap \mathcal{H}_-. \tag{8}$$

Proof. Let $x \in \mathcal{H}_0$ and $y = Tx$. Since $x \in \mathcal{H}_+$ and $\|y\| = \|x\|$, we have

$$\|T^n y\| = \|T^{n+1}x\| = \|x\| = \|y\| \quad (\forall n \in \mathbf{N})$$

and, hence, $y \in \mathcal{H}_+$. Moreover, according to Proposition 1, $T^* y = x$. But then (since $x \in \mathcal{H}_-$)

$$\|T^{*n} y\| = \|T^{*(n-1)}x\| = \|x\| = \|y\| \quad (\forall n \in \mathbf{N}),$$

i.e., $y \in \mathcal{H}_-$. Thus, $y \in \mathcal{H}_0$. This proves that the subspace $\mathcal{H}_0$ is invariant under the action of the operator T. Similarly, we establish that the subspace $\mathcal{H}_0$ is also invariant under the action of the operator T^*.

We consider the operator $V = T|_{\mathcal{H}_0}$. By virtue of (7), this is an isometric operator. To prove that this operator is unitary, it suffices to show that $\Delta_V = \mathcal{H}_0$.

Assume that $y \in \mathcal{H}_0$ and $x = T^* y$. Since $x \in \mathcal{H}_0$ and $\|x\| = \|y\|$, by virtue of

Proposition 1, we conclude that $y = Tx = Vx$. This proves that $\Delta_V = \mathcal{H}_0$.

■

The subspace $\mathcal{H}_1 = \mathcal{H} \ominus \mathcal{H}_0$, where $\mathcal{H}_0$ is defined by equality (8), is also invariant under the action of the operator T. Let us show that the operator $T_1 = T|_{\mathcal{H}_1}$ does not induce a unitary operator in any invariant subspace other than zero. Indeed, assume that the operator T_1 (and, hence, T) induces a unitary operator U in the subspace $\mathcal{H}_2 \subset \mathcal{H}_1$ ($\mathcal{H}_2 \neq \{0\}$). Let us show that $\mathcal{H}_2$ reduces the operator T. In fact, if $y \in \mathcal{H}_2$, then $y = Ux = Tx$, where $x \in \mathcal{H}_2$. Moreover, $\|y\| = \|x\|$ and, hence, by Proposition 1, $x = T^*y \in \mathcal{H}_2$. Thus, $T^*: \mathcal{H}_2 \to \mathcal{H}_2$. In this case, $T^*|_{\mathcal{H}_2}$ coincides with U^*.

This enables us to conclude that if $x \in \mathcal{H}_2$, then

$$\| T^n x \| = \| U^n x \| = \| x \|;$$

$$\| T^{*n} x \| = \| U^{*n} x \| = \| x \| \quad (\forall n \in \mathbb{N})$$

Consequently, $x \in \mathcal{H}_0$, which is possible only for $x = 0$.

A nonunitary operator is called *completely nonunitary* if it does not coincide with a unitary operator in any nontrivial reducing subspace. Therefore, in view of the reasoning presented above, the operator T_1 is completely nonunitary.

Thus, if T is a contraction in $\mathcal{H}$, then the space $\mathcal{H}$ is representable as the orthogonal sum of its subspaces $\mathcal{H}_0$ and $\mathcal{H}_1$ invariant with respect to the operator T, where the operator $V = T|_{\mathcal{H}_0}$ is unitary and the operator $T_1 = T|_{\mathcal{H}_1}$ is completely nonunitary. In this case, for $\mathcal{H}_0 = \{0\}$, the original operator T is completely nonunitary and, for $\mathcal{H}_1 = \{0\}$, it is unitary.

As a simple example, we consider the operator of right shift in the space $\mathcal{H} = l_2$ of sequences of the form $x = (x_0, x_1, x_2, ...)$. In this case, $\mathcal{H}_+ = \mathcal{H}$ and $\mathcal{H}_- = \{0\}$ (since $T^*x \to 0$ as $n \to \infty$) and, hence, $\mathcal{H}_0 = \{0\}$. This means that T is a completely nonunitary operator.

The decomposition $T = V \oplus T_1$ of the contraction T into the unitary and completely nonunitary parts is called the *canonical decomposition* of the operator T. The corresponding canonical decomposition of the operator T^* has the form $T^* = V^* \oplus T_1^*$. This implies that the contraction T is completely nonunitary if and only if the operator T^* is completely nonunitary.

Proposition 4. *The canonical decomposition of the contraction T is unique.*

Proof. Assume that, parallel with a decomposition $T = V \oplus T_1$, the operator T admits a decomposition $T = \tilde{V} \oplus \tilde{T}_1$, where $\tilde{V}$ is a unitary operator induced by the operator T in the subspace $\tilde{\mathcal{H}}_0$ and $\tilde{T}_1$ is a completely nonunitary operator induced by T in the subspace $\tilde{\mathcal{H}}_1 = \mathcal{H} \ominus \tilde{\mathcal{H}}_0$. Since

$$\| T^n x \| = \| \tilde{V}^n x \| = \| x \|, \quad \| T^{*n} x \| = \| x \|$$

for $x \in \tilde{\mathcal{H}}_0$ and $n \in \mathbf{N}$, we have $x \in \mathcal{H}_0$, i.e., $\tilde{\mathcal{H}}_0 \subset \mathcal{H}_0$. Moreover, the operator $T|_{\mathcal{H}'}$, where $\mathcal{H}' = \mathcal{H}_0 \ominus \tilde{\mathcal{H}}_0 \subset \tilde{\mathcal{H}}_1$, is also unitary. The fact that the operator $\tilde{T}_1$ is completely nonunitary implies that $\mathcal{H}' = \{0\}$, i.e., $\tilde{\mathcal{H}}_0 = \mathcal{H}_0$. But then $\tilde{\mathcal{H}}_1 = \mathcal{H}_1$. ∎

3.   Triangulation of Contractions

Assume that T is a contraction in $\mathcal{H}$, $\mathcal{H}_1$ is a subspace invariant under the action of the operator T, and $\mathcal{H}_2 = \mathcal{H} \ominus \mathcal{H}_1$. According to the decomposition $\mathcal{H} = \mathcal{H}_1 \oplus \mathcal{H}_2$, the operators T and T^* can be represented in the form

$$T = \begin{pmatrix} T_1 & \Gamma \\ 0 & T_2 \end{pmatrix}, \quad T^* = \begin{pmatrix} T_1^* & 0 \\ \Gamma^* & T_2^* \end{pmatrix}, \tag{9}$$

where $T_1 = T|_{\mathcal{H}_1}$ and $T_2 = \left(T^*|_{\mathcal{H}_2}\right)^*$ are contractions in $\mathcal{H}_1$ and $\mathcal{H}_2$, respectively, and Γ is a linear operator that maps $\mathcal{H}_2$ into $\mathcal{H}_1$.

Assume that the operators $\mathcal{D}$ and $\mathcal{D}_*$ are defined by equalities (2) and

$$\mathcal{D}_i = (I - T_i^* T_i)^{1/2}, \quad \mathcal{D}_{i*} = (I - T_i T_i^*)^{1/2}, \tag{10}$$

where $i = 1, 2$. In what follows, we also consider the corresponding defect spaces $\mathfrak{D}_i = \overline{\mathcal{D}_i \mathcal{H}_i}$ and $\mathfrak{D}_{i*} = \overline{\mathcal{D}_{i*} \mathcal{H}_i}$.

Theorem 5 *(Sz.-Nagy–Foias). The operator Γ in (9) can be represented in the form*

$$\Gamma = \mathcal{D}_{1*} L \, \mathcal{D}_2, \tag{11}$$

where L is a contraction from $\mathfrak{D}_2$ to $\mathfrak{D}_{1*}$.

Proof. Taking relation (9) into account, we obtain

$$\| x_2 \|^2 \geq \| T x_2 \|^2 = \| T_2 x_2 \|^2 + \| \Gamma x_2 \|^2 \quad (x_2 \in \mathcal{H}_2) \tag{12}$$

and

$$\| x_1 \|^2 \geq \| T^* x_1 \|^2 = \| T_1^* x_1 \|^2 + \| \Gamma^* x_1 \|^2 \quad (x_1 \in \mathcal{H}_1). \tag{13}$$

Then, by virtue of (12) and (10),

$$\| \Gamma x_2 \|^2 \leq (x_2, x_1) - (T_2^* T_2 x_2, x_2) = (\mathcal{D}_2^2 x_2, x_2) = \| \mathcal{D}_2 x_2 \|^2$$

and, thus,

$$\| \Gamma x_2 \| \leq \| \mathcal{D}_2 x_2 \| \quad (\forall x_2 \in \mathcal{H}_2). \tag{14}$$

Similarly, taking (13) and (10) into account, we get

$$\| \Gamma^* x_1 \| \leq \| \mathcal{D}_{1*} x_1 \| \quad (\forall x_1 \in \mathcal{H}_1). \tag{15}$$

By virtue of (15), the equality $\mathcal{D}_{1*} x_1 = 0$, implies that $\Gamma^* x_1 = 0$. Consequently, we can consider the linear operator N defined by the equality

$$N(\mathcal{D}_{1*} x_1) = \Gamma^* x_1 \quad (x_1 \in \mathcal{H}_1). \tag{16}$$

By virtue of (15) and (16),

$$\| N(\mathcal{D}_{1*} x_1) \| = \| \Gamma^* x_1 \| \leq \| \mathcal{D}_{1*} x_1 \|$$

and, thus, $\| N \| \leq 1$. By extending the operator N by continuity, we obtain the contraction also denoted by N which maps $\mathfrak{D}_{1*}$ into $\mathcal{H}_2$.

But then $N^* : \mathcal{H}_2 \to \mathfrak{D}_{1*}$.

Consider a linear operator L defined by the formula

$$L(\mathcal{D}_2 x_2) = N^* x_2. \tag{17}$$

Let us show that the definition of the operator L is correct. Let $\mathcal{D}_2 x_2 = 0$. Then $\Gamma x_2 = 0$ by virtue of (14). Relation (16) implies that $N \mathcal{D}_{1*} = \Gamma^*$ and, hence, $\mathcal{D}_{1*} N^* = \Gamma$. Therefore,

$$\mathcal{D}_{1*} N^* x_2 = \Gamma x_2 = 0.$$

Let $\varphi = N^* x_2$. Then

$$(\varphi, \mathcal{D}_{1*} y) = (\mathcal{D}_{1*} \varphi, y) = (\Gamma x_2, y) = 0$$

for any y from $\mathcal{H}_1$ and, hence, $\varphi \perp \mathfrak{D}_{1*}$. On the other hand, $\Delta_{N*} \subset \mathfrak{D}_{1*}$ and we have $\varphi \in \mathfrak{D}_{1*}$. Thus, $\varphi = 0$. This proves that the definition of the operator L is correct.

Relations (16) and (17) imply that $\Gamma = \mathfrak{D}_{1*} N^*$ and $N^* = L \mathfrak{D}_2$, whence we get relation (11).

Let us show that L is a contraction (from $\mathfrak{D}_2$ to $\mathfrak{D}_{1*}$). Indeed, by virtue of the equalities $T_1^* \mathfrak{D}_{1*} = \mathfrak{D}_1 T_1^*$ and (11), we find that

$$T_1^* \Gamma = \mathfrak{D}_1 T_1^* L \mathfrak{D}_2. \tag{18}$$

If in this case $x = x_1 \oplus x_2$ (this notation traditionally means that $x_1 \in \mathcal{H}_1$, $x_2 \in \mathcal{H}_2 = \mathcal{H} \ominus \mathcal{H}_1$ and, hence, $x_1 \perp x_2$), then, in view of relations (2) and (9) we obtain

$$\| \mathfrak{D}x \|^2 = ((I - T^* T)x, x) = \| x \|^2 - \| Tx \|^2$$

$$= (\| x_1 \|^2 + \| x_2 \|^2) - (\| T_1 x_1 + \Gamma x_2 \|^2 + \| T_2 x_2 \|^2)$$

$$= \| x_1 \|^2 - \| T_1 x_1 \|^2 + \| x_2 \|^2 - \| T_2 x_2 \|^2 - 2 \operatorname{Re}(T_1 x_1, \Gamma x_2) - \| \Gamma x_2 \|^2$$

Thus, by virtue of (11) and (18),

$$\| \mathfrak{D}x \|^2 = \| \mathfrak{D}_1 x_1 \|^2 + \| \mathfrak{D}_2 x_2 \|^2 - \| \mathfrak{D}_{1*} L \mathfrak{D}_2 x_2 \|^2 - 2 \operatorname{Re}(x_1, \mathfrak{D}_1 T_1^* L \mathfrak{D}_2 x_2)$$

$$= \| \mathfrak{D}_1 x_1 \|^2 + \| \mathfrak{D}_2 x_2 \|^2 - \| L \mathfrak{D}_2 x_2 \|^2$$

$$+ \| T_1^* L \mathfrak{D}_2 x_2 \|^2 - 2 \operatorname{Re}(\mathfrak{D}_1 x_1, T_1^* L \mathfrak{D}_2 x_2)$$

$$= \| \mathfrak{D}_1 x_1 - T_1^* L \mathfrak{D}_2 x_2 \|^2 + \| \mathfrak{D}_2 x_2 \|^2 - \| L \mathfrak{D}_2 x_2 \|^2 \geq 0. \tag{19}$$

Let $y_1 = \mathfrak{D}_1 x_1$ and $y_2 = \mathfrak{D}_2 x_2$. Then it follows from (19) that

$$\| L y_2 \|^2 \leq \| y_2 \|^2 + \| y_1 - T_1^* L y_2 \|^2. \tag{20}$$

Moreover,

$$T_1^* L y_2 = T_1^* L \mathfrak{D}_2 x_2 = T_1^* N^* x_2 \in T_1^* \mathfrak{D}_{1*} \subset \mathfrak{D}_1.$$

Since $y_1 \in \mathfrak{D}_1 \mathcal{H}_1$ and $\overline{\mathfrak{D}_1 \mathcal{H}_1} = \mathfrak{D}_1$, the second term on the right-hand side of inequality (20) can be made as small as desired by the proper choice of the vector x_1 (and, hence, y_1), . Consequently, $\| L y_2 \|^2 \leq \| y_2 \|^2$, i.e., L is a contraction from $\mathcal{D}_2$ to $\mathcal{D}_{1*}$. By extending the operator L by continuity, we arrive at the contraction L: $\mathcal{D}_2 \to \mathcal{D}_{1*}$.

∎

Let us show that the converse assertion is also true.

Theorem 6. *Assume that the operator T can be represented in the form (9), where T_1 is a contraction in $\mathcal{H}_1$, T_2 is a contraction in $\mathcal{H}_2$, and the operator Γ is defined by relation (11), where L is a contraction from $\mathcal{D}_2 = \overline{\mathcal{D}_2 \mathcal{H}_2}$ to $\mathcal{D}_{1*} = \overline{\mathcal{D}_{1*} \mathcal{H}_1}$. Then T is a contraction in $\mathcal{H} = \mathcal{H}_1 \oplus \mathcal{H}_2$.*

Proof. Let $x = x_1 \oplus x_2$. Then

$$\| Tx \|^2 = \| T_1 x_1 + \Gamma x_2 \|^2 + \| T_2 x_2 \|^2$$

$$= \| T_1 x_1 \|^2 + 2 \operatorname{Re} (T_1 x_1, \Gamma x_2) + \| \Gamma x_2 \|^2 + \| T_2 x_2 \|^2. \tag{21}$$

By using the equalities $\mathcal{D}_{1*} T_1 = T_1 \mathcal{D}_1$ and $\Gamma = \mathcal{D}_{1*} L \mathcal{D}_2$, we obtain

$$(T_1 x_1, \Gamma x_2) = (\mathcal{D}_1 x_1, T_1^* L \mathcal{D}_2 x_2) \tag{22}$$

and

$$\| \Gamma x_2 \|^2 = \| L \mathcal{D}_2 x_2 \|^2 - \| T_1^* L \mathcal{D}_2 x_2 \|^2. \tag{23}$$

Then, by virtue of equalities (21)–(23) and the inequality $\| L \| \leq 1$, we get

$$\| Tx \|^2 = \| T_1 x_1 \|^2 + \| \mathcal{D}_1 x_1 \|^2$$

$$- \| \mathcal{D}_1 x_1 - T_1^* L \mathcal{D}_2 x_2 \|^2 + \| L \mathcal{D}_2 x_2 \|^2 + \| T_2 x_2 \|^2$$

$$\leq \| T_1 x_1 \|^2 + \| \mathcal{D}_1 x_1 \|^2 + \| \mathcal{D}_2 x_2 \|^2 + \| T_2 x_2 \|^2$$

$$= \| x_1 \|^2 + \| x_2 \|^2 = \| x \|^2. \tag{24}$$

Thus, $\| T \| \leq 1$.

∎

4. Triangulation of Completely Nonunitary Contractions

If a completely nonunitary contraction T can be represented in the form (9), then it is easy to show that the contractions T_1 and T_2 are also completely nonunitary. The converse assertion holds not always. Indeed, the operators T_1 and T_2 in (9) can be completely nonunitary but this is not true for the contraction T.

We demonstrate this with the following example: Let $\mathcal{H} = l_2(-\infty, \infty)$ be a space of bilateral sequences. Consider the following subspaces of the space $\mathcal{H}$:

$$\mathcal{H}_1 = \langle e_0, e_1, e_2, \dots \rangle, \qquad \mathcal{H}_2 = \langle e_{-1}, e_{-2}, e_{-3}, \dots \rangle$$

where e_k is a unit vector in $\mathcal{H}$ whose kth component is equal to one while all the other components are equal to zero.

In the space $\mathcal{H}$, we introduce the operator of right shift by the formula

$$Te_k = e_{k+1} \quad (k \in \mathbb{Z}).$$

The space $\mathcal{H}_1$ is invariant under the action of this operator. Therefore, the operator T can be represented in the form (9), where T_1 is the operator of right shift in $\mathcal{H}_1$, T_2 is the operator adjoint to the operator of left shift in $\mathcal{H}_2$ (i.e., to the operator $T^* |_{\mathcal{H}_2}$), $\Gamma = P_0 L P_{-1}$, $P_0 = (I - T_1 T_1^*)^{1/2}$ is the orthoprojector onto the subspace $\langle e_0 \rangle$ in $\mathcal{H}$, $P_{-1} = (I - T_2^* T_2)^{1/2}$ is the orthoprojector onto the subspace $\langle e_{-1} \rangle$ in $\mathcal{H}$, and the operator L acting from $\langle e_{-1} \rangle$ into $\langle e_0 \rangle$ is defined by the equality $Le_{-1} = e_0$.

According to Subsection 2, the operators T_1 and T_2 are completely nonunitary, while the operator T is unitary.

At the same time, we have the following theorem:

Theorem 7. *Assume that T_1 and T_2 are completely nonunitary contractions in the subspaces $\mathcal{H}_1$ and $\mathcal{H}_2$, respectively, and the operator Γ is defined by equality (11). Also assume that, under these assumptions, the inequalities*

$$\|Lx_2\| < \|x_2\| \quad \text{and} \quad \|L^*x_1\| < \|x_1\| \tag{25}$$

hold for any nonzero vectors $x_1 \in \mathcal{D}_{1*}\mathcal{H}_1$ and $x_2 \in \mathcal{D}_2 \mathcal{H}_2$. Then the operator T defined by (9) is a completely nonunitary contraction.

Proof. Assume that $x = x_1 \oplus x_2 \in \mathcal{H}_0$, where $\mathcal{H}_0$ is defined by (8). Then by virtue of (5) and (6), we have

$$\|T^n x\| = \|x\|, \quad \|T^{*n} x\| = \|x\| \quad (\forall n \in \mathbb{N}). \tag{26}$$

In particular, we have $\|Tx\| = \|x\|$. This means that relation (24) turns into the equality and, hence,

$$\|L\mathcal{D}_2 x_2\| = \|\mathcal{D}_2 x_2\|, \quad \mathcal{D}_1 x_1 = T_1^* L \mathcal{D}_2 x_2. \tag{27}$$

According to (25), the first equality in (27) is possible only in the case where $\mathcal{D}_2 x_2 = 0$. Then, by virtue of (27) and (11), we have $\mathcal{D}_1 x_1 = 0$ and $\Gamma x_2 = 0$. Thus, $Tx = T_1 x_1 \oplus T_2 x_2$. Consequently,

$$T^n x = T_1^n x_1 \oplus T_2^n x_2 \quad (\forall n \in \mathbf{N})$$

and, therefore,

$$\mathcal{D}_1 T_1^n x_1 = 0, \quad \mathcal{D}_2 T_2^n x_2 = 0 \quad (\forall n \in \mathbf{N}). \tag{28}$$

By using the first equality in (28), we get

$$\| T_1^{n+1} x_1 \| = \| T_1^n x_1 \| \quad (\forall n \in \mathbf{N}),$$

whence

$$\| T_1^n x_1 \| = \| x_1 \| \quad (n \in \mathbf{N}). \tag{29}$$

Similarly, taking the second equality in (26) into account, we find that

$$\| T_1^{*n} x_1 \| = \| x_1 \| \quad (n \in \mathbf{N}). \tag{30}$$

Since T_1 is a completely nonunitary contraction, equalities (29) and (30) are possible only for $x_1 = 0$.

The same argument and the second equality in (28) imply that $x_2 = 0$.

Thus, it follows form relations (26) that $x = 0$. Consequently, T is a completely nonunitary contraction.

∎

Theorem 7 enables us to construct a new completely nonunitary contraction in terms of given completely nonunitary contractions T_1 and T_2.

For example, let the operators T_1 and T_2 be defined as at the beginning of Subsection 4. The operator $L: \langle e_{-1} \rangle \rightarrow \langle e_0 \rangle$ is defined by the formula $Le_{-1} = \alpha e_0$, where $\alpha \in \mathbf{C}$ and $|\alpha| < 1$. In the space $\mathcal{H} = \mathcal{H}_1 \oplus \mathcal{H}_2$, we introduce an operator

$$T_\alpha = \begin{pmatrix} T_1 & \Gamma \\ 0 & T_2 \end{pmatrix} \quad (\Gamma = P_0 L P_{-1}).$$

By virtue of Theorem 7, T_α is a completely nonunitary contraction. If in this case

$$x = (\ldots, x_{-2}, x_{-1}, \boxed{x_0}, x_1, x_2, \ldots) \in \mathcal{H}$$

(the frame denotes the location of the zero element), then

$$T_\alpha x = (\ldots, x_{-3}, x_{-2}, \boxed{\alpha x_{-1}}, x_0, x_1, \ldots),$$

i.e., T_α is an operator of right weighed shift.

Thus, by a slight "correction" of the given unitary operator (of right shift) T, we have constructed a completely nonunitary contraction T_α.

Remark. In Subsections 3 and 4, we followed Sz.-Nagy and Foias [13]. The canonical decomposition of the contraction T in Subsection 2 was obtained independently by Sz.-Nagy and Foias [5] and Langer [1].

Appendix 2. The Structure of J-Nonexpanding Operators

In Chapters 2–4, we frequently used the concepts of *J*-nonexpanding matrices and operators. Here, we study some properties of these operators.

1. Doubly J-Nonexpanding Operators

Assume that $\mathcal{H}$ is a Hilbert space with a scalar product $(\cdot, \cdot)$ and J is a self-adjoint unitary operator acting in $\mathcal{H}$. In $\mathcal{H}$, we define an indefinite scalar product $[x, y] = (Jx, y)$ (an indefinite metric). In what follows, we use standard terminology: *J*-metric, *J*-self-adjointness, etc. (see, e.g., Bognar [1] and Azizov and Iokhvidov [1]). In particular, the *J*-adjoint operator A^+ is connected with the adjoint operator A^* by the equality $A^+ = JA^*J$.

A bounded operator A is called *J-nonexpanding in a subspace* $\mathcal{H}_1$ of the space $\mathcal{H}$ provided that

$$[Ax, Ax] \leq [x, x] \quad (\forall x \in \mathcal{H}_1).$$

If, in addition, $\mathcal{H}_1 = \mathcal{H}$, then the operator A is called *J-nonexpanding (J-n.)*.

An operator A is called a *doubly J-nonexpanding in a subspace* $\mathcal{H}_1$ if, in this subspace, both the operator A and the operator A^* are *J*-nonexpanding. In particular, if $\mathcal{H}_1 = \mathcal{H}$, then the operator A is called *doubly J-nonexpanding* (d. *J*-n.).

Potapov [1] proved that an arbitrary *J*-n. operator acting in a finite-dimensional space is d. *J*-n. At the same time, one can easily construct a simple example of the operator of shift in the space l_2, which demonstrates that not all *J*-n. operators acting in an infinite-dimensional space are d. *J*-n.

Later, Ginzburg [1] showed that a *J*-n. operator A is d. *J*-n. if and only if one of the following equivalent conditions holds:

(a) $0 \in \rho(P_- A + P_+)$;

(b) $0 \in \rho(AP_- + P_+)$,

where the orthoprojectors P_+ and P_- are defined by the formulas

$$P_+ = \frac{1}{2}(I + J) \quad \text{and} \quad P_- = \frac{1}{2}(I - J). \tag{1}$$

We also mention that several other equivalent criteria of double J-nonexpandability can be found in Azizov and Iokhvidov [2], Chap. 2, § 1.8).

For an arbitrary J-n. operator S, Kuzhel [1, 3] described the set $\mathfrak{M}$ of all subspaces of the space $\mathcal{H}$, for which the operator A^* is J-n. In particular, the subspace $\mathcal{H}_0 = $ Ker $\tilde{P}(I + AA^*)$, where $\tilde{P}$ is the orthoprojector onto the subspace Ker $(P_- A^* + P_+)$ in $\mathcal{H}$, belongs to $\mathfrak{M}$ and is maximal in the given class. This enables us to obtain both the well-known criteria of double J-nonexpandability and some new criteria (e.g., the equality $\tilde{P}(I + AA^*) = 0$). It is worth noting that the property of "J-nonexpandability" of the operator A^* in the subspace $\mathcal{H} \ominus \mathcal{H}_0$ depends on the norm $\|A\|$. For example, if $\|A\| < \sqrt{c}$ (≈ 1.683), where c is the sole positive root of the polynomial $f(x) = 1 + 2x + 2x^2 - x^3$, then the operator A^* is J-expanding in the subspace $\mathcal{H} \ominus \mathcal{H}_0$. At the same time, there exist J-n. operators A such that the operator A^* is J-n. both in the subspace $\mathcal{H}_0$ and in the subspace $\mathcal{H} \ominus \mathcal{H}_0$. In this case, however, the operator A^* is not J-n. in the entire space $\mathcal{H}$ (it is clear that $\|A\| > \sqrt{c}$ for such operators).

2.   Regular Dilations of J-Nonexpandable Operators

As shown in Section 4 (Chapter 4), for an arbitrary bounded operator T, one construct a J-unitary dilation U, where the self-adjoint unitary operator J is defined in terms of the operators $V_T = \text{sign}(I - T^*T)$ and $V_{T^*} = \text{sign}(I - TT^*)$. The dependence of the operator J on T makes the study of the operator T in terms of its dilation U much more complicated. At the same time, it seems reasonable to extend to a wider class of bounded operators deep results obtained by Sz.-Nagy and Foias when investigating contracting operators by the method of unitary dilation.

Below, we show how the concepts of dilation and the characteristic function of a contracting operator can be generalized to the case of J-n. operators.

Assume that A is a bounded operator acting in a Hilbert space $\mathcal{H}$. A bounded operator U acting in the Hilbert space $\tilde{\mathcal{H}} \supset \mathcal{H}$ is called a dilation of the operator A if

$$A^n = P_{\mathcal{H}} U^n \quad (n \in \mathbb{N}), \tag{2}$$

where $P_{\mathcal{H}}$ is an orthoprojector onto $\mathcal{H}$ in $\tilde{\mathcal{H}}$.

By representing the operator U as an operator matrix $\left\| \tilde{U}_{ij} \right\|_{ij=1}^{2}$ that corresponds to the decomposition

$$\tilde{\mathcal{H}} = \mathcal{H} \oplus \mathcal{H}' \quad (\mathcal{H}' = (I - P_{\mathcal{H}})\tilde{\mathcal{H}}), \tag{3}$$

one can easily verify that condition (2) is equivalent to the following relations:

$$\tilde{U}_{11} = A, \quad \tilde{U}_{12}\tilde{U}_{22}^n\tilde{U}_{21} = 0 \quad (\forall n \in \mathbb{N} \cup \{0\}). \tag{4}$$

In what follows, the Hilbert space $\mathcal{H}$ with indefinite metric generated by a unitary self-adjoint operator J is called the *J-space* or the space $\langle \mathcal{H}, J \rangle$.

In the space $\tilde{\mathcal{H}}$, we consider a unitary self-adjoint operator $\tilde{J}$ that can be represented in the form

$$\tilde{J} = J \oplus I' \tag{5}$$

with respect to decomposition (3) (here, I' is the identity operator in $\mathcal{H}'$). If a $\tilde{J}$-unitary operator U acting in the space $\langle \tilde{\mathcal{H}}, \tilde{J} \rangle$ is a dilation of the operator A, it is called a *regular dilation* of the operator A.

Two regular dilations U_1 and U_2 of the operator A that act in the spaces $\langle \tilde{\mathcal{H}}_1, \tilde{J}_1 \rangle$ and $\langle \tilde{\mathcal{H}}_2, \tilde{J}_2 \rangle$, respectively, are called *isomorphic* if there exists a unitary mapping φ of the space $\tilde{\mathcal{H}}_2$ onto $\tilde{\mathcal{H}}_1$ such that $U_2 = \varphi^{-1}U_1\varphi$ and $\varphi|_{\mathcal{H}} = I$.

Proposition 1. *If there exists a regular dilation of an operator A, then A is a d.J-n. operator.*

Proof. Let U be a regular dilation of the operator A. Then, in view of (2), we have

$$[x, x] - [Ax, Ax] = [(I - P_{\mathcal{H}})Ux, (I - P_{\mathcal{H}})Ux].$$

for all $x \in \mathcal{H}$. By virtue of relations (3) and (5), the right-hand side of this equality is nonnegative and, hence, A is a *J-n.* operator. Similarly, it follows from the equality $A^+ = P_{\mathcal{H}}U^+$ that A^+ is a *J-n.* operator and, hence, A is a d. *J-n.* operator.

■

Consider the transformation

$$F_{\mathcal{H}}(X) = (P_+X + P_-)(P_-X + P_+)^{-1}$$

with the domain of definition

$$\mathfrak{D}_{F_{\mathcal{H}}} = \{X \in \mathcal{B}[\mathcal{H}] \mid 0 \in \rho(P_- X + P_+)\},$$

which is called the *Potapov–Ginzburg transformation* (PG-transformation).

Proposition 2. *Assume that* X *is a bounded operator in* $\langle \mathcal{H}, J \rangle$ *and* W *is its dilation in* $\langle \tilde{\mathcal{H}}, \tilde{J} \rangle$. *If* $X \in \mathfrak{D}_{F_{\mathcal{H}}}$ *then* $W \in \mathfrak{D}_{F_{\tilde{\mathcal{H}}}}$ *and the operator* $U = F_{\tilde{\mathcal{H}}}(W)$ *is a dilation of the operator* $Y = F_{\mathcal{H}}(X)$.

Proof. Let $x \in \tilde{\mathcal{H}}$. Then $x = x_1 + x_2$, where $x_1 \in \mathcal{H}$ and $x_2 \in \mathcal{H}'$. Taking relations (1) and (5) into account, we get

$$\tilde{P}_+ = \frac{1}{2}\left(\tilde{I} + \tilde{J}\right) = P_+ \oplus I', \quad \tilde{P}_- = \frac{1}{2}\left(\tilde{I} - \tilde{J}\right) = P_- \oplus 0. \tag{6}$$

Hence,

$$y = \left(\tilde{P}_- W + \tilde{P}_+\right) x = (P_- W_{11} + P_+) x_1 + P_- W_{12} x_2 + x_2, \tag{7}$$

where W_{ij} are operator elements of the matrix representation $\left\| W_{ij} \right\|_{ij=1}^{2}$ of W with respect to decomposition (3). The operator W is a dilation of the operator X. Therefore, by virtue of (4), we have $W_{11} = X$.

The operator equation (7) can be rewritten as a system of equations

$$\begin{cases} (P_- X + P_+) x_1 + P_- W_{12} x_2 = y_1, \\ x_2 = y_2, \end{cases}$$

where $y = y_1 + y_2$, $y_1 \in \mathcal{H}$, $y_2 \in \mathcal{H}'$. It is clear that this system is solvable for any $y_1 \in \mathcal{H}$ and $y_2 \in \mathcal{H}'$; its solution has the form

$$x = \left(\tilde{P}_- W + \tilde{P}_+\right)^{-1} y = (P_- X + P_+)^{-1}(y_1 - P_- W_{12} y_2) + y_2. \tag{8}$$

The last equality implies that $0 \in \rho\left(\tilde{P}_- W + \tilde{P}_+\right)$ and, hence, $W \in \mathfrak{D}_{F_{\tilde{\mathcal{H}}}}$.

By virtue of equalities (6) and (8), the operator

$$U = F_{\tilde{\mathcal{H}}}(W) = \left(\tilde{P}_+ W + \tilde{P}_-\right)\left(\tilde{P}_- W + \tilde{P}_+\right)^{-1}$$

can be represented with respect to decomposition (3) in the form

$$U = \begin{pmatrix} Y & -(YP_- - P_+)W_{12} \\ W_{21}(P_-Y + P_+) & W_{22} - W_{21}P_-YP_-W_{12} \end{pmatrix}. \tag{9}$$

The operator elements of this matrix satisfy relations (4) and, hence, the operator U is a dilation of the operator $Y = F_{\mathcal{H}}(X)$.

■

As is known (see Ginzburg [1]), the PG-transformation maps $\mathfrak{D}_F$ onto itself bijectively and $F^2 = I$. In particular, this transformation bijectively maps the class of d. J-n. (J-unitary) operators onto the intersection of the set of contracting (unitary) operators and the lineal $\mathfrak{D}_F$. By using this fact and Proposition 2, one can prove the following assertion:

Proposition 3. *Assume that* A *is a d.J-n. operator in the space* $\langle \mathcal{H}, J \rangle$ *and an operator* W *acting in a space* $\langle \tilde{\mathcal{H}}, \tilde{J} \rangle$ *is a unitary dilation of the contraction* $T = F_{\mathcal{H}}(A)$. *Then* $W \in \mathfrak{D}_{F_{\tilde{\mathcal{H}}}}$ *and the operator* $U = F_{\tilde{\mathcal{H}}}(W)$ *is a regular dilation of the operator* A. *The converse statement is also true, namely, if* U *is a regular dilation of the operator* A, *then* $W = F_{\tilde{\mathcal{H}}}(U)$ *is a unitary dilation of the contracting operator* T.

Thus, the study of regular dilations of the d.J-n. operator A is reduced to the study of unitary dilations of the contracting operator $T = F_{\mathcal{H}}(A)$. This enables us to prove the following statement that generalizes the well-known Sz.-Nagy–Foias theorem [4]:

Theorem 4. *An arbitrary d.J-n. operator* A *which acts in the space* $\langle \mathcal{H}, J \rangle$ *possesses a regular dilation* U *in a space* $\langle \tilde{\mathcal{H}}, \tilde{J} \rangle$ *and this dilation is minimal in the sense that*[9]

$$\tilde{\mathcal{H}} = \overset{\infty}{\underset{-\infty}{\vee}} U^n \mathcal{H}.$$

The minimal $\tilde{J}$*-unitary dilation of the operator* A *is determined uniquely to within an isomorphism.*

A subspace L is called $\tilde{J}$*-wandering* for a $\tilde{J}$-unitary operator U if, for all integer n and m $(n \neq m)$, the subspaces $U^n L$ and $U^m L$ are $\tilde{J}$-orthogonal.

---

[9] $\underset{\alpha}{\vee} L_\alpha$ is the closure of the linear span of the sets L_α.

Lemma 5. *Let* U *be the minimal regular dilation of the operator* A. *Then the subspaces*

$$L_+ = \overline{(U - A)\,\mathcal{H}}, \quad L_- = \overline{(U^+ - A^+)\,\mathcal{H}} \tag{10}$$

are $\tilde{J}$-*wandering for the operator* U *and*

$$\tilde{\mathcal{H}} = \ldots [\oplus]\, U^+ L_-[\oplus]\, L_-[\oplus]\, \mathcal{H}[\oplus]\, L_+[\oplus]\, U L_+[\oplus] \ldots, \tag{11}$$

where the sign $[\oplus]$ *denotes both* $\tilde{J}$-*orthogonality and the orthogonality of the elements of the direct sum.*

Proof. Parallel with the subspaces L_+ and L_-, we consider the subspaces

$$\hat{L}_+ = \overline{(W - T)\,\mathcal{H}} \quad \text{and} \quad \hat{L}_- = \overline{(W^* - T^*)\,\mathcal{H}},$$

where $T = F_{\mathcal{H}}(A)$ is a contracting operator acting in $\mathcal{H}$ and $W = F_{\mathcal{H}}(U)$ is its minimal unitary dilation acting in the space $\tilde{\mathcal{H}}$.

According to representation (9), we obtain

$$L_+ = \overline{(U - A)\,\mathcal{H}} = \overline{W_{21}(P_- A + P_+)\,\mathcal{H}} = \overline{W_{21}\,\mathcal{H}} = \overline{(W - T)\,\mathcal{H}} = \hat{L}_+,$$

$$L_- = \overline{(U^+ - A^+)\,\mathcal{H}} = \overline{W_{12}^*(P_- A^+ + P_+)\,\mathcal{H}} = \overline{W_{12}^*\,\mathcal{H}} = \overline{(W^* - T^*)\,\mathcal{H}} = \hat{L}_-$$

Similarly, taking into account relations (4) for the elements W_{ij} of the dilation W, we get

$$U^n L_+ = \overline{W_{22}^n W_{21}\,\mathcal{H}} = W^n \hat{L}_+ \quad (n \in \mathbf{N}),$$

$$(U^+)^n L_- = \overline{(W_{22}^*)^n W_{12}^*\,\mathcal{H}} = (W^*)^n \hat{L}_-. \tag{12}$$

As is known, the space $\tilde{\mathcal{H}}$ of the minimal unitary dilation W of the contracting operator T is representable in the form of an orthogonal sum

$$\tilde{\mathcal{H}} = \ldots \oplus W^* \hat{L}_- \oplus \hat{L}_- \oplus \mathcal{H} \oplus \hat{L}_+ \oplus W \hat{L}_+ \oplus \ldots. \tag{13}$$

By using equalities (12) and decomposition (13), we conclude that the subspaces L_+ and L_- are $\tilde{J}$-wandering for the $\tilde{J}$-unitary operator U and decomposition (11) takes place. By virtue of (5), the orthogonality of lineals in $\tilde{\mathcal{H}} \ominus \mathcal{H}$ is equivalent to their $\tilde{J}$-orthogonality.

■

Taking decomposition (11) into account, we obtain

$$U(L_-[\oplus] \mathcal{H}) = \mathcal{H}[\oplus] L_+. \tag{14}$$

Then, for any $x \in \mathcal{H}$ and $x_- \in L_-$,

$$Ux = Ax + U_{21}x, \qquad Ux_- = U_{12}x_- + U_{22}x_-, \tag{15}$$

where, in view of (14), $U_{21}: \mathcal{H} \rightarrow L_+$, $U_{12}: L_- \rightarrow \mathcal{H}$, and $U_{22}: L_- \rightarrow L_+$.

Consider the Hilbert space $\tilde{H}$ of vectors of the form

$$\tilde{x} = (\ldots, x_{-1}, \boxed{x_0}, x_1, \ldots) \qquad (x_0 \in \mathcal{H}, \ x_{\pm n} \in L_\pm)$$

with the norm

$$\|\tilde{x}\|^2 = \sum_{-\infty}^{\infty} \|x_i\|^2 < \infty$$

and the indefinite metric

$$[\tilde{x}, \tilde{y}] = (Jx_0, y_0) + \sum_{i \neq 0} (x_i, y_i) \tag{16}$$

where

$$\tilde{y} = (\ldots, y_{-1}, \boxed{y_0}, y_1, \ldots) \in \tilde{H}.$$

The space $\tilde{H}$ is a $\tilde{J}$-space with a self-adjoint unitary operator $\tilde{J}$ defined by the equality

$$\tilde{J}\tilde{x} = (\ldots, x_{-1}, \boxed{Jx_0}, x_1, \ldots).$$

In view of decomposition (11), the $\tilde{J}$-space $\tilde{\mathcal{H}}$ admits natural identification with the $\tilde{J}$-space $\tilde{H}$. In this case, the element $x \in \mathcal{H}$ is associated with the vector $\tilde{x} = (\ldots, 0, \boxed{x}, 0, \ldots) \in \tilde{H}$.

By virtue of (15), we conclude that

$$U\tilde{x} = (\dots, x_{-3}, x_{-2}, \boxed{Ax_0 + U_{12}x_{-1}}, U_{21}x_0 + U_{22}x_{-1}, x_1, x_2, \dots). \qquad (17)$$

Then, in view of (16), we get

$$U^+\tilde{y} = (\dots, y_{-1}, U_{12}^*Jy_0 + U_{22}^*y_1, \boxed{A^+y_0 + JU_{21}^*y_1}, y_2, y_3, \dots). \qquad (18)$$

Proposition 6. *If U is the minimal regular dilation of the d.J-n. operator A and $|\omega| > 1$, then $\omega \in \rho(U)$ if and only if $\omega \in \rho(A)$.*

Proof. We set

$$D_+ = \overset{\infty}{\underset{0}{[+]}} U^n \mathcal{L}_+ \quad \text{and} \quad D_- = \overset{\infty}{\underset{0}{[+]}} (U^+)^n \mathcal{L}_-.$$

By virtue of (17), we conclude that the operators $R_+ = P_{D_+}U|_{D_+}$ and $R_- = P_{D_-}U|_{D_-}$ are contracting in the spaces D_+ and D_-, respectively, and $\omega \in \rho(R_+) \cap \rho(R_-)$. In view of (17), this implies that the equation $(U - \omega I)\tilde{x} = \tilde{y}$ is equivalent to the following system of equations:

$$P_{D_-}\tilde{x} = (R_- - \omega I)^{-1} P_{D_-}\tilde{y},$$

$$(A - \omega I)x_0 = y_0 - U_{12}P_{\mathcal{L}_-}\tilde{x}, \qquad (19)$$

$$P_{D_+}\tilde{x} = (R_+ - \omega I)^{-1} (P_{D_+}\tilde{y} - U_{21}x_0 - U_{22}P_{\mathcal{L}_-}\tilde{x}).$$

System (19) implies that the equation $(U - \omega I)\tilde{x} = \tilde{y}$ is uniquely solvable for all $\tilde{y} \in \tilde{\mathcal{H}}$ if and only if the equation $(A - \omega I)x_0 = y_0$ is uniquely solvable for all $y_0 \in \mathcal{H}$. This assertion completes the proof.

■

3. Characteristic Functions of Doubly J-Nonexpanding Operators

Let U be the minimal regular dilation of a d.J-n. operator A.
The operator-valued function

$$\theta_A(\lambda) = P_{\mathcal{L}_-}(U - \lambda I)^{-1} P_{\mathcal{L}_+}, \qquad \lambda \in \rho(U), \quad |\lambda| < 1,$$

where $P_{\mathcal{L}_\pm}$ are the orthoprojectors onto the subspaces $\mathcal{L}_\pm$ in $\tilde{\mathcal{H}}$, is called the characteristic function of the operator A. For fixed λ, the value of the characteristic function $\theta_A(\lambda)$ is a bounded operator acting from L_+ to L_-.

It follows from the definition of characteristic function and relations (10) that

$$\theta_A^+(\lambda) = \theta_{A^+}(\bar{\lambda}).$$

Let $\lambda \in \rho(U)$ ($|\lambda| < 1$). For every $u \in L_+$, we set

$$\tilde{x} = U^+ u = \left(\ldots, 0, U_{22}^* u, \boxed{JU_{21}^* u}, 0 \ldots \right).$$

On solving the equation $(U^+ - \lambda^{-1} I)\tilde{y} = \tilde{x}$, we obtain

$$y_{-1} = P_{L_-}\tilde{y} = \lambda \left(U_{12}^* J(A^+ - \lambda^{-1} I)^{-1} JU_{21}^* - U_{22}^* \right) u,$$

where, by virtue of Proposition 6, $1/\lambda \in \rho(A^+)$.

Thus,

$$\theta_A(\lambda)u = -\frac{1}{\lambda} P_{L_-}(U^+ - \lambda^{-1} I)^{-1} U^+ u$$

$$= U_{22}^* u + \lambda U_{12}^* J(I - \lambda A^+)^{-1} JU_{21}^* u. \tag{20}$$

Note that in the case where $\mathcal{H}$ is a Hilbert space and the operator A is contracting in $\mathcal{H}$, the characteristic function defined above coincides with the characteristic function introduced by Sz.-Nagy and Foias [4].

It follows from the $\tilde{J}$-unitarity of the operator U and relations (17) and (18) that

$$I - U_{22}^* U_{22} = U_{12}^* JU_{12}, \quad I - U_{22} U_{22}^* = U_{21} JU_{21}^*, \tag{21}$$

$$I - A^+ A = JU_{21}^* U_{21}, \quad I - AA^+ = U_{12} U_{12}^* J, \tag{22}$$

$$U_{22}^* U_{21} = -U_{12}^* JA, \quad U_{12} U_{22}^* = -AJU_{21}^*. \tag{23}$$

By using equalities (20)–(23), we obtain

$$U_{22}\theta_A(\lambda) = I - U_{21}(I - \lambda A^+)^{-1} JU_{21}^*, \tag{24}$$

$$U_{12}\theta_A(\lambda) = (\lambda I - A)(I - \lambda A^+)^{-1} JU_{21}^*, \tag{25}$$

$$\theta_A(\lambda) U_{21} = U_{12}^* J(I - \lambda A^+)^{-1}(\lambda I - A). \tag{26}$$

By applying relations (24) and (25), we can immediately check that

$$I - \theta_A^+(\mu)\theta_A(\lambda) = (1 - \lambda\bar{\mu}) U_{21}(I - \bar{\mu}A)^{-1}(I - \lambda A^+)^{-1} JU_{21}^*.$$

Thus, for any $x \in \mathcal{L}_+$,

$$\|x\|^2 - \|\theta_A(\lambda)x\|^2 = (1 - |\lambda|^2)[(I - \lambda A^+)^{-1} JU_{21}^* x,\ (I - \lambda A^+)^{-1} JU_{21}^* x]$$

and, hence, unlike the case of Hilbert spaces, the values of the characteristic function $\theta_A(\lambda)$ are not, generally speaking, contracting operators.

Lemma 7. *If A is a d.J-n. operator, then $0 \in \sigma_z(A)$ ($z \in \{p,\ r,\ c\}$) if and only if $0 \in \sigma_z(U_{22}^*)$.*

Proof. Assume that $Ax = 0$, where $x \in \mathcal{H}$, $x \neq 0$. Taking the first equality in (22) into account, we conclude that $JU_{21}^* U_{21} x = x$ and, therefore, $y = U_{21} x \neq 0$. Moreover, according to (23), $U_{22}^* y = 0$ and, hence, $0 \in \sigma_p(U_{22}^*)$. Conversely, the fact that $U_{22}^* y = 0$ implies (by virtue of the second equalities in (21) and (23)) that $x = JU_{21}^* y \neq 0$ and $Ax = 0$. Therefore, $0 \in \sigma_p(A) \Leftrightarrow 0 \in \sigma_p(U_{22}^*)$.

If $0 \in \sigma_r(A)$, then there exists a nonzero vector $x \in \mathrm{Ker}\, A^+$. As above, we conclude that $U_{22} y = 0$, where $y = U_{12}^* Jx$ and, consequently, $0 \in \sigma_r(U_{22}^*)$. The case $0 \in \sigma_r(U_{22}^*)$ can be investigated similarly.

Now let $0 \in \sigma_c(A)$. Then there exists a sequence $\{x_n\}$, $\|x_n\| = 1$, such that

$$\lim_{n \to \infty} Ax_n = 0.$$

We set $y_n = U_{21} x_n$. Relation (22) yields

$$\lim_{n \to \infty} \|JU_{21}^* y_n\| = 1;$$

thus, without loss of generality, we can suppose that $\|y_n\| \geq \beta > 0$. Moreover, by virtue of (23), we have

$$\lim_{n \to \infty} \|U_{22}^* y_n\| = \lim_{n \to \infty} \|U_{12}^* JAx_n\| = 0$$

and, hence, $0 \in \sigma_c(U_{22}^*)$.

The case $0 \in \sigma_c(U_{22}^*)$ is considered analogously.

∎

We fix $a \in \rho(U)$ ($|a| < 1$). In view of Proposition 6, we have $(\bar{a})^{-1} \in \rho(A)$. Consider the operator $A_a = (A - aI)(I - \bar{a}A)^{-1}$ bounded in $\mathcal{H}$. One can directly check that A_a is a d.J-n. operator. Assume that U_a is the minimal regular dilation of the operator A_a and $\mathcal{L}_\pm(a)$ are the corresponding $\tilde{J}$-wandering subspaces (in the sense of relation (10)). By using equalities (22) and (26) and the same argument as in the case of Hilbert spaces (see Sz.-Nagy and Foias [4], Chap. 6, Sec. 1), one can easily prove the following assertion:

Lemma 8. *There are unitary mappings* $W_\pm: \mathcal{L}_\pm(a) \to \mathcal{L}_\pm$ *such that*

$$\theta_A\left(\frac{\lambda + a}{1 + \bar{a}\lambda}\right) = W_- \theta_{A_a}(\lambda) \bar{W}_+^{-1}.$$

Theorem 9. *Let* $\lambda \in \rho(U)$ *and* $|\lambda| < 1$. *Then* $0 \in \sigma_z(\theta_A(\lambda))$ *if and only if* $\lambda \in \sigma_z(A)$ ($z \in \{p, r, c\}$).

Proof. We fix arbitrary $\lambda = a \in \rho(U)$. It is clear that $a \in \sigma_z(A) \Leftrightarrow 0 \in \sigma_z(A_a)$. By virtue of Lemma 7, $0 \in \sigma_z(A_a) \Leftrightarrow 0 \in \sigma_z(U_{22}^*(a))$, where $U_{22}^*(a) = \theta_{A_a}(0)$ according to decomposition (20). Hence, $0 \in \sigma_z(A_a) \Leftrightarrow 0 \in \sigma_z(\theta_{A_a}(0))$. Then it follows from Lemma 8 that

$$0 \in \sigma_z(\theta_{A_a}(0)) \Leftrightarrow 0 \in \sigma_z(\theta_A(a)).$$

The proof of Theorem 9 is completed.

∎

4.   Remark

Subsections 2 and 3 are largely based on the results obtained by S. Kuzhel [2, 5].

Appendix 3. Lax–Phillips Abstract Scattering Scheme in Pontryagin Spaces

The Lax–Phillips abstract scheme developed in the early sixties (Lax and Phillips [1]) is one of the most important branches in scattering theory. This theory is based on the idea of incoming and outgoing subspaces, which enables one to obtain information on analytic properties of the scattering matrix in a natural way directly from the properties of the original unitary group. The necessary condition for the applicability of the Lax–Phillips abstract scheme to the study of individual systems is the positive definiteness of the corresponding energy metric. In the case where energy is negative, the original theory must be critically revised. This was done by Lax and Phillips [2] for some special classes of second-order operator differential equations.

In this appendix, the Lax–Phillips abstract scheme is generalized to the case of spaces with indefinite metric (the Pontryagin spaces). This construction is based on the ideas and methods of the theory of Pontryagin spaces. Note that the "indefinite" scheme considered below contains the results of Lax and Phillips ([3], Section 3) as a special case and substantially generalizes their results. In particular, it becomes clear that, as in the "definite" case, the generator of the associated semigroup determines spectral and analytic properties of the scattering matrix.

1. Pontryagin Spaces. Terminology and Auxiliary Statements

To make our presentation complete, we now recall some well-known facts from the theory of Pontryagin spaces whose complete proofs can be found, e.g., in the monographs by Bognar [1] and Azizov and Iokhvidov [1].

A J-space $\mathcal{H}$ is called a Pontryagin space if one of the orthoprojectors $P_+ = (I + J)/2$ or $P_- = (I - J)/2$ is finite-dimensional. In what follows we assume for definiteness that $\dim P_- \mathcal{H} = \kappa < \infty$ and use traditional terminology. Thus, the Pontryagin space $\mathcal{H}$ is denoted by Π_κ and instead of the terms "J-unitary", "J-orthogonal", etc., we use the terms "π-unitary", "π-orthogonal", etc.

A vector $x \in \Pi_\kappa$ is called *positive, negative,* or *neutral* if $[x, x] > 0$, $[x, x] < 0$, or $[x, x] = 0$, respectively.

Similarly, the lineal $\mathcal{L} \subset \Pi_\kappa$ is called *positive, negative,* or *neutral* if all its non-zero elements are, respectively, positive, negative, or neutral. Note that the indefinite metric determines a norm on positive lineals equivalent to the original norm.

Any closed lineal $\mathcal{L}$ in Π_κ can be represented in the form of a π-orthogonal direct sum of three subspaces (canonical expansion)

$$\mathcal{L} = \mathcal{L}^- \,[\dotplus]\, \mathcal{L}_0\,[\dotplus]\, \mathcal{L}^+,$$

where $\mathcal{L}_0$ is the neutral lineal and $\mathcal{L}^+$ and $\mathcal{L}^-$ are positive and negative lineals, respectively. Note that the lineal $\mathcal{L}_0$ in the canonical expansion is uniquely defined and called the *isotropic* part of the subspace $\mathcal{L}.$

The subspace L is called *nondegenerate* if $\mathcal{L}_0 = \{0\}$.

Proposition 1. *A subspace $\mathcal{L} \subset \Pi_\kappa$ is nondegenerate if and only if Π_κ is represented as the direct sum $\Pi_\kappa = \mathcal{L} \dotplus \mathcal{L}^{[\perp]}$ of the lineal $\mathcal{L}$ and its π-orthogonal complement $\mathcal{L}^{[\perp]}$.*

The operators acting in Π_κ admit a much more detailed description than their analogs in J-spaces. Thus, in the Pontryagin space, any π-nonexpanding operator is doubly π-nonexpanding.

The spectrum of a π-unitary operator U outside the unit circle consists of at most 2κ eigenvalues (counting multiplicities)

$$\lambda_1, \ldots, \lambda_n, \quad 1/\overline{\lambda}_1, \ldots, 1/\overline{\lambda}_n, \quad n \le \kappa \quad (|\lambda| < 1)$$

located symmetrically about the unit circle.

By $\mathcal{L}_\mu(U)$ we denote the root lineal of the operator U associated with an eigenvalue μ. We set

$$\mathcal{P}_- = \mathcal{L}_{\lambda_1}(U) \dotplus \ldots \dotplus \mathcal{L}_{\lambda_n}(U), \quad \mathcal{P}_+ = \mathcal{L}_{1/\overline{\lambda}_1}(U) \dotplus \ldots \dotplus \mathcal{L}_{1/\overline{\lambda}_n}(U).$$

The lineals $\mathcal{P}_-$ and $\mathcal{P}_+$ are neutral in Π_κ; their direct sum $\mathcal{P}_- \dotplus \mathcal{P}_+$ is a nondegenerate lineal.

Among the eigenvalues of a π-unitary operator U that lie on the unit circle, there are at most $m \le \kappa - n$ eigenvalues $\varepsilon_1, \ldots, \varepsilon_m$ such that the corresponding root lineals $\mathcal{L}_{\varepsilon_i}(U)$ contain nonpositive nonzero elements.

If the lineals $\mathcal{L}_{\varepsilon_i}(U)$ are nondegenerate, then the lineals $\mathcal{L}'_{\varepsilon_i} = \mathcal{L}_{\varepsilon_i}(U)\,[-]$ $\mathrm{Ker}_+\,(U - \varepsilon_i I)$ are also nondegenerate. Here, $\mathrm{Ker}_+\,(U - \varepsilon_i I)$ is the positive component of the canonical expansion of the space $\mathrm{Ker}\,(U - \varepsilon_i I)$. By construction, the lineals $\mathcal{L}'_{\varepsilon_i}$ are finite-dimensional and $U\mathcal{L}'_{\varepsilon_i} = \mathcal{L}'_{\varepsilon_i}$.

Furthermore, in view of relations $\mathcal{L}_\lambda(U)\,[\perp]\,\mathcal{L}_\mu(U)$ $(\lambda\overline{\mu} \ne 1)$, we conclude that the lineal

$$\mathfrak{M} = \mathcal{P}_- \dotplus \mathcal{P}_+ \,[\dotplus]\, (\mathcal{L}'_{\varepsilon_1}\,[\dotplus]\ldots[\dotplus]\,\mathcal{L}'_{\varepsilon_m})$$

is nondegenerate and $U\mathfrak{M} = \mathfrak{M}$.

Then, according to Proposition 1, the space Π_κ can be represented as the π-orthogonal direct sum

$$\Pi_\kappa = \mathcal{P}_- \dotplus \mathcal{P}_+ \, [\dotplus] \, (\mathcal{L}'_{\varepsilon_1} \, [\dotplus] \dots [\dotplus] \, \mathcal{L}'_{\varepsilon_m}) \, [\dotplus] \, \Pi_0 \tag{1}$$

of the finite-dimensional subspace $\mathfrak{M}$ and the positive definite subspace $\Pi_0 = \mathfrak{M}^{[\perp]}$, which are invariant under the action of the operator U.

2. Outgoing and Incoming Subspaces

For simplicity, we first consider the discrete case of the Lax–Phillips "indefinite" scheme.

Let U be a π-unitary operator acting in Π_κ. The subspace D_+ is called *outgoing* for the operator U if

(a) $UD_+ \subset D_+$;

(b) $\displaystyle\bigcap_{n=0}^{\infty} U^n D_+ = \{0\}$.

The subspace D_- is called *incoming* for the operator U if D_- is outgoing for the operator U^+.

Suppose that all root lineals $\mathcal{L}_{\varepsilon_i}(U)$, $i = 1, \dots, m$, *are nondegenerate.*

Then Π_κ can be represented in the form (1). By P_0 we denote the π-orthogonal projector onto the subspace Π_0 in Π_κ. We set

$$D'_+ = P_0 D_+, \quad D'_- = P_0 D_-. \tag{2}$$

Theorem 2. *The lineals* D'_+ *and* D'_- *are, respectively, the outgoing and incoming subspaces for the operator* U. *Furthermore,*

$$D_+ = \{d_+ = d'_+ + s \mid d'_+ \in D'_+, \, s \in \mathcal{P}_-\} \tag{3}$$

and

$$D_- = \{d_- = d'_- + p \mid d'_- \in D'_-, \, p \in \mathcal{P}_+\}, \tag{4}$$

where the vectors s *and* p *are uniquely determined by* d'_+ *and* d'_-, *respectively.*

Proof. The inclusion $U D'_+ \subset D'_+$ follows from the invariance of the subspaces $\mathfrak{M}$ and Π_0 under the action of the operator U and condition (a) for the subspace D_+. If

$$d'_0 \in \bigcap_{n=0}^{\infty} U^n D'_+,$$

then, for all $n \in \mathbb{N} \cup \{0\}$, there exists $d'_n \in D'_+$ such that $d'_0 = U^n d'_n$. By virtue of (2), $d'_n = d_n - y_n$, where $d_n \in D_+$ and $y_n \in \mathfrak{M}$. Then $U^n d_n - d_0 \stackrel{.}{=} U^n y_n - y_0$ The left-hand side of this equality belongs to D_+, while its right-hand side belongs to $\mathfrak{M}$. Taking into account the fact that $\mathfrak{M}$ is finite-dimensional and property (b) for the subspace D_+, we conclude that

$$D_+ \cap \mathfrak{M} = \{0\}. \tag{5}$$

Then $U^n d_n = d_0$ and, hence,

$$d_0 \in \bigcap_{n=0}^{\infty} U^n D_+.$$

Thus, $d_0 = 0$ and $d'_0 = 0$. Hence, condition (b) is satisfied for the lineal D'_+.

The fact that the lineal D'_+ is close follows from the closeness of the lineal D_+ and the finite-dimensionality of the space $\mathfrak{M}$.

By virtue of (1), an arbitrary element $d_+ \in D_+$ can be represented in the form $d_+ = d'_+ + y$, where $d'_+ \in D'_+$ and $y \in \mathfrak{M}$. Assume that $y \notin \mathcal{P}_-$. Without loss of generality, we can suppose that $y \in \mathrm{Ker}\,(U - \mu I)$, where $\mu \in \{1/\bar{\lambda}_j,\ \varepsilon_i\}$, $j = 1, \dots, n$; $i = 1, \dots, m$. Then

$$\frac{1}{\mu^n}[U^n d_+, v] = \frac{1}{\mu^n}[U^n d'_+, v] + [y, v] \tag{6}$$

for all $v \in \Pi_\kappa$. The space Π_0 is positive and, hence, the subspace

$$M_+ = \bigvee_{-\infty}^{\infty} U^n D'_+$$

is a Hilbert space with respect to the indefinite metric and the unitary operator $U' = U/M_+$ is isomorphic to a certain right shift in the outgoing translation representation constructed according to the subspace D'_+, which is outgoing for U'. Then

$$\lim_{n \to \infty} [U^n d'_+, v] = 0.$$

This equality together with relation (6) implies that the sequence $\{1/\mu^n U^n d_+\}$ of elements from D_+ weakly converges to the element $y \in \mathfrak{M}$. By virtue of (5), this is possible only in the case where $y = 0$. We have thus proved the validity of representation (3), where the element $y = s \in \mathcal{P}_-$ is uniquely determined for given d'_+.

For the subspace D'_-, the assertion of the theorem is proved similarly.

∎

By virtue of the positivity of Π_0 and neutrality of $\mathcal{P}_+$ and $\mathcal{P}_-$, relations (3) and (4) mean that the subspaces D_+ and D_- are positive.

Proposition 3. *The subspaces*

$$\overset{\infty}{\underset{-\infty}{\mathbf{V}}} U^n D_+ \quad and \quad \overset{\infty}{\underset{-\infty}{\mathbf{V}}} U^n D_-$$

are nonnegative in Π_κ *and their isotropic parts coincide with the lineals* $(I - P_0)D_+$ *and* $(I - P_0)D_-$, *respectively.*

Proof. The nonnegativity of the subspace

$$L = \overset{\infty}{\underset{-\infty}{\mathbf{V}}} U^n D_+$$

follows from the positivity of the subspaces $U^n D_+$ $(n \in \mathbb{Z})$. Furthermore, according to relation (3), we have $U^n D_+ \subset \Pi_0 [\dotplus] \mathcal{P}_-$ and, therefore, $L \subset \Pi_0 [\dotplus] \mathcal{P}_-$. This inclusion means that the isotropic part L_0 of the subspace L has the form $L_0 = L \cap \mathcal{P}_-$.

Let us show that $(I - P_0)D_+ \subset L_0$. Assume that $s \in (I - P_0)D_+$. Without loss of generality, we can take $s \in L_\lambda(U) \cap (I - P_0)D_+$. Assume that the condition

$$(U - \lambda I)s = \tilde{s} \in L_0 \tag{7}$$

is satisfied for the element s. By using the same reasoning as in the proof of Theorem 2 and taking the almost evident identity

$$(\lambda U^+)^n s = s - \sum_{k=1}^{n} (U^+)^k \lambda^{k-1} \tilde{s}$$

into account, we conclude that the sequence

$$\left\{ (\lambda U^+)^n d_+ + \sum_{k=1}^{n} (U^+)^k \lambda^{k-1} \bar{s} \right\}_{n=1}^{\infty} \qquad (d_+ = d'_+ + s)$$

of elements from L weakly converges to s and, hence, $s \in L \cap \mathcal{P}_- = L_0$. It is clear that all elements $s \in \mathrm{Ker}\,(U - \lambda I) \cap (I - P_0)D_+ = L'$ satisfy condition (7); therefore, $L' \subset L_0$. Then it follows from the inclusion $(U - \lambda I)(I - P_0)D_+ \subset (I - P_0)D_+$ that

$$(U - \lambda I)(\mathrm{Ker}\,(U - \lambda I)^2 \cap (I - P_0)D_+) \subset L' \subset L_{\bar{0}}$$

and, hence,

$$\mathrm{Ker}\,(U - \lambda I)^2 \cap (I - P_0)D_+ \subset L_0.$$

Acting as above, after finitely many steps we arrive at the inclusion $L_\lambda(U) \cap (I - P_0)D_+ \subset L_0$ and, consequently, $(I - P_0)D_+ \subset L_0$. Let us prove the inverse inclusion. Let $s \in L_0$ and let $\left\{ U^{n_k} d_{n_k} \right\}_{k=1}^{\infty}$ $(d_{n_k} \in D_+,\ n_k \in \mathbb{Z})$ be a sequence of elements from L strongly convergent to s. Then the sequence of elements $(I - P_0)U^{n_k} d_{n_k} = U^{n_k}(I - P_0)d_{n_k}$ is also strongly convergent to s. According to (3) and the finite-dimensionality of the lineal $(I - P_0)D_+$, this is possible only in the case where $s \in (I - P_0)D_+$. Thus, $(I - P_0)D_+ \supset L_0$ and, therefore, $(I - P_0)D_+ = L_0$.
    The case where

$$L = \overset{\infty}{\underset{-\infty}{\mathsf{V}}} U^n D_-$$

can be investigated similarly.

■

    Note that, in the general case, the translation representation of the operator U cannot be constructed for given positive subspaces D_+ and D_- because the subspaces

$$\overset{\infty}{\underset{-\infty}{\mathsf{V}}} U^n D_+ \quad \text{and} \quad \overset{\infty}{\underset{-\infty}{\mathsf{V}}} U^n D_-$$

are not Hilbert spaces with respect to the indefinite metric. This is a consequence of Proposition 3.
    Consider the Hilbert spaces

$$M_+ = \overset{\infty}{\underset{-\infty}{\bigvee}} U^n D'_+$$

and

$$M_- = \overset{\infty}{\underset{-\infty}{\bigvee}} U^n D'_-.$$

By virtue of Proposition 3, we obtain

$$\overset{\infty}{\underset{-\infty}{\bigvee}} U^n D_- = M_- \ [\dotplus] \ (I - P_0) D_-,$$

$$(8)$$

$$\overset{\infty}{\underset{-\infty}{\bigvee}} U^n D_+ = M_+ \ [\dotplus] \ (I - P_0) D_+.$$

We set

$$D''_- = D_- \cap \Pi_0 \quad \text{and} \quad D''_+ = D_+ \cap \Pi_0. \qquad (9)$$

Note that $D''_\pm \subset D'_\pm$ and, hence, the subspace D''_+ (D''_-) is also an outgoing (incoming) subspace for the operator U.

Proposition 4. *The following equalities hold:*

$$M_- = \overset{\infty}{\underset{-\infty}{\bigvee}} U^n D''_- \quad \text{and} \quad M_+ = \overset{\infty}{\underset{-\infty}{\bigvee}} U^n D''_+.$$

Proof. It is obvious that

$$\overset{\infty}{\underset{-\infty}{\bigvee}} U^n D''_+ \subset M_+.$$

Let us prove the inverse inclusion. Consider the operator

$$T = (U - \lambda_1 I)^\kappa \dots (U - \lambda_n I)^\kappa.$$

Since $0 \in \rho(T/M_+)$ and $T d'_+ \in D_+ \cap \Pi_0 = D''_+$ for all $d'_+ \in D'_+$ by virtue of (3), we have

$$M_+ = TM_+ = T \bigvee_{-\infty}^{\infty} U^n D_+' \subset \bigvee_{-\infty}^{\infty} U^n D_+''$$

and, thus,

$$M_+ = \bigvee_{-\infty}^{\infty} U^n D_+''.$$

The equality

$$M_- = \bigvee_{-\infty}^{\infty} U^n D_-''$$

is proved similarly.

■

3. Scattering Operator. Definition and Simple Properties

In what follows, we assume that the subspaces D_+ and D_- are π-orthogonal. By virtue of Theorem 2, the subspaces D_-' and D_-'' are incoming for the operator U/M_- unitary (with respect to the indefinite metric) in M_-. The theorem on translation representation (Lax and Phillips [1]) and Proposition 4 imply that there are two incoming translation representations of the operator U/M_-, namely $l_2(\mathbb{Z}, UN_-')$, where $UN_-' = UD_-' [-] D_-'$, and $l_2(\mathbb{Z}, UN_-'')$, where $UN_-'' = UD_-'' [-] D_-''$. The first of these representations is constructed according to D_-', while the second one is constructed according to D_-''. We denote the images of a vector f from M_- in these translation representations by f' and f'', respectively.

Similarly, there are two outgoing translation representations of the operator U/M_+ unitary in M_+: $l_2(\mathbb{Z}, N_+')$ and $l_2(\mathbb{Z}, N_+'')$ ($N_+' = D_+' [-] UD_+'$, $N_+'' = D_+'' [-] UD_+''$) constructed according to the outgoing subspaces D_+' and D_+'', respectively.

Consider a vector $g = P_{M_+} f$, where $f \in M_-$ and P_{M_+} is the π-orthoprojector onto M_+ in Π_0. By g' and g'', we denote the images of g in these outgoing representations.

An operator that establishes correspondence between any incoming representative of the element f and any outgoing representative of the element g can be regarded as a scattering operator. At the same time, Lax and Phillips showed that the most natural definition of this operator has the form

$$S: f' \to g'.$$

In view of the evident equality $P_{M_+} U = U P_{M_+}$, we conclude that the scattering operator S is a contracting operator from $l_2(\mathbb{Z}, U N'_-)$ into $l_2(\mathbb{Z}, N'_+)$ commuting with the operators of right shift in these spaces. In the spectral representation, the operator S defines the contraction

$$\mathcal{B} = F_{N'_+} S F_{U N'_-}^{-1}$$

(F_N is the Fourier transformation of the space $l_2(\mathbb{Z}, N)$ onto $L_2((0, 2\pi), N))$ acting from $L_2((0, 2\pi), U N'_-)$ into $L_2((0, 2\pi), N'_+)$ and commuting with the operators of multiplication by bounded measurable functions.

Generally speaking, the operator $\mathcal{B}$ is not causal because the π-orthogonality of D_+ and D_- does not always imply the π-orthogonality of the subspaces D'_+ and D'_-.

Parallel with the scattering operator S, we consider the operators

$$S_- : f'' \to f' \quad \text{and} \quad S_-^0 : f'' \to g'. \tag{10}$$

Each of these operators can be regarded as a scattering operator with respect to an orthogonal pair of incoming and outgoing subspaces. For example, the unitary operator S_- is a scattering operator with respect to the orthogonal pair D''_- and $M_-[-] D'_-$ of subspaces of the Hilbert space M_-. In this case, the equality

$$\bigvee_{-\infty}^{\infty} U^n D''_- = \bigvee_{-\infty}^{\infty} U^n (M_-[-] D'_-)$$

holds for this pair of subspaces. Therefore, all results of the Lax–Phillips classical scattering scheme hold for the scattering operator S_-.

In its turn, the contracting operator S_-^0 is a scattering operator in the generalized Lax–Phillips scheme (i.e., under the condition that

$$\bigvee_{-\infty}^{\infty} U^n D''_- \neq \bigvee_{-\infty}^{\infty} U^n D'_+;$$

for more details, see Adamjan and Arov [3] and Kuzhel and Tretyakov [1]) with respect to the orthogonal pair of the incoming D''_- and outgoing D'_+ subspaces of the Hilbert space Π_0.

Thus, the operator S_- in the spectral representation determines the causal operator $\mathcal{B}_-$, which can be realized as an operator of multiplication by the function $\mathcal{B}_-(e^{i\theta})$ whose values, for almost all θ, are unitary operators from $U N''_-$ into $U N'_-$.

Similarly, the spectral representation $\mathcal{B}_-^0$ of the operator S_-^0 can be realized as a multiplication by the function $\mathcal{B}_-^0(e^{i\theta})$ whose values are contracting operators from

$U N_-''$ into $U N_+'$.

By virtue of relations (10), we have

$$\mathcal{B} \mathcal{B}_-(e^{i\theta}) = \mathcal{B}_-^0(e^{i\theta}).$$

In view of the unitarity of the operator $\mathcal{B}_-(e^{i\theta})$, this implies that the operator $\mathcal{B}$ can be represented as an operator of multiplication by the function

$$\mathcal{B} = \mathcal{B}(e^{i\theta}) = \mathcal{B}_-^0(e^{i\theta}) \mathcal{B}_-^{-1}(e^{i\theta}), \tag{11}$$

whose values are contracting operators from $U N_-'$ into N_+'.

The function $\mathcal{B}(e^{i\theta})$ is called the *scattering matrix.*

Let us show that the properties of the scattering matrix are uniquely determined in terms of the spectral properties of the following π-nonexpanding operator A:

$$A = P_{K_\kappa} U |_{K_\kappa} \tag{12}$$

acting in the Pontryagin space $K_\kappa = \Pi_\kappa [-] \, (D_+ [+] D_-)$. For this purpose, we first study the structure of the operator A.

4. Spectral Analysis of the Operator A

Lemma 5. *The π-unitary operator U is a regular dilation of the operator A.*

Proof. By virtue of the Sarason criterion (Lemma 5.1, Chapter 4) and relations $U D_+ \subset D_+$ and $U^+ D_- \subset D_-$, we get $A^n = P_{K_\kappa} U^n$ $(n \in \mathbb{N})$. Moreover, since the indefinite metric determines a norm equivalent to the original one on the positive subspace $D_+ [+] D_-$, we can regard without loss of generality that $\pi|_{K_\kappa} = \pi$ and $\pi|_{D_+[+]D_-} = I$. Hence, U is a regular dilation of the operator A.

∎

Generally speaking, the dilation U is not minimal. In other words, the following assertion is true:

Proposition 6. *The space Π_κ can be represented as a π-orthogonal direct sum*

$$\Pi_\kappa = \Pi_\kappa^{\min} [+] \mathfrak{N}$$

of the Pontryagin space $\Pi_\kappa^{\min}$ and the Hilbert space $\mathfrak{N}$ invariant with respect to the operator U. The restriction of the operator U to $\Pi_\kappa^{\min}$ determines the minimal regular dilation of the operator A, and the restriction U to the Hilbert space $\mathfrak{N}$ is a bilateral shift.

Proof. We set $\mathcal{L}_+ = \overline{(U-A)K_\kappa}$ and $\mathcal{L}_- = \overline{(U^+ - A^+)K_\kappa}$. Let us show that $\mathcal{L}_+ \subset N_+$, where $N_+ = D_+[-]UD_+$. In fact, the decomposition $\Pi_\kappa = D_-[+] K_\kappa[+] D_+$ and the relations $UD_- \supset D_-$ and $UD_+ \subset D_+$ yield

$$UK_\kappa \subset K_\kappa[+]D_+. \tag{13}$$

Then $\mathcal{L}_+ \subset D_+$ by virtue of (12). In its turn, taking the equality

$$[(U-A)x, Ud_+] = [x, d_+] - [Ax, Ud_+] = 0 \quad (x \in K_\kappa, \, d_+ \in D_+)$$

into account, we conclude that $\mathcal{L}_+ [\perp] UD_+$ and, hence, $\mathcal{L}_+ \subset N_+$. Similarly, we can show that $\mathcal{L}_- \subset N_-$, where $N_- = D_-[-]U^+D_-$.

We set $\mathcal{R} = N_+[-]\mathcal{L}_+$ and $\mathcal{R}_1 = N_-[-]\mathcal{L}_-$. It is easy to show that the expansion

$$D_+ = \overset{\infty}{\underset{0}{[+]}} U^n N_+ := N_+[+] UN_+[+] \ldots$$

holds for the outgoing subspace D_+. Then

$$D_+ = \left(\overset{\infty}{\underset{0}{[+]}} U^n \mathcal{L}_+ \right) [+] \left(\overset{\infty}{\underset{0}{[+]}} U^n \mathcal{R} \right). \tag{14}$$

Similarly,

$$D_- = \overset{\infty}{\underset{0}{[+]}} (U^+)^n N_- = \left(\overset{\infty}{\underset{0}{[+]}} (U^+)^n \mathcal{L}_- \right) [+] \left(\overset{\infty}{\underset{0}{[+]}} (U^+)^n \mathcal{R}_+ \right). \tag{15}$$

We set

$$\Pi_\kappa^{\min} = \left(\overset{\infty}{\underset{0}{[+]}} (U^+)^n \mathcal{L}_- \right) [+] K_\kappa [+] \left(\overset{\infty}{\underset{0}{[+]}} U^n \mathcal{L}_+ \right)$$

and

$$\mathfrak{N} = \overset{\infty}{\underset{0}{[+]}} (U^+)^n \mathcal{R}_1 [+] \overset{\infty}{\underset{0}{[+]}} U^n \mathcal{R}. \tag{16}$$

By virtue of relations (14) and (15), we obtain

$$\Pi_{\kappa} = \Pi_{\kappa}^{\min} [\dot{+}] \mathfrak{N}.$$

In view of Lemma 5 in Appendix 2, we conclude that the restriction U to the Pontryagin space $\Pi_{\kappa}^{\min}$ is the minimal regular dilation. Moreover, $U\Pi_{\kappa}^{\min} = \Pi_{\kappa}^{\min}$ and, hence, $U\mathfrak{N} = \mathfrak{N}$. Therefore, by using (16), we can show that $U\mathcal{R}_1 = \mathcal{R}$ and, hence,

$$\mathfrak{N} = \overset{\infty}{\underset{-\infty}{[\dot{+}]}} U^n \mathcal{R}.$$

The operator U is thus a bilateral shift in the Hilbert space $\mathfrak{N}$ (with respect to indefinite metric) with the generating space $\mathcal{R}$.

∎

Proposition 7. *Outside the unit disk, the spectrum of a π-nonexpanding operator A coincides with the set $1/\overline{\lambda}_1, \dots, 1/\overline{\lambda}_n$ of the eigenvalues of the operator U. Moreover,*

$$\mathcal{L}_{1/\overline{\lambda}_i}(A) = P_{K_{\kappa}} \mathcal{L}_{1/\overline{\lambda}_i}(U), \quad i = 1, \dots, n \tag{17}$$

and $\dim \mathcal{L}_{1/\overline{\lambda}_i}(A) = \dim \mathcal{L}_{1/\overline{\lambda}_i}(U).$

Proof. According to Proposition 6 (Appendix 2) and Proposition 6 given above, we conclude that, outside the unit disk, the spectrum of the operator A coincides with the set $1/\overline{\lambda}_1, \dots, 1/\overline{\lambda}_n$ of the eigenvalues of the operator U.

Let $x \in \mathcal{L}_{1/\overline{\lambda}}(U)$ and $(U - 1/\overline{\lambda}I)^p x = 0$ $(p \in \mathbb{N})$. Then, for any $d_- \in D_-$, we have

$$0 = [(U - 1/\overline{\lambda}I)^p x, d_-] = [x, (U^+ - 1/\lambda I)^p d_-]. \tag{18}$$

Since $U^+|_{D_-}$ is a π-isometric operator, the lineal D_- is positive, and $|\lambda| < 1$, we have $(U^+ - 1/\lambda I)^p D_- = D_-$.

Thus, taking (18) into account, we conclude that $x [\perp] D_-$ and, hence, $(I - P_{K_{\kappa}})x \in D_+$. In this case, in view of the fact that D_+ is a outgoing subspace for U, we have

$$P_{K_{\kappa}}(U - 1/\overline{\lambda}I)^p (I - P_{K_{\kappa}})x = 0$$

and, hence,

$$P_{K_\kappa}(U - 1/\overline{\lambda}I)^p x = P_{K_\kappa}(U - 1/\overline{\lambda}I)^p P_{K_\kappa} x = (A - 1/\overline{\lambda}I)^p P_{K_\kappa} x.$$

Thus,

$$P_{K_\kappa} \mathcal{L}_{1/\overline{\lambda}}(U) \subset \mathcal{L}_{1/\overline{\lambda}}(A).$$

Conversely, assume that $x_0 \in \mathcal{L}_{1/\overline{\lambda}}(A)$ and, therefore, $(A - 1/\overline{\lambda}I)^p x_0 = 0$ for some natural p.

Since $U_1 = U|_{D_+}$ is a π-isometric operator on the positive lineal D_+ and $|\lambda| < 1$, we find that the operator $(U_1 - 1/\overline{\lambda}I)^{-1}$ exists and is defined in the entire space D_+. Consider a vector $x = x_0 + d_+$, where the vector $d_+ \in D_+$ is given by the equality

$$d_+ = -(U_1 - 1/\overline{\lambda}I)^{-p} P_{D_+}(U - 1/\overline{\lambda}I)^p x_0.$$

Then relation (13) and the definition of the operator A imply that

$$(U - 1/\overline{\lambda}I)^p x = (A - 1/\overline{\lambda}I)^p x_0 = 0.$$

Thus, $P_{K_\kappa} \mathcal{L}_{1/\overline{\lambda}}(U) = \mathcal{L}_{1/\overline{\lambda}}(A)$ and the dimensionalities of these lineals are equal.
∎

In view of Proposition 7, we can show that the space K_κ can be represented as

$$K_\kappa = \mathcal{L}_{1/\overline{\lambda}_1}(A) \dotplus \ldots \dotplus \mathcal{L}_{1/\overline{\lambda}_n}(A) \dotplus \mathfrak{H}, \tag{19}$$

where $\mathfrak{H}$ is a subspace of K_κ invariant for A such that $\sigma(A|\mathfrak{H}) = \sigma(A)\setminus\{1/\overline{\lambda}_1, \ldots, 1/\overline{\lambda}_n\}$.

By virtue of (19),

$$\mathfrak{H} = (A - 1/\overline{\lambda}_1 I)^\kappa \ldots (A - 1/\overline{\lambda}_n I)^\kappa K_\kappa \tag{20}$$

Then taking an equality similar to (17) for the operators U^+ and A^+ into account, we obtain

$$\mathfrak{H}^{[\perp]} = \mathcal{L}_{1/\lambda_1}(A^+) \dotplus \ldots \dotplus \mathcal{L}_{1/\lambda_n}(A^+) = P_{K_\kappa} \mathcal{P}_-. \tag{21}$$

Assume that ε_i ($|\varepsilon_i| = 1$) is an eigenvalue of the operator U such that the root lineal $\mathcal{L}_{\varepsilon_i}(U)$ contains nonpositive elements (see, Subsection 1). In view of expansion

(15), we conclude that the operator U^+ is isomorphic in D_- to the operator of right shift and, hence, all points of the unit circle belong to the continuous spectrum of the operator $U^+|_{D_-}$. Therefore, $\overline{(U^+ - \bar{\varepsilon}_i I)^n D_-} = D_-$. If we now replace $1/\lambda$ by ε_i in (18), then, by using the same reasoning as in the proof of Proposition 7, we obtain

$$\mathcal{L}_{\varepsilon_i}(U)\,[\perp]\,D_-.$$

By analogy, it follows from the equality $\overline{(U - \varepsilon_i I)^n D_+} = D_+$ that

$$\mathcal{L}_{\bar{\varepsilon}_i}(U^+)\,[\perp]\,D_+.$$

Finally, the equality $\mathcal{L}_{\varepsilon_i}(U) = \mathcal{L}_{\bar{\varepsilon}_i}(U^+)$ implies that $\mathcal{L}_{\varepsilon_i}(U) \subset K_\kappa$ and, therefore, for any $x \in \mathcal{L}_{\varepsilon_i}(U)$,

$$U^n x = A^n x \quad \text{and} \quad (U^+)^n x = (A^+)^n x \quad (n \in \mathbf{N}). \tag{22}$$

Therefore, according to relation (20), $\mathcal{L}_{\varepsilon_i}(U) \subset \mathfrak{H}$. Since the lineal

$$\overset{m}{\underset{1}{[+]}}\,\mathcal{L}_{\varepsilon_i}(U)$$

is nondegenerate, the lineal

$$\overset{m}{\underset{1}{[+]}}\,\mathcal{L}'_{\varepsilon_i}$$

is also nondegenerate. Consequently,

$$\mathfrak{H} = \mathfrak{H}_0\,[+]\,\overset{m}{\underset{1}{[+]}}\,\mathcal{L}'_{\varepsilon_i}, \tag{23}$$

where $\mathfrak{H}_0$ is a subspace of K_κ.

In the Hilbert space Π_0, we consider the subspace

$$K_+^0 = \Pi_0\,[-]\,(D_-' \,[+]\, D_+'').$$

Lemma 8. *A π-orthoprojector P_{K_κ} is a bijection of the subspace K_+^0 onto the subspace $\mathfrak{H}_0$.*

Proof. The π-orthogonality of the subspaces K_+^0 and D'_- and the inclusion $K_+^0 \subset \Pi_0$ imply that $K_+^0 [\perp] D_-$. Thus, an arbitrary element $t \in K_+^0$ can be represented in the form

$$t = t_0 + d_+, \tag{24}$$

where $t_0 \in K_\kappa$ and $d_+ \in D_+$. Let us show that $t_0 \in \mathfrak{H}_0$. First, we note that $D_\pm [\perp] \mathcal{P}_\mp$ by virtue of (3) and (4) and, hence,

$$\mathcal{P}_- \subset D_- [\dot{+}] K_\kappa, \quad \mathcal{P}_+ \subset K_\kappa [\dot{+}] D_+. \tag{25}$$

Since $t \in \Pi_0$, we have $t [\perp] \mathcal{P}_-$ which, in view of relations (24) and (25) implies that $t_0 [\perp] P_{K_\kappa} \mathcal{P}_-$. By virtue of (21), this means that $t_0 \in \mathfrak{H}^{[\perp][\perp]} = \mathfrak{H}$.

Similarly, it follows from the condition $t \in \Pi_0$ that

$$t_0 [\perp] \overset{m}{\underset{1}{[\dot{+}]}} \mathcal{L}'_{\varrho_i},$$

whence, by virtue of (23), we conclude that $t_0 \in \mathfrak{H}_0$.

For any $t_0 \in \mathfrak{H}_0$, we set $t = t_0 - (I - P_{K_\kappa}) \hat{p}$, where $\hat{p} \in \mathcal{P}_+$ and satisfies the condition

$$[t_0, p] + [P_{K_\kappa} \hat{p}, p] = 0 \quad (\forall p \in \mathcal{P}_+). \tag{26}$$

We note that equality (26) uniquely determines the element $d_+ = -(I - P_{K_\kappa}) \hat{p}$. Indeed, if there exist elements $\hat{p}_1$ and $\hat{p}_2$ which satisfy (26), then $[P_{K_\kappa}(\hat{p}_1 - \hat{p}_2), p] = 0$. In view of the neutrality of the lineal $\mathcal{P}_+$, this relation means that $\hat{p}_1 - \hat{p}_2 \in K_\kappa$ and, consequently, $(I - P_{K_\kappa})(\hat{p}_1 - \hat{p}_2) = 0$ (unlike d_+, the vector $\hat{p}$ is determined to within a vector from $\mathcal{P}_+ \cap K_\kappa$).

By virtue of relations (1), (9), (23), (25), and (26), it can be directly verified that $t \in K_+^0$. Thus, the operator $P_{K_\kappa} t = t_0$ maps K_+^0 onto $\mathfrak{H}_0$. Furthermore, in equality (24), we have $t_0 \neq 0$ since, otherwise, $d_+ \in D''_+$ and $d_+ [\perp] D''_+$. Consequently, the mapping $P_{K_\kappa} | K_+^0$ is a bijection.

∎

In the Hilbert space (with respect to the indefinite metric) K_+^0, we consider the contracting operator

$$A_+^0 = P_{K_+^0} U\big|_{K_+^0},$$

where $P_{K_+^0}$ is the π-orthoprojector onto K_+^0 in Π_0.

Theorem 9. *For all* $|\mu| \leq 1$, $\mu \neq \varepsilon_i$, $(i = 1, \ldots, m)$, *the number* μ *lies in* $\sigma_\alpha(A)$ *if and only if* $\mu \in \sigma_\alpha(A_+^0)$, $\alpha \in \{p, r, c\}$.

Proof. The invariance of the subspace $\mathfrak{H}$ for the operator A and relations (22) and (23) imply that $A: \mathfrak{H}_0 \to \mathfrak{H}_0$. Then, according to decomposition (19) and the definition of the subspace $\mathfrak{H}_0$, we get

$$\mu \in \sigma_\alpha(A) \Leftrightarrow \mu \in \sigma_\alpha(A\big|_{\mathfrak{H}_0}), \quad |\mu| = 1, \quad \mu \neq \varepsilon_i. \tag{27}$$

Taking the decomposition $\Pi_0 = D'_- [+] K_+^0 [+] D''_+$ and relations $U D'_- \supset D'_-$ and $U D''_+ \subset D''_+$ into account, we find that $U K_+^0 \subset K_+^0 [+] D''_+$; therefore, $(I - P_{K_+^0}) U t \in D''_+$ for all $t \in K_+^0$. Then, in view of the inclusion $D''_+ \subset D_+$, we have $P_{K_\kappa}(I - P_{K_+^0}) U t = 0$, i.e.,

$$P_{K_\kappa} U t = P_{K_\kappa} A_+^0 t \quad (t \in K_+^0).$$

In this case, this equality and (24) yield

$$P_{K_\kappa} A_+^0 t = P_{K_\kappa} U(t_0 + d_+) = P_{K_\kappa} U t_0 = A P_{K_\kappa} t$$

and, thus,

$$P_{K_\kappa}(A_+^0 - \mu I) t = (A - \mu I) P_{K_\kappa} t \quad (t \in K_+^0).$$

By combining this relation, Lemma 8, and relation (27), we complete the proof. ∎

The same reasoning as in Lemma 8 and Theorem 9 applied to the operator

$$A_-^0 = P_{K_-^0} U^+\big|_{K_-^0} \tag{28}$$

acting in the Hilbert space

$$K_-^0 = \Pi_0 [-] (D''_- [+] D'_+)$$

yields the following assertion "dual" to Theorem 9:

Theorem 10. *For all* $|\mu| \leq 1$, $\mu \neq \bar{\varepsilon}_i$, $(i = 1, \ldots, m)$, *the number* μ *lies in* $\sigma_\alpha(A^+)$ *if and only if* $\mu \in \sigma_\alpha(A_-^0)$, $\alpha \in \{p, r, c\}$.

By using the operators A_+^0 and A_-^0, we can describe the important situation where the scattering operator S defined in Subsection 3 is unitary.

Theorem 11. *The scattering operator S is unitary if and only if the completely nonunitary part of the operator* A_-^0 (A_+^0) *belongs[10] to the class* C_{00}.

Proof. It follows from the definition of the scattering operator that the operator S is a unitary mapping of the space $l_2(\mathbb{Z}, UN_-')$ onto $l_2(\mathbb{Z}, N_+')$ if and only if $M_+ = M_-$.

The definition of the operator A^0 and Lemma 5.1 (Chapter 4) imply that the operator U^+ acting in the Hilbert space (with respect to the indefinite metric) Π_0 is a unitary dilation of the contraction A^0. Acting as in the proof of Proposition 6, we conclude that the space Π_0 can be represented in the form $\Pi_0 = \Pi_0^{\min} [\dotplus] \mathfrak{N}$, where

$$\Pi_0^{\min} = \overset{\infty}{\underset{0}{[\dotplus]}} (U^+)^n L_-'' [\dotplus] K_-^0 [\dotplus] \overset{\infty}{\underset{0}{[\dotplus]}} U^n L_+',$$

$$L_+' = \overline{\left(U - (A_-^0)^+\right) K_-^0}, \quad L_-'' = \overline{\left(U^+ - A_-^0\right) K_-^0},$$

is the space of minimal unitary dilation and

$$\mathfrak{N} = \overset{\infty}{\underset{-\infty}{[\dotplus]}} U^n \mathcal{R}' = \overset{\infty}{\underset{-\infty}{[\dotplus]}} U^n \mathcal{R}'',$$

$$\mathcal{R}'' = N_-'' [-] L_-'', \quad \mathcal{R}' = N_+' [-] L_+', \quad U^+ \mathcal{R}' = \mathcal{R}'',$$

is the space of bilateral shift. Then it is easy to see that

$$M_- = \overset{\infty}{\underset{-\infty}{\bigvee}} U^n D_-'' = \overset{\infty}{\underset{-\infty}{[\dotplus]}} U^n N_-'' = \overset{\infty}{\underset{-\infty}{[\dotplus]}} U^n L_-'' [\dotplus] \mathfrak{N},$$

$$M_+ = \overset{\infty}{\underset{-\infty}{\bigvee}} U^n D_+' = \overset{\infty}{\underset{-\infty}{[\dotplus]}} U^n N_+' = \overset{\infty}{\underset{-\infty}{[\dotplus]}} U^n L_+' [\dotplus] \mathfrak{N}.$$

---

[10] The operator $T \in \mathcal{B}[\mathcal{H}]$ belongs to the class C_{00} if $T^n x \to 0$ and $(T^*)^n x \to 0$ $(\forall x \in \mathcal{H})$ as $n \to \infty$.

Thus,

$$M_- = M_+ \Leftrightarrow [\overset{\infty}{+}] U^n L''_- = [\overset{\infty}{+}] U^n L'_+ .$$
$$ _{-\infty} _{-\infty}$$

This equality holds only in the case where the completely nonunitary part of the operator A^0_- belongs to the class C_{00} (see Sz.-Nagy and Foias [4], Chapter 2, Theorem 1.2 and Proposition 1.4).

The case of the operator A^0_+ is similar.

∎

Theorem 9 demonstrates that the part of spectrum of the π-nonexpanding operator A that lies in the unit disk coincides (with the exception of m critical points ε_i) with the spectrum of the contracting operator A^0_+. To characterize the part of the spectrum of the operator A that lies outside the unit disk, we consider the Hilbert spaces

$$K_- = D'_- [-] D''_-, \quad K_+ = D'_+ [-] D''_+ ,$$

and the contracting operators

$$A_- = P_{K_-} U|_{K_-}, \quad A_+ = P_{K_+} U^+|_{K_+}$$

acting in K_- and K_+, respectively.

Proposition 12. *The operator A_- is completely nonunitary and its spectrum contains only those eigenvalues λ_i of the operator U for which the lineal $\mathcal{L}_{1/\lambda_i}(A^+)$ contains negative elements.*

Proof. By virtue of the first equality in (25), we have $(I - P_{K_-}) \mathcal{P}_- \subset D_-$. One can easily show that $(I - P_{K_-}) \mathcal{P}_- [\perp] D''_-$ and, hence,

$$P_0 (I - P_{K_-}) \mathcal{P}_- \subset D'_- [-] D''_- = K_- .$$

Suppose that $f \in K_-$ and $f [\perp] P_0 (I - P_{K_-}) \mathcal{P}_-$. In view of (1), we get $f [\perp] P_{K_-} \mathcal{P}_-$. By virtue of (4), $f = d_- - p$, where $d_- \in D_-$ and $p \in \mathcal{P}_+$. Therefore,

$$0 = [f, P_{K_-} s] = -[p, s] \quad (\forall s \in \mathcal{P}_-).$$

By virtue of the fact that the lineal $\mathcal{P}_- \dotplus \mathcal{P}_+$ is nondegenerate, this relation holds only in

the case where $p = 0$. Then $f \in D''_-$ and, hence, $f = 0$.
   Thus,

$$K_- = P_0(I - P_{K_\chi}) \mathcal{P}_-. \tag{29}$$

Relation (29) immediately implies that an arbitrary element $f \in K_-$ can be represented in the form

$$f = (I - P_{K_\chi}) \tilde{s} - \tilde{p} \quad (\tilde{s} \in \mathcal{P}_-), \tag{30}$$

where the element $\tilde{p}$ belongs to $\mathcal{P}_+$ and is uniquely defined by the equality

$$[\tilde{p}, s] + [P_{K_\chi} \tilde{s}, s] = 0 \quad (\forall s \in \mathcal{P}_-). \tag{31}$$

   By virtue of (29), the space K_- is finite-dimensional and, hence, the operator A_- has only point spectrum.
   The decomposition

$$M_- = D''_- [\dotplus] K_- [\dotplus] (M_- [-] D'_-)$$

and the inclusion $U D''_- \supset D''_-$ imply that

$$U K_- \subset K_- [\dotplus] (M_- [-] D'_-)$$

and, therefore,

$$A_-^n = P_{K_-} U^n |_{K_-} = P_{D'_-} U^n |_{K_-} \quad (n \in \mathbb{N})$$

($P_{D'_-}$ is a π-orthoprojector onto D'_- in Π_χ).
   According to equalities (4) and (30), for arbitrary $f \in K_-$ and $d_- \in D_-$, we have

$$[A_-^n f, d_-] = [f, (U^+)^n P_{D'_-} d_-] = [f, (U^+)^n d_-] = [U^n \tilde{s}, d_-]$$

and, consequently,

$$K_- = \mathcal{L}_{\lambda_1}(A_-) \dotplus \ldots \dotplus \mathcal{L}_{\lambda_n}(A_-), \tag{32}$$

where

$$\mathcal{L}_{\lambda_i}(A_-) = P_0(I - P_{K_\chi}) \mathcal{L}_{\lambda_i}(U).$$

In this case, by virtue of Proposition 7 and decomposition (30), we have $\mathcal{L}_{\lambda_i}(A_-) \neq \{0\}$ (and, hence, $\lambda_i \in \sigma(A_-)$) only if the lineal $\mathcal{L}_{1/\lambda_i}(A^+)$ contains nonnegative elements.

The complete nonunitariness of the contraction A_- follows from the fact that the spectrum of the operator A_- lies inside the unit circle.

∎

Note that the π-orthogonal elements $(I - P_{K_\chi}) \tilde{s}$ and $\tilde{p}$ in equality (30) are equivalent in the sense that each of them is uniquely determined by the other by equality (31). Moreover, in view of the decomposition $D'_- = K_- [+] D''_-$ and equality (4), the set of all vectors $\tilde{p}$ from decomposition (30) obviously coincides with the lineal $(I - P_0) D_-$.

It follows from the proof of Proposition 12 that the root lineals of the operator A_- are, in fact, determined by the term $(I - P_{K_\chi}) \tilde{s}$ in decomposition (30). At the same time, it can be shown that the root lineals of the operator A_-^+ depend on the component $\tilde{p}$ of the element f.

Lemma 13. *If $f' \in K_-$ and $f' = (I - P_{K_\chi}) s' - p'$, then the equality $(A_-^+ - \bar{\lambda} I)^n f' = 0$ is equivalent to the equality $(U^+ - \bar{\lambda} I)^n p' = 0$.*

Proof. Let us show that

$$A_-^+ f' = r = (I - P_{K_\chi}) s'' - U^+ p', \tag{33}$$

where the element $(I - P_{K_\chi}) s''$ is determined by equality of type (31) for the element $U^+ p'$.

By virtue of equalities (25) and (30) and the fact that the lineals $\mathcal{P}_+ = U^+ \mathcal{P}_+$ and $\mathcal{P}_- = U^+ \mathcal{P}_-$ are neutral, we get

$$[r, f] = -[P_{K_\chi} s'', \tilde{s}] = [U^+ p', \tilde{s}] \tag{34}$$

for arbitrary $f \in K_-$.

On the other hand,

$$[A_-^+ f', f] = -[U^+ (I - P_{K_\chi}) s', P_{K_\chi} \tilde{s}] + [U^+ (I - P_{K_\chi}) s', \tilde{s}].$$

The first term on the right-hand side of this equality is equal to zero because, by virtue of (25), $(I - P_{K_\chi}) s' \subset D_-$. For the second term, taking (31) into account, we obtain

$$[U^+(I-P_{K_\varkappa})s', \tilde{s}] = -[P_{K_\varkappa}s', U\tilde{s}] = [p', U\tilde{s}].$$

Thus, $[A_-^+ f', f] = [p', U\tilde{s}]$. It now follows from (34) that equality (33) holds. Moreover, the element $p' \in (I-P_{\Pi_0})D_-$ determines the element $f' \in K_-$ uniquely. The lemma is proved.

■

Assertions similar to Proposition 12 and Lemma 13 can also be formulated for the operator A_+ acting in the Hilbert space K_+. Here, we restrict ourselves only to the statements which will be used below.

By analogy with the proof of Proposition 12, we conclude that an arbitrary vector $g \in K_+$ can be represented in the form

$$g = (I-P_{K_\varkappa})\hat{p} - \hat{s} \quad (\hat{p} \in \mathcal{P}_+), \tag{35}$$

where the Π-orthogonal elements $(I-P_{K_\varkappa})\hat{p}$ and $\hat{s} \in (I-P_{\Pi_0})D_+ \subset \mathcal{P}_-$ uniquely determine each other by the equality

$$[\hat{s},p] + [P_{K_\varkappa}\hat{p}, p] = 0 \quad (\forall p \in \mathcal{P}_+). \tag{36}$$

Furthermore,

$$A_+^n = P_{D_+'}\left(U^+\right)^n\big|_{K_+} \quad (n \in \mathbf{N})$$

and, for arbitrary $g \in K_+$ and $d_+ \in D_+$, the following equality holds:

$$[A_+^n g, d_+] = \left[\left(U^+\right)^n \hat{p}, d_+\right]. \tag{37}$$

This yields

$$K_+ = L_{\bar{\lambda}_1}(A_+) + ... + L_{\bar{\lambda}_n}(A_+),$$

where

$$L_{\bar{\lambda}_i}(A_+) = P_0(I-P_{K_\varkappa}) L_{\bar{\lambda}_i}(U^+).$$

Proposition 14. *For all $f \in K_-$, the following equality holds:*

$$A_+ P_{K_+} f = P_{K_+} A_-^+ f,$$

where P_{K_+} is a π-orthoprojector onto K_+ in Π_0.

Proof. Let us show that, for any $f = (I - P_{K_\kappa})\tilde{s} - \tilde{p}$ from K_-,

$$P_{K_+}f = g = -((I - P_{K_\kappa})\tilde{p} - \hat{s}(\tilde{p})), \tag{38}$$

where the element $\hat{s}(\tilde{p})$ belongs to $\mathcal{P}_-$ and is determined via $(I - P_{K_\kappa})\tilde{p}$ by equality (36).

Indeed, $g \in K_+$ by construction. In this case,

$$f - g = \tilde{s} - \hat{s}(\tilde{p}) - P_{K_\kappa}(\tilde{s} + \tilde{p}).$$

The vector $\tilde{s} - \hat{s}(\tilde{p})$ is π-orthogonal to the space $\overline{\Pi_0} \supset K_+$. Furthermore, one can easily verify that the element $P_{K_\kappa}(\tilde{s} + \tilde{p})$ is π-orthogonal to the subspace K_+ (see equalities (31) and (35)). Therefore, $(f - g)[\perp] K_+$, which proves equality (38).

Equality (37) with $g = P_{K_+}f$ yields

$$[A_+ P_{K_+}f, d_+] = -[U^+\tilde{p}, d_+] \quad (\forall\, d_+ \in D_+)$$

On the other hand, by virtue of equalities (33) and (38), we have

$$[P_{K_+}A^+_\kappa f, d_+] = -[(I - P_{K_\kappa})U^+\tilde{p}, d_+] = -[U^+\tilde{p}, d_+].$$

Then

$$[A_+ P_{K_+}f, d_+] = [P_{K_+}A^+_\kappa f, d_+] \quad (\forall\, d_+ \in D_+).$$

Thus, Proposition 14 is proved.

∎

To conclude, we prove one more assertion necessary for what follows.

Lemma 15. *If $g \in L^\kappa_{\bar\lambda}(A_+)$, then its image $\hat{g}$ in the outgoing spectral representation $L_2((0, 2\pi), N'_+)$ has the form*

$$\hat{g} = \sum_{k=0}^{\kappa} \frac{e^{ik\theta}}{\left(1 - \bar\lambda e^{i\theta}\right)^{k+1}} P_{N'_+}(A_+ - \bar\lambda I)^k g.$$

Proof. The outgoing spectral representation $L_2((0, 2\pi), N'_+)$ is the Fourier transform of the space

$$M_+ = \overset{\infty}{\underset{-\infty}{V}} U^n D'_+ = \overset{\infty}{\underset{-\infty}{[+]}} U^n N'_+,$$

where $N'_+ = D'_+ [-] U D'_+$ is a wandering space with respect to the operator U. Since

$$D'_+ = \overset{\infty}{\underset{0}{[+]}} U^n N'_+$$

and $g \in K_+ \subset D'_+$, we have

$$g = \sum_{n=0}^{\infty} U^n g_n \quad (g_n \in N'_+). \tag{39}$$

In view of the relation $A_+ = P_{D'_+} U^+ \big|_{K_+}$, one can conclude that if $(A_+ - \bar{\lambda} I)g = 0$, then

$$\sum_{n=1}^{\infty} U^{n-1} g_n = \bar{\lambda} \sum_{n=0}^{\infty} U^n g_n$$

and, hence, $g_n = \bar{\lambda}^n g_0 = \bar{\lambda}^n P_{N'_+} g$.

Similarly, if $(A_+ - \bar{\lambda} I)^p g = 0$ $(p \in \mathbf{N})$, then it is easy to check that the elements g_n in decomposition (39) are determined by the relation

$$g_n = \sum_{k=0}^{n} c_n^k \bar{\lambda}^{n-k} P_{N'_+} \left(A_+ - \bar{\lambda} I\right)^k g.$$

Therefore, the spectral representation of the element g has the form

$$\hat{g} = \sum_{n=0}^{\infty} e^{i\theta n} g_n = \sum_{n=0}^{\infty} e^{i\theta n} \sum_{k=0}^{n} c_n^k \bar{\lambda}^{n-k} P_{N'_+} \left(A_+ - \bar{\lambda} I\right)^k g$$

$$= \sum_{k=0}^{\kappa} e^{i\theta k} \left(\sum_{n \geq k} c_n^k \bar{\lambda}^{n-k} e^{i\theta(n-k)} \right) P_{N'_+} \left(A_+ - \bar{\lambda} I\right)^k g.$$

To complete the proof, it suffices to take the identity

$$\sum_{n \geq k} c_n^k \bar{\lambda}^{n-k} e^{i\theta(n-k)} = \frac{1}{\left(1 - \bar{\lambda} e^{i\theta}\right)^{k+1}}$$

into account.

■

5.   Properties of Scattering Matrix

It follows from equality (11) that the properties of the scattering matrix $\mathcal{B}(e^{i\theta})$ are determined by the properties of the scattering matrices $\mathcal{B}_-^0(e^{i\theta})$ and $\mathcal{B}_-(e^{i\theta})$ of the operators S_-^0 and S_-, respectively.

The well-known properties of these functions are summarized in the two theorems presented below.

Theorem 16. *The scattering matrix $\mathcal{B}_-^0(e^{i\theta})$ is a boundary value of the operator-valued function $\mathcal{B}_-^0(\omega)$ analytic in the domain $|\omega| > 1$, whose values are contracting operators from $U N_-''$ to N_+'.. Furthermore,*

(a) $0 \in \sigma_\alpha(\mathcal{B}_-^0(\omega))$ if and only if $\omega^{-1} \in \sigma_\alpha(A_-^0)$ ($\alpha \in \{p, r, c\}, |\omega| > 1$), where the operator A_-^0 is defined by equality (28);

(b) the scattering matrix $\mathcal{B}_-^0(e^{i\theta})$ is analytic[11] and unitary on the set $C \setminus \sigma\left(\left(A_{BH}^0\right)^+\right)$, where C is the unit circle in the complex plane and A_{BH}^0 is completely nonunitary part of the operator A_-^0.

To make our presentation complete, we indicate the principal steps of the proof. By virtue of equality (10), the operator S_-^0 is, in fact, a translation representation of the orthoprojector P_{M_+} from M_- to M_+. In view of the orthogonality of the incoming subspace D_-'' and the outgoing subspace D_+', we get

$$P_{M_+} U N_-'' \subset \overset{0}{\underset{-\infty}{[+]}} U^n N_+'$$

Following the proof of Lemma 3.1 in the monograph of Sz.-Nagy and Foias [4, Chapter 5], we can conclude that the scattering matrix $\mathcal{B}_-^0(e^{i\theta})$ is the boundary value

---

[11] We say that a scattering matrix is analytic at the point $e^{i\theta}$ if the function $\mathcal{B}(\omega)$ can be analytically extended outside the unit circle in the neighbourhood of this point.

of the function

$$\mathcal{B}_-^0(\omega) \;=\; P_{N_+'} \sum_{n=0}^{\infty} \left(\frac{U}{\omega}\right)^n \;=\; P_{N_+'}(U^+ - 1/\omega I)^{-1} U^{-1}.$$

This function is analytic in the domain $|\omega| > 1$ and its values are contracting operators from $U N_-''$ to N_+'. By analogy with the proof of Theorem 11, we get

$$\mathcal{B}_-^0(\omega) \;=\; \left(\theta_{A_-^0}(\bar{\lambda}) \oplus U P_{\mathcal{R}''}\right) U^{-1}, \tag{40}$$

where

$$\theta_{A_-^0}(\bar{\lambda}) \;=\; P_{L_+'}(U^+ - \bar{\lambda} I)^{-1} P_{L_-''} \qquad (\bar{\lambda} = 1/\omega)$$

is the characteristic function of the operator A_-^0 and $P_{\mathcal{R}''}$ is an orthoprojector onto $\mathcal{R}''$ in $\mathfrak{N}$.

The validity of assertion (a) follows from decomposition (40) and Theorem 9 in Appendix 2. Assertion (b) is a consequence of Theorem 4.1 in the monograph of Sz.-Nagy and Foias [4, Chapter 6].

■

Theorem 16 shows that the properties of the scattering matrix $\mathcal{B}_-^0(e^{i\theta})$ are determined by the properties of the contracting operator A_-^0. By analogy, one can establish a similar relationship between the scattering matrix $\mathcal{B}_-(e^{i\theta})$ and the spectrum of the contracting operator A_-^+. Proposition 12 enables us to strengthen assertions (a) and (b) in Theorem 16.

Theorem 17. *The scattering matrix $\mathcal{B}_-(e^{i\theta})$ is a boundary value of the operator-valued function $\mathcal{B}_-(\omega)$ analytic in the domain $|\omega| > 1$, whose values are contracting operators from $U N_-''$ to $U N_-'$. Furthermore,*

(a) *the operator $\mathcal{B}_-(\omega)$ is completely invertible (i.e., $0 \in \rho(\mathcal{B}_-(\omega))$) for all $|\omega| > 1$ except the finite set of points $\{\omega_i = 1/\bar{\lambda}_i\}$ for which the lineal $\mathcal{L}_{1/\lambda_i}(A^+)$ contains negative elements;*

(b) *the scattering matrix $\mathcal{B}_-(e^{i\theta})$ is analytic and unitary on the unit circle.*

We set

$$\mathcal{B}(\omega) \;=\; \mathcal{B}_-^0(\omega) \mathcal{B}_-^{-1}(\omega). \tag{41}$$

By virtue of Theorems 16 and 17, we can conclude that the operator-valued function $\mathcal{B}(\omega)$ is meromorphic in the domain $|\omega| > 1$ and the set of its singular points coincides with the set of points $\{\omega_i\}$ at which the operator $\mathcal{B}_-(\omega)$ is not invertible. Equality (11) and Theorems 10, 16, and 17 yield the following assertion:

Theorem 18. *If all root lineals* $\mathcal{L}_{\varepsilon_i}(U)$, $i = 1, \ldots, m$, *are nondegenerate, then, for almost all* θ, *the scattering matrix* $\mathcal{B}(e^{i\theta})$ *is a boundary value of the operator-valued function* $\mathcal{B}(\omega)$ *meromorphic in the domain* $|\omega| > 1$, *whose set of singular points coincides with the set* $\{\omega_i = 1/\bar{\lambda}_i\}$ *of eigenvalues of the operator* A *for which the lineal* $\mathcal{L}_{1/\bar{\lambda}_i}(A^+)$ *contains negative elements. If the point* ω *is not singular, then* $0 \in \sigma_\alpha(\mathcal{B}(\omega))$ *if and only if* $1/\omega \in \sigma_\alpha(A^+)$ $(\alpha \in \{p, r, c\})$.

The scattering matrix $\mathcal{B}(e^{i\theta})$ *is analytic and unitary on the set* $C \setminus \sigma\left((A_{BH}^0)^+\right)$, *where* A_{BH}^0 *is a completely nonunitary part of the operator* A_-^0.

6. Scattering Matrix and Characteristic Function
of π-Nonexpanding Operator A^+

In this subsection, we show that properties of the scattering matrix $\mathcal{B}(e^{i\theta})$ are determined by properties of the characteristic function $\theta_{A^+}(\bar{\lambda}) = P_{\mathcal{L}_+}(U^+ - \bar{\lambda}I)^{-1}P_{\mathcal{L}_-}$ of the π-nonexpanding operator A^+. Note that this correspondence is not so evident as in the definite case where the scattering matrix is, in fact, a boundary value of the characteristic function of some contraction,.

Theorem 19. *Let* $\omega \notin \sigma(A)$. *Then* $0 \in \sigma_\alpha(\mathcal{B}(\omega))$ *if and only if*

$$0 \in \sigma_\alpha(\theta_{A^+}(1/\omega)) \quad (\alpha \in \{p, r, c\}).$$

Proof. This assertion is a consequence of Theorem 9 in Appendix 2 and Theorem 18. ∎

Assume that $P_{\bar{\lambda}_i}$ is a projector onto $\mathcal{L}_{\bar{\lambda}_i}(U^+)$ in Π_κ defined by decomposition (1). By using the Laurent expansion of the resolvent $(U^+ - \bar{\lambda}I)^{-1}$ in the neighborhood of the singular point $\bar{\lambda}_i$, we get

$$\theta_{A^+}(\bar{\lambda}) = P_{L_+}\left[-\sum_{l=0}^{\kappa} \frac{\left(U^+ - \bar{\lambda}_i I\right)^l}{\left(\bar{\lambda} - \bar{\lambda}_i\right)^{l+1}} P_{\bar{\lambda}_i} + \sum_{k=0}^{\infty} \frac{\left(U^+ - \bar{\lambda}_i I\right)^{-(k+1)}}{\left(\bar{\lambda} - \bar{\lambda}_i\right)^{-k}} \left(I - P_{\bar{\lambda}_i}\right)\right] P_{L_-}.$$

$$(42)$$

It is clear from (42) that the singular point $\bar{\lambda}_i$ of the function $\theta_{A^+}(\bar{\lambda})$ has the order

$$\gamma = \max_{p_i \in P_{\bar{x}_i} L_-} \max_{0 \le l < \kappa} \{l + 1 \,|\, P_{L_+}(U^+ - \bar{\lambda}_i I)^l p_i \ne 0\}.$$

In view of relations (18) (Appendix 2) and (25), this yields

$$\gamma = \max_{p_i \in P_{\bar{x}_i} L_-} \max_{0 \le k < \kappa} \{k \,|\, (U^+ - \bar{\lambda}_i I)^k p_i \in K_\kappa\}. \qquad (43)$$

Theorem 20. *The point* ω_i *is a pole of order* γ $(\gamma > 0)$ *of the function* $\mathcal{B}(\omega)$ *if and only if the point* $\bar{\lambda}_i = 1/\omega_i$ *is a pole of the same order of the characteristic function.* $\theta_{A^+}(\bar{\lambda})$.

Proof. For any $y \in UN'_-$, we have

$$P_{M_+} y = (I - P_{UD'_+})y + P_{UD'_+} y, \qquad (44)$$

where $P_{UD'_+}$ is a π-orthoprojector onto UD'_+ in Π_0. Furthermore,

$$(I - P_{UD'_+})y \in \overset{0}{\underset{-\infty}{[+]}} U^n N'_+$$

and

$$(I - P_{UD'_+})y = \sum_{-\infty}^{0} U^n P_{N'_+} U^{-n} y. \qquad (45)$$

On the other hand, by setting $y = Ud'_-$ (here, $d'_- \in N'_-$, $d'_- = d''_- + f$, $d''_- \in D''_-$, and $f \in K_-$), we get $P_{UD'_+} y = P_{UD'_+} Uf = UP_{D'_+} f$. Since the element f is π-orthogonal to D''_+, we have $UP_{D'_+} f = UP_{K_+} f$. By virtue of equality (32) and Proposition 14, we conclude that

$$f = \sum_{i=1}^{n} f_i \qquad (f_i \in L_{\bar{\lambda}_i}(A^+_-))$$

and

$$P_{K_+}f = \sum_{i=1}^{n} g_i,$$

where $P_{K_+}f_i = g_i \in L_{\bar{\lambda}_i}(A_+).$
  Thus,

$$P_{UD'_+}y = UP_{K_+}f = U \sum_{i=1}^{n} g_i. \tag{46}$$

By passing to the spectral representation in equality (44) and taking equalities (45) and (46) and Lemma 15 into account, we obtain

$$\mathcal{B}(e^{i\theta})y = P_{N'_+} \sum_{-\infty}^{n=0} e^{in\theta} U^{-n} y + \sum_{i=1}^{n} \sum_{k=1}^{\kappa} \frac{e^{ik\theta}}{\left(1-\bar{\lambda}_i e^{i\theta}\right)^k} P_{N'_+} (A_+ - \bar{\lambda}_i I)^{k-1} g_i.$$

Hence,

$$\mathcal{B}(\omega)y = P_{N'_+} \left(I - \frac{U}{\omega}\right)^{-1} y + \sum_{i=1}^{n} \sum_{k=1}^{\kappa} \frac{\omega^k}{\left(1-\bar{\lambda}_i \omega\right)^k} P_{N'_+} (A_+ - \bar{\lambda}_i I)^{k-1} g_i. \tag{47}$$

It follows from the proof of Lemma 15 that if $g_i \in \text{Ker}\,(A_+ - \bar{\lambda}_i I)$ and $g_i \neq 0$, then $P_{N'_+} g_i \neq 0$. Therefore, by virtue of decomposition (47) and Proposition 14, the point $\omega_i = 1/\bar{\lambda}_i$ is a pole of order

$$\gamma_1 = \max_{f \in P_{K_-} N'_-} \min_k \{k | P_{K_+}(A_-^+ - \bar{\lambda}_i I)^k f_i = 0\} \tag{48}$$

of the function $\mathcal{B}(\omega)$. Equalities (30), (33), and (38) imply that the vector f_i can be represented in the form $f_i = (I - P_{K_+}) \bar{s}_i - \tilde{p}_i$, where $\tilde{p}_i \in L_{\bar{\lambda}_i}(U^+)$; furthermore, the equality $P_{K_+}(A_-^+ - \bar{\lambda}_i I)^k f_i = 0$ holds if and only if $(U^+ - \bar{\lambda}_i I)^k \tilde{p}_i \in K_\kappa$.
  It follows from (48) that

$$\gamma_1 = \max_{\tilde{p}_i \in P_{\bar{\lambda}_i}(I - P_{D_-})P_{K_-} N'_-} \min_k \{k | (U^+ - \bar{\lambda}_i I)^k \tilde{p}_i \in K_\kappa\}. \tag{49}$$

According to equality (4), any $d'_- \in N'_-$ can be represented in the form $d'_- = d_- - \tilde{p}$, where $d_- \in N_- = D_-[-]U^+D_-$. In this case, $d'_- = d''_- + f$, where $f =$

$(I - P_{K_\kappa}) \tilde{s} - \tilde{p} \in K_-$. Then

$$P_{\tilde{\lambda}_i} (I - P_0) P_{K_-} d'_- = P_{\tilde{\lambda}_i} (I - P_{D_-}) f = - P_{\tilde{\lambda}_i} \tilde{p} = - P_{\tilde{\lambda}_1} d_-$$

and, hence,

$$P_{\tilde{\lambda}_i} (I - P_{D_-}) P_{K_-} N'_- = P_{\tilde{\lambda}_i} N_-,$$

where $P_{\tilde{\lambda}_i} N_- = P_{\tilde{\lambda}_i} L_-$ by virtue of Proposition 6. Taking equalities (43) and (49) into account, we get $\gamma_1 = \gamma$. Theorem 20 is proved.

∎

7.   Individual Case

Within the framework of the classical Lax–Phillips scheme, the most interesting results can be obtained under the condition that the orthogonal pair of subspaces D_- and D_+ of the Hilbert space $\mathcal{H}$ satisfies the following additional condition:

$$\overset{\infty}{\underset{-\infty}{V}} U^n D_- = \overset{\infty}{\underset{-\infty}{V}} U^n D_+ = \mathcal{H}. \tag{50}$$

It follows from Proposition 3 that equality (50) is not valid in the Pontryagin space Π_κ ($\kappa > 0$).

Below, we consider the following "weakened" version of condition (50) for the π-orthogonal pair of the subspaces D_- and D_+:

$$\overset{\infty}{\underset{-\infty}{V}} U^n (D_- [+] D_+) = \Pi_\kappa \tag{51}$$

Lemma 21. *If condition (51) is satisfied, the operator U has no eigenvalues on the unit circle.*

Proof. Let us show that if $Ux = \varepsilon x$ ($|\varepsilon| = 1$), then $x [\perp] U^m D_+$ ($m \in \mathbb{Z}$). Indeed, by virtue of conditions (a) and (b) (see Subsection 2) and Proposition 2.9 in the monograph of Azizov and Iokhvidov [1, Chapter 3], the subspace D_+ is positive. Therefore, the subspace D_+ is a Hilbert space with respect to the indefinite metric and the operator $U|_{D_+}$ is isomorphic to right shift in D_+. Then, for any $d_+ \in D_+$, we get

$$[x, U^m d_+] = \lim_{n \to \infty} \varepsilon^{n-m} [x, U^n d'_+] = 0$$

and, hence, $x \, [\perp] \, U^m D_+$.

Similarly, we establish that $x \, [\perp] \, U^m D_-$. In view of (51), this implies that $x \, [\perp] \, \Pi_\kappa$ and, consequently, $x = 0$.

■

By virtue of Lemma 21, decomposition (1) takes the following form for the operator U:

$$\Pi_\kappa = \Pi_0 \, [\dotplus] \, (\mathcal{P}_- \dotplus \mathcal{P}_+).$$

At the same time, by virtue of equalities (8) and (51), we have

$$\Pi_\kappa = M_+ \, V M_- \, [\dotplus] \, (I - P_{\Pi_0})(D_+ \dotplus D_-).$$

Taking (3) and (4) into account, we obtain

$$M_+ \, V M_- = \Pi_0 \tag{52}$$

and

$$(I - P_{\Pi_0})D_- = \mathcal{P}_+, \quad (I - P_{\Pi_0})D_+ = \mathcal{P}_-. \tag{53}$$

Theorem 22. *Suppose that condition (51) is satisfied. Then the following assertions are true:*

(a) *For almost all θ, the scattering matrix $\mathcal{B}(e^{i\theta})$ is a boundary value of the operator-valued function $\mathcal{B}(\omega)$ meromorphic in the domain $|\omega| > 1$, whose set of singular points coincides with the set $\{\omega_i = 1/\bar{\lambda}_i\}$ of eigenvalues of the operator A lying outside the unit circle. The order of the pole ω_i is equal to the index[12] of the eigenvalue $1/\bar{\lambda}_i$ of the operator A.*

(b) *The scattering matrix $\mathcal{B}(e^{i\theta})$ is analytic and unitary on the set $C \setminus \sigma(A)$, where C is the unit circle.*

Proof. Let $x \in \mathcal{P}_- \cap K_\kappa$. Then $x \, [\perp] \, D_-$ and, by virtue of equalities (4) and (53), $x \, [\perp] \, \mathcal{P}_+$. This is possible only for $x = 0$ because the lineal $\mathcal{P}_- \dotplus \mathcal{P}_-$ is non-degenerate. Therefore, $\mathcal{P}_- \cap K_\kappa = \{0\}$. According to Proposition 7, the lineals $\mathcal{L}_{1/\lambda_i}(A^+)$ $(i = 1, \dots, n)$ are negative and, hence, the set of singular points of the

---

[12] If $\dim \mathcal{L}_\mu(A) < \infty$, then the index of the eigenvalue μ of the operator A is the least minimal natural n such that $\mathrm{Ker}\,(A - \mu I)^n = \mathcal{L}_\mu(A)$.

function $\mathcal{B}(\omega)$ coincides with the set of eigenvalues $1/\bar{\lambda}_1, \ldots, 1/\bar{\lambda}_n$ of the operator A (see Theorem 18).

It follows from the proof of Theorem 20 that the function $\mathcal{B}(\omega)$ has a pole of order

$$\gamma = \max_{p_i \in P_{\bar{\lambda}_i} N_-} \min_k \{k \,|\, (U^+ - \bar{\lambda}_i I)^k p_i \in K_\kappa\}.$$

at the point $\omega_i = 1/\bar{\lambda}_i$.

Similarly, we establish that $\mathcal{P}_+ \cap K_\kappa = \{0\}$. Therefore, the lineals $\mathcal{L}_{1/\bar{\lambda}_i}(A)$ are also negative. By virtue of equality (17), the element $(U^+ - \bar{\lambda}_i I)^k p_i$ $(p_i \in \mathcal{L}_{1/\bar{\lambda}_i}(U))$ belongs to K_κ only if $(U^+ - \bar{\lambda}_i I)^k p_i = 0$. Hence,

$$\gamma = \max_{p_i \in P_{\bar{\lambda}_i} N_-} \min_k \{k \,|\, (U^+ - \bar{\lambda}_i I)^k p_i = 0\}$$

$$= \max_{p_i \in P_{\bar{\lambda}_i} V_{n=0}^\kappa (U^+)^n N_-} \min_k \{k \,|\, (U^+ - \bar{\lambda}_i I)^k p_i = 0\}. \tag{54}$$

Since $\dim \mathcal{L}_{1/\bar{\lambda}_i}(U) \leq \kappa$, we have

$$P_{\bar{\lambda}_i} \bigvee_{n=0}^\kappa (U^+)^n N_- = (I - P_{\Pi_0}) D_-.$$

Together with the first equality in (53) and equality (54), this proves statement (a).

By virtue of Theorems 10 and 18, to prove statement (b), it suffices to show that the operator A_-^0 is completely nonunitary.

Indeed, it follows from the proof of Proposition 11 and equality (52) that

$$\left[\dotplus \right]_{-\infty}^\infty U^n \mathcal{L}_-'' \bigvee \left[\dotplus \right]_{-\infty}^\infty U^n \mathcal{L}_+' = \Pi_0^{\min},$$

which is equivalent to the complete nonunitariness of the operator A_-^0 (see Sz.-Nagy and Foias [4, Chapter 2, Proposition 1.4]).

∎

8.   Continuous Case

Assume that $W(t)$ is a π-unitary semigroup of the class C_0 in Π_κ

The subspace D_+ is called outgoing for the semigroup $W(t)$ if the following conditions are satisfied:

(a) $W(t)D_+ \subset D_+$ $(t \geq 0)$,

(b) $\bigcap\limits_{t \geq 0} W(t)D_+ = \{0\}$.

The subspace D_- is called incoming for the semigroup $W(t)$ if D_- is outgoing for the semigroup $W^+(t)$.

The generator R of the semigroup $W(t)$ is a π-skewself-adjoint operator in Π_κ. Therefore, there exists $\xi_0 > 0$ such that all ξ with $\mathrm{Re}\,\xi \geq \xi_0$ belong to the resolvent set $\rho(R)$. The π-unitary operator

$$U = (R + \xi_0 I)(R - \xi_0 I)^{-1}$$

is called a cogenerator of the semigroup $W(t)$.

Following the proof of Lemma 3.1 in the monograph of Lax and Phillips [4, Chapter 2], one can easily establish the following assertion:

Lemma 23. *A subspace D is outgoing (incoming) for the semigroup $W(t)$ if and only if it is outgoing (incoming) for the cogenerator U.*

Assume that $D_+ [\perp] D_-$. Then the π-nonexpanding operators $Z(t) = P_{K_\kappa} W(t)|_{K_\kappa}$ form a semigroup of the class C_0, which acts in the Pontryagin space $K_\kappa = \Pi_\kappa[-]$ $(D_+[+]D_-)$.

Assume that an operator B is a generator of the semigroup $Z(t)$. Then its cogenerator

$$A = (B + \xi_0 I)(B - \xi_0 I)^{-1}$$

is a π-nonexpanding operator for which the π-unitary operator U is a regular dilation.

By using Lemma 23 and the transformation

$$z = i\xi_0 \frac{1 + \omega}{1 - \omega},$$

which maps the domain $|\omega| \geq 1$ into the lower half-plane $\mathrm{Im}\,z \leq 0$ (as in the Lax–Phillips method [1]), we reduce the investigation of the scattering matrix $\mathcal{B}_W(\delta)$ $(\delta \in \mathbb{R})$ for the semigroup $W(t)$ to studying the corresponding scattering matrix $\mathcal{B}(e^{i\theta})$ for the cogenerator U of this semigroup.

To conclude, we summarize the principal results.

If all root lineals $\mathcal{L}_{i\gamma}(R)$, $\gamma \in \mathbb{R}$, are nondegenerate, then, for almost all $\delta \in \mathbb{R}$, the scattering matrix $\mathcal{B}_W(\delta)$ is a boundary value of the operator-valued function $\mathcal{B}_W(z)$ meromorphic in the domain $\text{Im}\, z < 0$, whose set of singular points coincides with the set $\{z_i\}$ of eigenvalues of the operator $-iB$ for which the lineal $\mathcal{L}_{\bar{z}_i}(iB^+)$ contains negative elements.

The order of the singular point z_i coincides with the order of the pole of the characteristic function $\theta_{A^+}(\bar{\lambda})$ at the point $\bar{\lambda}_i = (z_i + i\xi_0)(z_i - i\xi_0)^{-1}$.

If the point z is not singular, then $0 \in \sigma_\alpha(\mathcal{B}_W(z))$ if and only if $z \in \sigma_\alpha(iB^+)$, $\alpha \in \{p, r, c\}$.

If condition (51) is satisfied for the subspaces D_+ and D_-, then

(a) the scattering matrix $\mathcal{B}_W(\delta)$ is analytic and unitary on the set

$$\mathbb{R} \setminus \left\{ i\xi_0 \frac{1+\gamma}{1-\gamma} \,\middle|\, \gamma \in \sigma(A) \right\};$$

(b) the set of singular points of the operator function $\mathcal{B}_W(z)$ coincides with the set of eigenvalues $\{z_i\}$ of the operator $-iB$ in the lower half-plane. The order of the pole z_i is equal to the index of the eigenvalue z_i of the operator $-iB$.

9.   Remark

This appendix was written by S. Kuzhel on the basis of his paper [4].

REFERENCES

Adamjan, V. M. and Arov, D. Z.

[1] *About one class scattering operators and characteristic operator functions contractions,* Dokl. Akad. Nauk SSSR **160** (1965), No. 1, 9–12. *(Russian)*

[2] *On scattering operators and contraction semigroups in Hilbert spaces,* Dokl. Akad. Nauk SSSR **165** (1965), No. 1, 9–12. *(Russian)*

[3] *Unitary couplings of semi-unitary operators,* Mat. Issled. **1** (1966), No. 2, 3–64. *(Russian)*

[4] *A general solution of a certain problem in the linear prediction of stationary processes,* Teor. Veroyatn. i Primen. **13** (1968), No. 3, 419–423. *(Russian)*

Azizov, T. Ya. and Iokhvidov, I. S.

[1] Foundations of the Theory of Linear Operators in the Spaces with Indefinite Metrics, Nauka, Moscow, 1986; *English transl.* Wiley, 1989.

[2] *Linear operators in spaces with indefinite metrics and their applications,* VINITI Series on Contemporary Problems in Mathematics. The Mathematical Analysis, vol. 17, VINITI, Moscow, 1979, pp. 113–205. *(Russian)*

Akhiezer, N. I.

[1] The Classic Problem of Moments, Nauka, Moscow, 1961; *English transl.* Oliver and Boyl, Edinburgh, 1965.

Akhiezer, N. I. and Glazman, I. M.

[1] Theory of Linear Operators in Hilbert Space, vol. 1, Vyshcha Shkola, Kharkov, 1977; *English transl.* Pitman (APP), 1981.

[2]   Theory of Linear Operators in Hilbert Space, vol. 2, Vyshcha Shkola, Kharkov, 1978; *English transl.* Pitman (APP), 1981.

Ball, J. A.

[1]   *Models for non-contractions*, J. Math. Anal. and Appl. **52** (1975), 235–254.

[2]   *Factorization and model theory for contraction operators with unitary part,* Memoirs of the AMS **13** (1978), No. 198, 1–68.

Berezanskii, Yu. M.

[1]   Expansions in Eigenfunctions of Selfadjoint Operators, Naukova Dumka, Kiev, 1965; *English transl.*: Amer. Math. Soc. Transl. vol. **17**, Providence, R.I., 1968.

Berezanskii, Yu. M., Us, G. F., and Sheftel, Z. G.

[1]   Functional Analysis, Vyshcha Shkola, Kiev, 1990. *(Russian)*

Bognar, J.

[1]   Indefinite Inner Product Spaces, Springer, Berlin, 1974.

de Branges, L. and Rovnyak, J.

[1]   *Canonical models in quantum scattering theory: appendix on square summable power series,* Perturbation Theory and Its Applications in Quantum Mechanics, Wiley, New York, 1966.

[2]   Square Summable Power Series, Holt: Reinhart and Wilson, New York, 1966.

Brodskii, M. S.

[1]   Triangular and Jordan Representations of Linear Operators, Nauka, Moscow, 1969. *(Russian)*

Brodskii, M. S. and Kisilevskii, G. E.

[1]   *Criterion for unicellularity of dissipative Volterra operators with nuclear imaginary components,* Izv. Akad. Nauk SSSR. Ser. Mat. **30** (1966), No. 6, 1213–1228. *(Russian)*

Brodskii, M. S. and Livsic, M. S.

[1] *Spectral analysis of nonself-adjoint operators and intermediate systems,* Uspekhi Mat. Nauk **13** (1958), No. 1, 3–85. *(Russian)*

Brodskii, V. M.

[1] *On multiplication and division theorems of characteristic function of invertible operator,* Acta Sci Math. **32** (1971), 161–171. *(Russian)*

Brodskii, V. M., Gohberg, I. C., and Krein, M. G.

[1] *Definition and basic properties of characteristic function of J-coupling,* Funk. Anal. i Prilozh. **4** (1970), No. 1, 88–90. *(Russian)*

van Casteren, J. A.

[1] *Operators similar to unitary of selfadjoint ones,* Pacific. J. Math. **104** (1983), No. 1, 241–255.

Clark, D. N.

[1] *Concrete model theory for a class of operators,* J. Funct. Anal. **14** (1973), 269–280.

[2] *On models for non-contractions,* Acta Sci. Math. **36** (1974), 5–16.

Davis, Ch.

[1] *J-uniraty dilation of a general operator,* Acta Sci. Math. **31** (1970), 75–86.

Davis, Ch. and Foias, C.

[1] *Operators with bounded characteristic function and their J-unitary dilation,* Acta Sci. Math. **32** (1971), 127–139.

Derkach, V. A. and Malamud, M. M.

[1] *Veyl function of Hermitian operators and its relation to characteristic function,* Preprint No. 85–9 (104), Fiz.-Tekhn. Inst., Donetsk, 1985. *(Russian)*

254   *References*

Do Kong Han

[1]   *On multiplication theorem of characteristic functions of unbounded operators,* Teor. Funktsij Funk. Anal. i Prilozh. (1969), No. 9, 65–74. *(Russian)*

Douglas, R. G.

[1]   *Canonical models,* Math. Surveys, AMS, Providence, 1974, vol. 13, 161–218.

Dunford, N. and Schwartz, J.T.

[1]   Linear Operators, vol. I: General Theory, Interscience, New York – London, 1958.
[2]   Linear Operators, vol. II: Spectral Theory. Selfadjoint Operators in Hilbert Spaces, Interscience, New York – London, 1963.

Faddeev, L. D.

[1]   *On Friedrichs model in the theory of perturbations of the continuous spectrum,* Trudy Mat. Inst. Steklov **73** (1964), 293–313. *(Russian)*

Foias, C.

[1]   *Modeles fonctionnels, liaison entreles theories de la pridiction, de la fonction caracteristique et de la dilatation unitaire,* Deuxieme Colloque Surl'analyse Fonctionnele, Liegi, 4–6 mai, 63–76 (1964).

[2]   *On the Lax–Phillips nonconservative scattering theory,* J. Funct. Anal. **19** (1975), 273–301.

Friedrichs, K.

[1]   *Über die Spectralzerlegungeines integral operators,* Match. Ann. **115** (1938), No. 2, 249–272.

[2]   *On the perturbations of continuous spectra,* Comm. Pure Appl. Math. **1** (1948), No. 4, 361–406.

Garnett, J. B.

[1]   Bounded Analytic Functions, Academic Press, New York, 1981.

Ginzburg, Yu. P.

[1]   *On J-nonexpanding operator function,* Dokl. Akad. Nauk SSSR **117** (1957), No. 2, 171–173. *(Russian)*

Gohberg, I. C. and Krein, M. G.

[1]   Theory of Volterra Operators in Hilbert Space and Its Applications, Nauka, Moscow, 1967. *(Russian)*

[2]   *On a description of contraction operators similar to unitary ones,* Funk. Anal. i Prilozh. **1** (1967), 38–60.

Gorbachuk, M. L. and Gorbachuk, V. I.

[1]   Boundary-Value Problems for Operator-Differential Equations, Naukova Dumka, Kiev, 1984; *English transl.* Kluwer, 1991.

Gorbachuk, M. L. Gorbachuk, V. I., and Kochubei, A. N.

[1]   *The theory of extensions of symmetric operators and boundary-value problems for differential equations,* Ukrain. Mat. Zh. **41** (1989), No. 10, 1299–1313. *(Russian)*

Gubreev, G. M.

[1]   *Definition and fundamental properties of the characteristic function of a W-coupling,* Dokl. Akad. Nauk Ukrain. SSR. Ser. A (1978), No. 1, 3–6. *(Russian)*

Helson, H.

[1]   Lectures on Invariant Subspaces, Academic Press, New York, 1964.

Helton, J. W.

[1]   *Discrete time systems, operator model and scattering theory,* J. Funct. Anal. **16** (1974), No. 1, 15–38.

Hoffman, K.

[1]   Banach Spaces of Analytic Functions, Prentice Hall, Englewood Cliffs, 1962.

Karpenko, I. I.

[1] *Characteristic operator functions of* K_1^r *operators,* Dokl. Akad. Nauk Ukrain. SSR. Ser. A (1983), No. 9, 3–5. *(Russian)*

[2] Characteristic Functions and Models of Regular Extension of Hermitian Operators, Candidate Degree Thesis, Simferopol, 1991. *(Russian)*

Karpenko, I. I. and Tikhonov, A. S.

[1] *A functional model of unbounded dissipative operators,* Dokl. Akad. Nauk Ukrain. SSR. Ser. A. (1985), No. 6, 7–10. *(Russian)*

Kisilevskii, G. E.

[1] *On the unicellularity of dissipative Volterra operators,* Ukrain. Mat. Zh. **16** (1964), No. 5, 689–696. *(Russian)*

[2] *Cyclic subspaces of dissipative operators,* Dokl. Akad. Nauk SSSR **173** (1967), No. 5, 1006–1009. *(Russian)*

Kochubei, A. N.

[1] *On extensions and characteristic functions of symmetric operators,* Izv. Akad. Nauk Armenian SSR **15** (1980), No. 3, 219–232. *(Russian)*

[2] *About symmetric operators commuting with a family of unitary operators,* Funk. Anal. i Prilozh. **13** (1979), No. 4, 77–78. *(Russian)*

Koosis, P.

[1] Introduction to H_p Spaces, Cambridge University Press, Cambridge, 1980.

Kostjucenko, A. G. and Sargsjan, I. S.

[1] Distribution of Eigenvalues, Nauka, Moscow, 1979. *(Russian)*

Krasnosel'skii, M. A.

[1] *On a selfadjoint extensions of Hermitian operators,* Ukrain. Mat. Zh. **1** (1949), No. 1, 21–28. *(Russian)*

Krein, M. G.

[1] *Resolvents of Hermitian operators with deficiency index* (m, m), Dokl. Akad. Nauk SSSR **52** (1946), No. 8, 657–660. *(Russian)*

Kriete, T. L.

[1] *Canonical models and the selfadjoint parts of dissipative operators*, J. Funct. Anal. **23** (1976), 39–94.

Kudryashov, Yu. L.

[1] *Symmetric and selfadjoint dilation of dissipative operators*, Teor. Funktsij Funk. Anal. i Prilozh. (1982), No. 37, 51–54. *(Russian)*

[2] *J-Hermitian and J-selfadjoint dilations of linear operators*, Dynamical Systems (1984), No. 3., 94–98. *(Russian)*

Kuzhel, A. V.

[1] *Reduction of unbounded nonself-adjoint operators to triangular form*, Dokl. Akad. Nauk SSSR **119** (1958), No. 5, 868–871. *(Russian)*

[2] *Spectral analysis of unbounded nonself-adjoint operators*, Dokl. Akad. Nauk SSSR **125** (1959), No. 1, 35–37. *(Russian)*

[3] *Multiplication theorem for characteristic matrix functions of nonunitary operators*, Nauch. Dokl. Vysh. Shk. (1959), No. 3, 33–41. *(Russian)*

[4] *Perturbation of a self-adjoint operator by a finite-dimensional one and the condition of completeness*, Ukrain. Mat. Zh. **13** (1961), No. 1, 106–111. *(Russian)*

[5] *Spectral analysis of quasiunitary operators of rank one in a space with indefinite metric*, Dokl. Akad. Nauk Ukrain. SSR (1961), No. 8, 1001–1003. *(Ukrainian)*

[6] *On a class of linear operators*, Dokl. Akad. Nauk Ukrain. SSR (1961), No. 11, 1412–1414. *(Ukrainian)*

[7] *Conditions under which* $D_A = D_{A^*}$ *for unbounded operators*, Uspekhi Mat. Nauk **16** (1961), No. 3 (99), 189–190. *(Russian)*

[8] *A triangular model of* K_1*-operators in a space with indefinite metric*, Dokl. Akad. Nauk Ukrain. SSR (1962), No. 5, 572–574. *(Ukrainian)*

[9] *Characteristic matrix functions of quasiunitary operators of arbitrary rank in a space with indefinite metric*, Dokl. Akad. Nauk Ukrain. SSR (1962), No. 9, 1135–1138. *(Ukrainian)*

[10] *Spectral decomposition of quasiunitary operators of arbitrary rank in a space with indefinite metric*, Dokl. Akad. Nauk Ukrain. SSR (1963), No. 4, 430–434. *(Ukrainian)*

[11] *Spectral analysis of bounded nonself-adjoint operators in a space with indefinite metric*, Dokl. Akad. Nauk SSSR **151** (1963), No. 4, 772–774. *(Russian)*

[12] *On nonself-adjoint operators generated by Jacobian matrices*, Dokl. Akad. Nauk SSSR **154** (1964), No. 5, 1027–1029 *(Russian)*

[13] *On the spectrum of a regular quasidifferential operator*, Dokl. Akad. Nauk SSSR **156** (1964), No. 4, 731–733 *(Russian)*

[14] *On the spectrum of a nonself-adjoint Sturm–Liouville operator on a semiaxis*, Dokl. Akad. Nauk Ukrain. SSR (1965), No. 2, 157–160. *(Ukrainian)*

[15] *On a case of existence of invariant subspaces of quasiunitary operators in a space with an indefinite metrix*, Dokl. Akad. Nauk Ukrain. SSR (1966), No. 5, 583–585. *(Ukrainian)*

[16] *Spectral analysis of quasiunitary operators in a space with indefinite metric*, Teor. Funktsij Funk. Anal. i Prilozh. (1967), No. 4, 3–27.

[17] *A characteristic operator function of an arbitrary bounded operator*, Dokl. Akad. Nauk Ukrain. SSR. Ser. A (1968), No. 3, 233–236. *(Ukrainian)*

[18] *Spectral analysis of unbounded nonself-adjoint operator in a space with indefinite metric*, Dokl. Akad. Nauk SSSR **178** (1968), No. 1, 31–34. *(Russian)*

[19] *A generalization of the Sz.-Nagy and Foias theorem on the factorization of a characteristic operator function*, Dokl. Akad. Nauk Ukrain. SSR. Ser. A (1969), No. , 588–591. *(Ukrainian)*

[20] *A generalization of the Sz.-Nagy and Foias theorem on the factorization of the characteristic operator-function*, Acta Sci. Math. **30** (1969), 225–234 *(Russian)*

[21] *An analogue of the Sz.-Nagy and Foias theorem for dissipative operators*, Dokl. Akad. Nauk SSSR **215** (1974), No. 2, 253–254. *(Russian)*

[22] *Regular extensions of Hermitian operators*, Dokl. Akad. Nauk SSSR **251** (1980), No. 1, 30–33. *(Russian)*

[23] *An analogue of Krein formula for resolvents of nonself-adjoint extensions of a Hermitian operator*, Teor. Funktsij Funk. Anal. i Prilozh. (1981), No. 36, 49–55. *(Russian)*

[24] *Regular extension of Hermitian operators in a space with indefinite metric*, Dokl. Akad. Nauk SSSR **265** (1982), No. 5, 1059–1061. *(Russian)*

[25] *Selfadjoint and J-selfadjoint dilations of linear operators*, Teor. Funktsij Funk. Anal. i Prilozh. (1982), No. 37, 54–62. *(Russian)*

[26] *J-selfadjoint and J-unitary dilations of linear operators*, Funk. Anal. i Prilozh. **17** (1983), No. 1 75–76. *(Russian)*

[27] *Classification of the spectrum in algebras*, Dokl. Akad. Nauk Ukrain. SSR. Ser. A (1983), No. 10, 11–14. *(Russian)*

[28] Extensions of Hermitian Operators, Vyshcha Shkola, Kiev, 1989. *(Russian)*

[29] *Evolution of the concept of the characteristic function of a linear operator*, Ukrain. Mat. Zh. **45** (1993), No. 6, 731–744. *(Russian)*

Kuzhel, A. V. and Kudryashov, Ju. L.

[1] *Symmetric and selfadjoint dilations of dissipative operators*, Dokl. Akad. Nauk SSSR **253** (1980), No. 4, 812–815. *(Russian)*

Kuzhel, A. V. and Rudenko, L. I.

[1] *Regular extension of Hermitian and isometric operators*, Ukr. Math. Zh. **33** (1981), No. 6, 810–814. *(Russian)*

[2] *Description of regular extensions of Hermitian operators*, Funk. Anal. i Prilozh. **16** (1982), No. 1, 74–75. *(Russian)*

Kuzhel, A. V. and Tretyakov, D. V.

[1] *On a generalization of the Lax–Phillips scheme in scattering theory*, Dokl. Akad. Nauk Ukrain. SSR. Ser. A (1982), No. 2, 19–21. *(Russian)*

[2] *Generalized Lax–Phillips scheme in scattering theory*, Dynamical Systems (1983), No. 2, 115–121. *(Russian)*

Kuzhel, S. A.

[1] *J-nonexpanding operators*, Teor. Funktsij Funk. Anal. i Prilozh. (1986), No. 43, 82–87. *(Russian)*

[2] *On the stability of doubly J-nonexpanding operators*, Teor. Funktsij Funk. Anal. i Prilozh. (1989), No. 54, 68–74. *(Russian)*

[3] Spectral Analysis on Nonself-Adjoint Extensions of Hermitian Operators and Characteristics of Contractions, Candidate Degree Thesis, Kiev, 1990. *(Russian)*

[4] *About Lax–Phillips scattering theory in Pontryagin spaces*, Dokl. Ukrain. Akad. Nauk (1993), No. 4, 19–22. *(Ukrainian)*

[5] *Spectral analysis of doubly J-nonexpanding operators,* Ukrain. Mat. Zh. **45** (1993), No. 3, 384–388. *(Russian)*

Langer, H.

[1] *Ein zerspaltungsats für operatoren im Hilbertraum,* Acta. Math. Hung. **12** (1961), 441–445. *(German)*

Lax, P. D. and Phillips, R. S.

[1] Scattering Theory, Academic Press, New York, 1967.

[2] *The acoustic equation with an indefinite energy form and the Schrödinger equation,* J. Funct. Anal. **1** (1967), 37–85.

[3] Scattering Theory for Automorphic Functions, Princeton University Press, Princeton, 1976.

Levitan, B. M. and Sargsjan, I. S.

[1] Sturm–Liouville and Dirac operators, Nauka, Moscow, 1988. *(Russian)*

Livsic, M. S.

[1] *On one class of linear operators in Hilbert space,* Mat. Sbornik **19** (1946), 239–260. *(Russian)*

[2] *Isometric operators with equal defect numbers and quasiunitary operators,* Mat. Sbornik **26** (1950), 247–264. *(Russian)*

[3] *On a spectral decomposition of linear self-adjoint operators,* Mat. Sbornik **34** (1954), 144–199. *(Russian)*

[4] Operators, Oscillations, Waves. Open Systems, Nauka, Moscow, 1966; *English transl.* AMS 34, 1973.

Livsic, M. S. and Potapov, V. P.

[1] *A theorem on the multiplication of characteristic matrix functions,* Dokl. Akad. Nauk SSSR **72** (1950), No. 4, 625–628. *(Russian)*

Livsic, M. S. and Yantsevich, A. A.

[1] The Theory of Operator Colligations in Hilbert Spaces, Kharkov University, Kharkov, 1971. *(Russian)*

Makarov, N. G.

[1] *Stability of the essential spectrum on operators that are close to unitary operators,* Funk. Anal. i Prilozh. **16** (1982), No. 3, 72–73. *(Russian)*

Makarov, N. G. and Vasyunin, V. I.

[1] *A model for noncontractions and stability of the continuous spectrum,* Lect. Notes Math. (1981), No. 864, 365–412.

Malamud, M. M.

[1] *A criterion for a closed operator to be similar to a self-adjoint operator,* Ukrain. Mat. Zh. **37** (1985), No. 1, 49–56. *(Russian)*

[2] *On new approach to the theory of extensions of nondensely defined Hermitian operators,* Dokl. Akad. Nauk Ukrain. SSR (1990), No. 3, 20–25. *(Russian)*

McEnnis, B. W.

[1] *Purely contractive analytic functions and characteristic functions of noncontractions,* Acta Sci. Math. **41** (1979), No. 1–2, 161–172.

[2] *Characteristic functions and dilations of noncontractions,* J. Operator Theory **3** (1980), 71–78.

[3] *Models for operators with bounded characteristic function,* Acta Sci. Math. **43** (1981), No. 1–2, 71–90.

[4] *Uniformly continuous semigroups with bounded characteristic functions,* J. Operator Theory **22** (1989), No. 2, 209–232.

Mukminov, B. R.

[1] *On expansion in eigenfunctions of dissipative kernels,* Dokl. Akad. Nauk SSSR **99** (1954), No. 4, 499–502. *(Russian)*

Naboko, S. N.

[1] *Absolutely continuous spectrum of a nondissipative operator and a functional model. I,* Zap. LOMI **65** (1977), 90–102. *(Russian)*

[2] *Absolutely continuous spectrum of a nondissipative operator and a functional model. II,* Zap. LOMI **73** (1977), 118–135. *(Russian)*

[3] *Wave operators for nonself-adjoint operators and a functional model,* Zap. LOMI **69** (1977), 129–135. *(Russian)*

[4] *Functional model of perturbation theory and its application to scattering theory,* Trudy Mat. Inst. Steklov **147** (1980), 86–114. *(Russian)*

[5] *Conditions of similarity of operators to unitary or selfadjoint ones,* Funk. Anal. i Prilozh. **18** (1984), No. 1, 16–27. *(Russian)*

Naimark, M. A.

[1] *On a self-adjoint extensions of the second kind of symmetric operator,* Izv. Akad. Nauk SSSR **4** (1940), No. 1, 53–104. *(Russian)*

von Neumann, J.

[1] *Allgemeine eigenwer theorie Hermitesher Funktional operatoren,* Math. Ann. **102** (1926), 49–131. *(German)*

Nikolsky, N. K.

[1] Lectures on the Shift Operator, Nauka, Moscow, 1980. *(Russian)*

Nikolsky, N. K. and Khrushchev, S. V.

[1] *A functional model and some problems of the spectral theory of functions,* Trudy Mat. Inst. Steklov **176** (1987), 97–210. *(Russian)*

Nikolsky, N. K. and Vasyunin, V. I.

[1] *General approach to function models and the transcription problem,* Preprint No. E.5–86, LOMI, Leningrad, 1986.

Pavlov, B. S.

[1] *Conditions for separation of spectral components of a dissipative operator,* Izv. Akad. Nauk SSSR **39** (1975), No. 1, 123–148. *(Russian)*

[2] *Self-adjoint dilation of a dissipative Schrödinger operator and expansion its eigenfunction,* Mat. Sbornik **102** (1977), No. 4, 511–536. *(Russian)*

[3] *The theory of extensions, and explicitly solvable models,* Uspekhi Mat. Nauk **42** (1987), No. 6. 99–131. *(Russian)*

Petrov, A. M.

[1] Spectral Projectors of Dissipative Operators, Dynamical Systems (1987), No. 6, 109–114. *(Russian)*

Phillips, R. S.

[1] *Dissipative operators and hyperbolic systems of partial differential equations,* Trans. AMS **90** (1959), 195–254.

[2] *The extension of dual subspaces invariant under an algebra,* Proceedings of the International Symposium (Jerusalem, 1960), Pergamon Press, 1961.

Polyakov, V. N.

[1] *On one problem in the theory of characteristic functions of linear operators,* Mat. Zametki **9** (1971), No. 2, 171–180. *(Russian)*

Polyatskii, V. T.

[1] *Reduction of quasiunitary operators to triangular form,* Dokl. Akad. Nauk SSSR **113** (1957), No. 4, 756–759. *(Russian)*

Potapov, V. P.

[1] *The multiplicative structure of J-contractive matrix functions,* Trudy Mosk. Mat. Obshch. **4** (1955), 125–236. *(Russian)*

[2] Collected Papers, Hokkaido University, Research Institute of Applied Electricity, Supporo, 1982.

Richtmyer, R. D.

[1] Principles of Advanced Mathematical Physics, Berlin, Springer, 1978.

Rofe–Beketov, F. S.

[1] *Self-adjoint extensions of differential operators in a space of vector-valued functions,* Teor. Funktsij Funk. Anal. i Prilozh. (1969) No. 8, 3–24. *(Russian)*

Rota, G. C.

[1] *On models for linear operators,* Comm. Pure Appl. Math. **13** (1960), 468–472.

Rutkas, A. G.

[1] *Characteristic function and a model of a linear operator pencil*, Teor. Funktsij Funk. Anal. i Prilozh. (1986) No. 45, 98-111. *(Russian)*

Sahnovich, L. A.

[1] *Reduction on nonself-adjoint operators with continuous spectrum to triangular form*, Mat. Sbornik **44** (1958), No. 4, 509–548. *(Russian)*

[2] *Operators similar to unitary with absolutely continuous spectrum*, Funk. Anal. i Prilozh. **2** (1968), No. 1, 51–63. *(Russian)*

[3] *Dissipative operators with absolutely continuous spectrum*, Trudy Mosk. Mat. Obshch. **19** (1968), 211–270. *(Russian)*

[4] *On J-unitary dilations of bounded operators*, Funk. Anal. i Prilozh. **8** (1974), No. 3, 83–84. *(Russian)*

Sarason, D. E.

[1] *On spectral sets with connected complements*, Acta Sci. Math. **26** (1965), 289–299.

Shmulyan, Yu. L.

[1] *Operators with degenerate characteristic functions*, Dokl. Akad. Nauk SSSR **93** (1953), No. 6, 985–988. *(Russian)*

Shtraus, A. V.

[1] *On theory of Hermitian operators*, Dokl. Akad. Nauk SSSR **67** (1949), No. 4, 211–214. *(Russian)*

[2] *On a class of regular operator functions*, Dokl. Akad. Nauk SSSR **70** (1950), No. 4, 577–580. *(Russian)*

[3] *On generalized resolvents of symmetric operators*, Dokl. Akad. Nauk SSSR **71** (1950), No. 2, 241–244. *(Russian)*

[4] *On the theory of generalized resolvents of symmetric operators*, Dokl. Akad. Nauk SSSR **78** (1951), No. 2, 217–220. *(Russian)*

[5] *On characteristic functions of linear operators*, Dokl. Akad. Nauk SSSR **126** (1959), No. 3, 514–516. *(Russian)*

[6] *Characteristic functions of linear operators,* Izv. Akad. Nauk SSSR. Ser. Mat. **24** (1960), No. 1, 43–74. *(Russian)*

[7] *On selfadjoint operators in ortogonal sum of Hilbert spaces,* Dokl. Akad. Nauk SSSR **144** (1962), No. 3, 512–515. *(Russian)*

[9] *On extensions and characteristic functions of symmetric operators,* Izv. Akad. Nauk SSSR. Ser. Mat. **32** (1968), No. 1, 186–207. *(Russian)*

[10] *Extensions and generalized resolvents of nondensely defined symmetric operators,* Izv. Akad. Nauk SSSR. Ser. Mat. **34** (1970), No. 1, 175–202. *(Russian)*

Shvartsman, Ya. S.

[1] *Invariant subspaces of dissipative operators and the divisor of their characteristic functions,* Funk. Anal. i Prilozh. **4** (1970), No. 4, 85–86. *(Russian)*

Solomyak, B. M.

[1] *A functional model for dissipative operators. Coordinate-free approach,* Zap. LOMI **178** (1989), 57–91. *(Russian)*

[2] *Scattering theory for almost unitary operators and functional model,* Zap. LOMI **178** (1989), 92–119. *(Russian)*

Storozh, O. G.

[1] *Extensions of symmetric operators with different defect numbers,* Mat. Zametki **36** (1984), No. 5, 791–795. *(Russian)*

Szökefalvi–Nagy, B.

[1] *Unitary dilations of Hilbert space operators and related topics,* in: Lectures from the CBMS, (Regional Conference, University of New Hampshire, 1979), AMS, Providence, 1979.

Szökefalvi–Nagy, B. and Foias, C.

[1] *Models fonctionnels des contractions de lespace de Hilbert, La fonction caŕacteristique,* C. R. **256** (1963), 3236–3239.

[2] *Proprietes des fonctions characteristiques, modeles trianqularires et une classification des contractions,* C. R. **258** (1963), 3413–3415.

[3] *Forme trianqulaire d'une contractionet factorisation de la fonction carac-teristique,* Acta Sci. Math. **28** (1967), 201–212.

[4] Analyse Harmonique des Operateurs de l'Espace de Hilbert, Academiai Kiadó, 1967.

[5] *Sur les contractions de l'espace de Hilbert. IV,* Acta Sci. Math. **21** (1960), 251–259.

Taljush, M. O.

[1] *The typical structure of dissipative operators,* Dokl. Akad. Nauk Ukrain. SSR. Ser. A (1973), No. 11, 993–996. *(Ukrainian)*

Titchmarsh, E. C.

[1] Eigenfunction Expansions Associated with Second-Order Differential Equations, Clarendon Press, Oxford, 1946.

Tretyakov, D. V.

[1] *Wave operators and associated semigroup in generalized Lax–Phillips sheme,* Dynamical Systems (1985), No. 4, 125–130. *(Russian)*

[2] *Outgoing and incoming representations. Spectral analysis of the generator of as-sociated semigroup,* Dynamical Systems (1985), No. 5, 120–127. *(Russian)*

Tsekanovskii, E. R.

[1] *Triangular models of unbounded accretive operators and the regular factor-ization of their characteristic operator functions,* Dokl. Akad. Nauk SSSR **297** (1987), No. 3, 552–556.

Tsekanovskii, E. R. and Shmulyan, Yu. L.

[1] *The theory of biextensions of operators in rigged Hilbert spaces. Unbounded operator couplings and characteristic functions,* Uspekhi Mat. Nauk **32** (1977), No. 5, 69–124. *(Russian)*

Vasyunin, V. I.

[1] *The construction of the Sz.-Nagy–Foias functional model,* Zap. LOMI **73** (1977), 16–23. *(Russian)*

[2] *Two classical theorems on a model in coordinate-free presentation,* Zap. LOMI
     178 (1989), 5–22. *(Russian)*

Vernik, A. N. and Chunaeva, M. S.

[1] *Symmetric dilation of linear relations in a space with indefinite metric,* in: Funk.
     Anal. Spectral Theory (1982), No. 18, 36–43. *(Russian)*

SUBJECT INDEX

AUTHOR INDEX

NOTATION

Other *Mathematics and Its Applications* titles of interest:

A.F. Filippov: *Differential Equations with Discontinuous Righthand Sides*. 1988, 320 pp. ISBN 90-277-2699-X

A.T. Fomenko: *Integrability and Nonintegrability in Geometry and Mechanics.* 1988, 360 pp. ISBN 90-277-2818-6

G. Adomian: *Nonlinear Stochastic Systems Theory and Applications to Physics.* 1988, 244 pp. ISBN 90-277-2525-X

A. Tesar and Ludovt Fillo: *Transfer Matrix Method.* 1988, 260 pp.
ISBN 90-277-2590-X

A. Kaneko: *Introduction to the Theory of Hyperfunctions.* 1989, 472 pp.
ISBN 90-277-2837-2

D.S. Mitrinovic, J.E. Pecaric and V. Volenec: *Recent Advances in Geometric Inequalities.* 1989, 734 pp. ISBN 90-277-2565-9

A.W. Leung: *Systems of Nonlinear PDEs: Applications to Biology and Engineering.* 1989, 424 pp. ISBN 0-7923-0138-2

N.E. Hurt: *Phase Retrieval and Zero Crossings: Mathematical Methods in Image Reconstruction.* 1989, 320 pp. ISBN 0-7923-0210-9

V.I. Fabrikant: *Applications of Potential Theory in Mechanics. A Selection of New Results.* 1989, 484 pp. ISBN 0-7923-0173-0

R. Feistel and W. Ebeling: *Evolution of Complex Systems. Selforganization, Entropy and Development.* 1989, 248 pp. ISBN 90-277-2666-3

S.M. Ermakov, V.V. Nekrutkin and A.S. Sipin: *Random Processes for Classical Equations of Mathematical Physics.* 1989, 304 pp. ISBN 0-7923-0036-X

B.A. Plamenevskii: *Algebras of Pseudodifferential Operators.* 1989, 304 pp.
ISBN 0-7923-0231-1

N. Bakhvalov and G. Panasenko: *Homogenisation: Averaging Processes in Periodic Media. Mathematical Problems in the Mechanics of Composite Materials.* 1989, 404 pp. ISBN 0-7923-0049-1

A.Ya. Helemskii: *The Homology of Banach and Topological Algebras.* 1989, 356 pp. ISBN 0-7923-0217-6

M. Toda: *Nonlinear Waves and Solitons.* 1989, 386 pp. ISBN 0-7923-0442-X

M.I. Rabinovich and D.I. Trubetskov: *Oscillations and Waves in Linear and Nonlinear Systems.* 1989, 600 pp. ISBN 0-7923-0445-4

A. Crumeyrolle: *Orthogonal and Symplectic Clifford Algebras. Spinor Structures.* 1990, 364 pp. ISBN 0-7923-0541-8

V. Goldshtein and Yu. Reshetnyak: *Quasiconformal Mappings and Sobolev Spaces.* 1990, 392 pp. ISBN 0-7923-0543-4

Other *Mathematics and Its Applications* titles of interest:

I.H. Dimovski: *Convolutional Calculus.* 1990, 208 pp.   ISBN 0-7923-0623-6

Y.M. Svirezhev and V.P. Pasekov: *Fundamentals of Mathematical Evolutionary Genetics.* 1990, 384 pp.   ISBN 90-277-2772-4

S. Levendorskii: *Asymptotic Distribution of Eigenvalues of Differential Operators.* 1991, 297 pp.   ISBN 0-7923-0539-6

V.G. Makhankov: *Soliton Phenomenology.* 1990, 461 pp.   ISBN 90-277-2830-5

I. Cioranescu: *Geometry of Banach Spaces, Duality Mappings and Nonlinear Problems.* 1990, 274 pp.   ISBN 0-7923-0910-3

B.I. Sendov: *Hausdorff Approximation.* 1990, 384 pp.   ISBN 0-7923-0901-4

A.B. Venkov: *Spectral Theory of Automorphic Functions and Its Applications.* 1991, 280 pp.   ISBN 0-7923-0487-X

V.I. Arnold: *Singularities of Caustics and Wave Fronts.* 1990, 274 pp.
ISBN 0-7923-1038-1

A.A. Pankov: *Bounded and Almost Periodic Solutions of Nonlinear Operator Differential Equations.* 1990, 232 pp.   ISBN 0-7923-0585-X

A.S. Davydov: *Solitons in Molecular Systems. Second Edition.* 1991, 428 pp.
ISBN 0-7923-1029-2

B.M. Levitan and I.S. Sargsjan: *Sturm-Liouville and Dirac Operators.* 1991, 362 pp.   ISBN 0-7923-0992-8

V.I. Gorbachuk and M.L. Gorbachuk: *Boundary Value Problems for Operator Differential Equations.* 1991, 376 pp.   ISBN 0-7923-0381-4

Y.S. Samoilenko: *Spectral Theory of Families of Self-Adjoint Operators.* 1991, 309 pp.   ISBN 0-7923-0703-8

B.I. Golubov A.V. Efimov and V.A. Scvortsov: *Walsh Series and Transforms.* 1991, 382 pp.   ISBN 0-7923-1100-0

V. Laksmikantham, V.M. Matrosov and S. Sivasundaram: *Vector Lyapunov Functions and Stability Analysis of Nonlinear Systems.* 1991, 250 pp.
ISBN 0-7923-1152-3

F.A. Berezin and M.A. Shubin: *The Schrödinger Equation.* 1991, 556 pp.
ISBN 0-7923-1218-X

D.S. Mitrinovic, J.E. Pecaric and A.M. Fink: *Inequalities Involving Functions and their Integrals and Derivatives.* 1991, 588 pp.   ISBN 0-7923-1330-5

Julii A. Dubinskii: *Analytic Pseudo-Differential Operators and their Applications.* 1991, 252 pp.   ISBN 0-7923-1296-1

V.I. Fabrikant: *Mixed Boundary Value Problems in Potential Theory and their Applications.* 1991, 452 pp.   ISBN 0-7923-1157-4

Other *Mathematics and Its Applications* titles of interest:

A.M. Samoilenko: *Elements of the Mathematical Theory of Multi-Frequency Oscillations.* 1991, 314 pp.   ISBN 0-7923-1438-7

Yu.L. Dalecky and S.V. Fomin: *Measures and Differential Equations in Infinite-Dimensional Space.* 1991, 338 pp.   ISBN 0-7923-1517-0

W. Mlak: *Hilbert Space and Operator Theory.* 1991, 296 pp.   ISBN 0-7923-1042-X

N.Ja. Vilenkin and A.U. Klimyk: *Representation of Lie Groups and Special Functions. Volume 1: Simplest Lie Groups, Special Functions, and Integral Transforms.* 1991, 608 pp.   ISBN 0-7923-1466-2

N.Ja. Vilenkin and A.U. Klimyk: *Representation of Lie Groups and Special Functions. Volume 2: Class I Representations, Special Functions, and Integral Transforms.* 1992, 630 pp.   ISBN 0-7923-1492-1

N.Ja. Vilenkin and A.U. Klimyk: *Representation of Lie Groups and Special Functions. Volume 3: Classical and Quantum Groups and Special Functions.* 1992, 650 pp.   ISBN 0-7923-1493-X

(Set ISBN for Vols. 1, 2 and 3: 0-7923-1494-8)

K. Gopalsamy: *Stability and Oscillations in Delay Differential Equations of Population Dynamics.* 1992, 502 pp.   ISBN 0-7923-1594-4

N.M. Korobov: *Exponential Sums and their Applications.* 1992, 210 pp.
                    ISBN 0-7923-1647-9

Chuang-Gan Hu and Chung-Chun Yang: *Vector-Valued Functions and their Applications.* 1991, 172 pp.   ISBN 0-7923-1605-3

Z. Szmydt and B. Ziemian: *The Mellin Transformation and Fuchsian Type Partial Differential Equations.* 1992, 224 pp.   ISBN 0-7923-1683-5

L.I. Ronkin: *Functions of Completely Regular Growth.* 1992, 394 pp.
                    ISBN 0-7923-1677-0

R. Delanghe, F. Sommen and V. Soucek: *Clifford Algebra and Spinor-valued Functions. A Function Theory of the Dirac Operator.* 1992, 486 pp.
                    ISBN 0-7923-0229-X

A. Tempelman: *Ergodic Theorems for Group Actions.* 1992, 400 pp.
                    ISBN 0-7923-1717-3

D. Bainov and P. Simenov: *Integral Inequalities and Applications.* 1992, 426 pp.
                    ISBN 0-7923-1714-9

I. Imai: *Applied Hyperfunction Theory.* 1992, 460 pp.   ISBN 0-7923-1507-3

Yu.I. Neimark and P.S. Landa: *Stochastic and Chaotic Oscillations.* 1992, 502 pp.
                    ISBN 0-7923-1530-8

H.M. Srivastava and R.G. Buschman: *Theory and Applications of Convolution Integral Equations.* 1992, 240 pp.   ISBN 0-7923-1891-9

Other *Mathematics and Its Applications* titles of interest:

A. van der Burgh and J. Simonis (eds.): *Topics in Engineering Mathematics.* 1992, 266 pp. ISBN 0-7923-2005-3

F. Neuman: *Global Properties of Linear Ordinary Differential Equations.* 1992, 320 pp. ISBN 0-7923-1269-4

A. Dvurecenskij: *Gleason's Theorem and its Applications.* 1992, 334 pp.
ISBN 0-7923-1990-7

D.S. Mitrinovic, J.E. Pecaric and A.M. Fink: *Classical and New Inequalities in Analysis.* 1992, 740 pp. ISBN 0-7923-2064-6

H.M. Hapaev: *Averaging in Stability Theory.* 1992, 280 pp. ISBN 0-7923-1581-2

S. Gindinkin and L.R. Volevich: *The Method of Newton's Polyhedron in the Theory of PDE's.* 1992, 276 pp. ISBN 0-7923-2037-9

Yu.A. Mitropolsky, A.M. Samoilenko and D.I. Martinyuk: *Systems of Evolution Equations with Periodic and Quasiperiodic Coefficients.* 1992, 280 pp.
ISBN 0-7923-2054-9

I.T. Kiguradze and T.A. Chanturia: *Asymptotic Properties of Solutions of Non-autonomous Ordinary Differential Equations.* 1992, 332 pp. ISBN 0-7923-2059-X

V.L. Kocic and G. Ladas: *Global Behavior of Nonlinear Difference Equations of Higher Order with Applications.* 1993, 228 pp. ISBN 0-7923-2286-X

S. Levendorskii: *Degenerate Elliptic Equations.* 1993, 445 pp.
ISBN 0-7923-2305-X

D. Mitrinovic and J.D. Kečkić: *The Cauchy Method of Residues, Volume 2.* Theory and Applications. 1993, 202 pp. ISBN 0-7923-2311-8

R.P. Agarwal and P.J.Y Wong: *Error Inequalities in Polynomial Interpolation and Their Applications.* 1993, 376 pp. ISBN 0-7923-2337-8

A.G. Butkovskiy and L.M. Pustyl'nikov (eds.): *Characteristics of Distributed-Parameter Systems.* 1993, 386 pp. ISBN 0-7923-2499-4

B. Sternin and V. Shatalov: *Differential Equations on Complex Manifolds.* 1994, 504 pp. ISBN 0-7923-2710-1

S.B. Yakubovich and Y.F. Luchko: *The Hypergeometric Approach to Integral Transforms and Convolutions.* 1994, 324 pp. ISBN 0-7923-2856-6

C. Gu, X. Ding and C.-C. Yang: *Partial Differential Equations in China.* 1994, 181 pp. ISBN 0-7923-2857-4

V.G. Kravchenko and G.S. Litvinchuk: *Introduction to the Theory of Singular Integral Operators with Shift.* 1994, 288 pp. ISBN 0-7923-2864-7

A. Cuyt (ed.): *Nonlinear Numerical Methods and Rational Approximation II.* 1994, 446 pp. ISBN 0-7923-2967-8

Other *Mathematics and Its Applications* titles of interest:

G. Gaeta: *Nonlinear Symmetries and Nonlinear Equations.* 1994, 258 pp.
ISBN 0-7923-3048-X

V.A. Vassiliev: *Ramified Integrals, Singularities and Lacunas.* 1995, 289 pp.
ISBN 0-7923-3193-1

N.Ja. Vilenkin and A.U. Klimyk: *Representation of Lie Groups and Special Functions.* Recent Advances. 1995, 497 pp. ISBN 0-7923-3210-5

Yu. A. Mitropolsky and A.K. Lopatin: *Nonlinear Mechanics, Groups and Symmetry.* 1995, 388 pp. ISBN 0-7923-3339-X

R.P. Agarwal and P.Y.H. Pang: *Opial Inequalities with Applications in Differential and Difference Equations.* 1995, 393 pp. ISBN 0-7923-3365-9

A.G. Kusraev and S.S. Kutateladze: *Subdifferentials: Theory and Applications.* 1995, 408 pp. ISBN 0-7923-3389-6

M. Cheng, D.-G. Deng, S. Gong and C.-C. Yang (eds.): *Harmonic Analysis in China.* 1995, 318 pp. ISBN 0-7923-3566-X

M.S. Livšic, N. Kravitsky, A.S. Markus and V. Vinnikov: *Theory of Commuting Nonselfadjoint Operators.* 1995, 314 pp. ISBN 0-7923-3588-0

A.I. Stepanets: *Classification and Approximation of Periodic Functions.* 1995, 360 pp. ISBN 0-7923-3603-8

C.-G. Ambrozie and F.-H. Vasilescu: *Banach Space Complexes.* 1995, 205 pp.
ISBN 0-7923-3630-5

E. Pap: *Null-Additive Set Functions.* 1995, 312 pp. ISBN 0-7923-3658-5

C.J. Colbourn and E.S. Mahmoodian (eds.): *Combinatorics Advances.* 1995, 338 pp. ISBN 0-7923-3574-0

V.G. Danilov, V.P. Maslov and K.A. Volosov: *Mathematical Modelling of Heat and Mass Transfer Processes.* 1995, 330 pp. ISBN 0-7923-3789-1

A. Laurinčikas: *Limit Theorems for the Riemann Zeta-Function.* 1996, 312 pp.
ISBN 0-7923-3824-3

A. Kuzhel: *Characteristic Functions and Models of Nonself-Adjoint Operators.* 1996, 283 pp. ISBN 0-7923-3879-0